Lecture Notes in Bioinformatic

T0250674

Edited by S. Istrail, P. Pevzner, and M. Waterman

Subseries of Lecture Notes in Computer Science

Michael R. Berthold   Robert Glen
Ingrid Fischer (Eds.)

Computational Life Sciences II

Second International Symposium, CompLife 2006
Cambridge, UK, September 27-29, 2006
Proceedings

 Springer

Series Editors

Sorin Istrail, Brown University, Providence, RI, USA
Pavel Pevzner, University of California, San Diego, CA, USA
Michael Waterman, University of Southern California, Los Angeles, CA, USA

Volume Editors

Michael R. Berthold
Ingrid Fischer
University of Konstanz
Department of Computer and Information Science
Box M712, 78457 Konstanz, Germany
E-mail: berthold@ieee.org, Ingrid.Fischer@inf.uni-konstanz.de

Robert Glen
Unilever Center
University of Cambridge
Lensfield Road, Cambridge CB2 1EW, UK
E-mail: rcg28@cam.ac.uk

Library of Congress Control Number: 2006932818

CR Subject Classification (1998): H.2, H.3, H.4, J.3

LNCS Sublibrary: SL 8 – Bioinformatics

ISSN   0302-9743
ISBN-10   3-540-45767-4 Springer Berlin Heidelberg New York
ISBN-13   978-3-540-45767-1 Springer Berlin Heidelberg New York

Springer is a part of Springer Science+Business Media

springer.com

© Springer-Verlag Berlin Heidelberg 2006

Typesetting: Camera-ready by author, data conversion by Scientific Publishing Services, Chennai, India
Printed on acid-free paper   SPIN: 11875741   06/3142   5 4 3 2 1 0

Preface

Since our first CompLife symposium last year, we have seen the predicted trends in the life and computer science areas continue with ever-increasing production of high-quality data mated to novel analysis methods. The integration of the most advanced computational methods into experimental design and in particular the validation of these methods will remain a challenge. However, there is increasing appreciation between the different scientific communities in computer science and biology that each has substantial goals in common and much to gain by collaboration on complex problems. Providing a forum for an open and lively exchange between computer scientists, biologists, and chemists remains our goal. To encourage precisely this type of exchange, crossing the borders of the sciences, we organized the First Symposium on Computational Life Science in Konstanz, Germany in September 2005 (the proceedings were published in this series as LNBI 3695). Due to the success of the symposium, especially in bringing together scientists with diverse backgrounds, a second symposium was held in Cambridge (September 27-29, 2006).

The conference program shows that the scientific mix worked out very well again. We received higher quality submissions (56 this time) and selected 23 for oral presentation. As a supplement to the normal conference program we arranged for a "Free Software Session," where a dozen open source tools and toolkits were presented. Due to the nature of such software projects it seemed inappropriate to cover them in printed form but the conference Web site will continue to link to the respective pages (www.complife.org). Adding this session to the symposium also educated attendees on how to use some of the methods presented and shed some light on the wealth of free tools available already.

Selecting the papers included in this volume would not have been possible without the help of our Area Chairs and an international Program Committee that put in countless hours to create a minimum of three detailed reviews for each paper! And, of course, a successful conference relies on many individuals working hard behind the scenes. We would like to thank first and foremost Susan Begg and Heather Fyson for conference and local organization and keeping everybody on track. Peter Burger worked on the Web pages promoting the conference and Thorsten Meinl was the man behind the free software session and, together with Andreas Bender, also took care of publicity. Last, but certainly not least, thanks go to Ingrid Fischer and Richard van de Stadt for putting together this volume!

July 2006

Michael R. Berthold
Robert Glen

Organization

General Chair   Michael R. Berthold
University of Konstanz, Germany
Michael.Berthold@uni-konstanz.de

Program Chair   Robert Glen
Unilever Center, University of Cambridge, UK
rcg28@cam.ac.uk

Publication Chair   Ingrid Fischer
University of Konstanz, Germany
Ingrid.Fischer@uni-konstanz.de

Publicity Chairs   Andreas Bender
Novartis Institutes for BioMedical Research, USA
andreas.bender@complife.org

Thorsten Meinl
University of Konstanz, Germany
Thorsten.Meinl@uni-konstanz.de

Conference Chair   Heather Fyson
University of Konstanz, Germany
heather.fyson@uni-konstanz.de

Local Chair   Susan Begg
Unilever Center, University of Cambridge, UK
smb28@cam.ac.uk

Webmaster   Peter Burger
University of Konstanz, Germany
Peter.Burger@uni-konstanz.de

Program Committee

Area Chairs

Giuseppe Di Fatta, University of Konstanz, Germany and CNR Palermo, Italy
Aldo Faisal, University of Cambridge, UK
Hans-Christian Hege, Zuse Institute Berlin, Germany
Janette Jones, Unilever Research, UK
Oliver Kohlbacher, Tübingen University, Germany
Peter Murry-Rust, University of Cambridge, UK
Jeremy Nicholson, Imperial College, UK
Gisbert Schneider, Frankfurt University, Germany
Brian Shoichet, UC San Francisco, USA
Arno Siebes, Utrecht University, NL
Hong Yan, City University, Hong Kong
Daniel Zaharevitz, NIH, USA

Program Committee

Alexander Bockmayr, FU Berlin, Germany
Sebastian Böcker, Friedrich Schiller University Jena, Germany
Tim Clark, Friedrich Alexander University Erlangen-Nuremberg, Germany
Thomas Exner, University of Konstanz, Germany
Peter Haebel, ALTANA Pharma Konstanz, Germany
Lawrence Hall, University of South Florida Tampa, USA
Joost Kok, Leiden University, The Netherlands
Hans-Peter Lenhof, Saarland University Saarbrücken, Germany
Xiaohui Liu, Brunel University, UK
Vladimir Marik, Czech Technical University Prague, Czech Republic
Srinivasan Parthasarathy, The Ohio State University, USA
David Patterson, Vistamont Consultancy, USA
Matthias Rarey, University of Hamburg, Germany
Knut Reinert, FU Berlin, Germany
Klaus Schäfer, ALTANA Pharma Konstanz, Germany
Hannu Toivonen, University of Helsinki, Finland
Allan Tucker, Brunel University, UK
Alexandre Urzhumtsev, University H. Poincare Nancy, France
Peter Willet, The University of Sheffield, UK
Mohammed Zaki, Rensselaer Polytechnic Institute, USA
Ralf Zimmer, LMU Munich, Germany
Albert Y. Zomaya, The University of Sydney, Australia

Additional Reviewers

Fatih Altiparmak, Sitaram Asur, Robert Banfield, Daniel Baum, Fabian Bir-
zele, Andreas Döring, Caroline C. Friedel, Jan Gewehr, Clemens Gröpl, Volk-
hard Helms, Andreas Hildebrandt, Wilhelm Huisinga, Hans-Michael Kaltenbach,
Andreas Keller, Robert Kueffner, Stefan Kurtz, Jan Küntzer, Hans Lamecker,
Abdelhalim Larhlimi, Zsuzsanna Liptak, Andreas Moll, Ozgur Ozturk, Hei-
ke Pospisil, Sven Rahmann, Alexander Rurainski, Johannes Schmidt-Ehrenberg,
Larry Shoemaker, Selina Sommer, Jens Stoye, Wiebke Timm, Duygu Ucar,
Chao Wang, Hui Yang

Sponsoring Institutions

We thank our sponsors for their support in making the second International
Symposium on Computational Life Science a successful event: Molecular Gra-
phics Modelling Society (MGMS), ALTANA Pharma AG, AstraZeneca R&D,
Boehringer Ingelheim, Inpharmatica, Unilever R&D Port Sunlight and Unilever
Corporate Research.

Table of Contents

Molecular Simulation

Molecular Informatics

Systems Biology

Biological Networks / Metabolism

Computational Neuroscience

Improved Robustness in Time Series Analysis of Gene Expression Data by Polynomial Model Based Clustering

Michael Hirsch[1,*], Allan Tucker[1], Stephen Swift[1], Nigel Martin[2], Christine Orengo[3], Paul Kellam[4], and Xiaohui Liu[1]

[1] School of Information Systems Computing and Mathematics, Brunel University, Uxbridge UB8 3PH, UK
[2] School of Computer Science and Information Systems Birkbeck, University of London, Malet Street, London, WC1E 7HX, UK
[3] Department of Biochemistry and Molecular Biology, University College London, Gower Street, London, WC1E 6BT, UK
[4] Department of Infection, University College London, Gower Street, London, WC1E 6BT, UK

Abstract. Microarray experiments produce large data sets that often contain noise and considerable missing data. Typical clustering methods such as hierarchical clustering or partitional algorithms can often be adversely affected by such data. This paper introduces a method to overcome such problems associated with noise and missing data by modelling the time series data with polynomials and using these models to cluster the data. Similarity measures for polynomials are given that comply with commonly used standard measures. The polynomial model based clustering is compared with standard clustering methods under different conditions and applied to a real gene expression data set. It shows significantly better results as noise and missing data are increased.

1 Introduction

Microarray experiments are widely used in medical and life science research [11]. This technology makes it possible to examine the behaviour of thousands of genes simultaneously. Moreover, microarray time series experiments provide an insight into the dynamics of gene activity as an essential part of cell processes.

Despite efforts to produce high quality microarray data, such data is often burdened with a considerable amount of noise. Attempts to reduce the noise are manifold, including intelligent experimental design, multiple repeats of the experiment and noise reduction techniques in the data preprocessing [13]. In addition to the noise problem, parts of the data often can not be retrieved properly so that the dataset contains missing values. For example, a dataset of several experiments with yeast (about 500,000 values) [10] has more than 11% missing values.

* This work is in part supported by the BBSRC in UK (BB/C506264/1).

M.R. Berthold, R. Glen, and I. Fischer (Eds.): CompLife 2006, LNBI 4216, pp. 1–10, 2006.

With decreasing quality the direct clustering (DC) of the data with standard methods [5] becomes less reliable. If the data has considerable missing data, the straightforward calculation of the score functions homogeneity and separation [4] for the cluster quality becomes impossible. To overcome these problems this paper suggests the modelling of the data with continuous functions. The model based clustering is done not on the original dataset directly, but on models learnt from it. The models reduce random noise and interpolate missing values, thereby increasing the robustness of clustering.

In this paper the polynomial model based clustering (PMC) is introduced. In contrast to the DC of the data, which calculates the similarity matrix directly from the data, PMC comprised of three steps: the modelling, the calculation of the similarity matrix from the models and the grouping.

2   Methods

The application of continuous functions in time series modelling is motivated by some specific assumptions. Time series result from measurements of a quantity at different time points (TP) over a certain time period. The quantity changes continuously if it could be measured at any time in the presumed time period. Measurement restrictions are due to extrinsic factors such as technical restrictions. Moreover, if a continuous quantity has the value x at TP a and the value y at TP b, then the quantity has any value between x and y at some TP between a and b. Often time series or functions have no sharp edges in the time response, i.e. they are differentiable or smooth.

Any smooth function can be approximated by the Taylor expansion, i.e. by a polynomial. Polynomials are easy to handle since basic operations can be done by simple algebraic manipulations on the parameters. Therefore polynomials are a natural choice in time series modelling. Nevertheless, other classes of functions might be used as well. Previously, polynomials have also been used in other applications of gene expression data modelling [8,12].

2.1   Modelling

Consider series of observations, $y_l(t_i)$ ($l = 1 \ldots N$, $i \in I = \{1, \ldots, T\}$), of N quantities at T TPs. The time elapsed between two measurements at t_i and t_{i+1} might be different through the series. A sub-series of $y_l(t_i)$ in which the missing values are omitted is denoted by $\tilde{y}_l(t_i)$ $i \in J$, where the index set J is the subset of I that contains these time-indices, where a value is available. If J is equal to I, then $\tilde{y}_l(t_i) = y_l(t_i)$.

Polynomials have the general form

$$P(t) = \sum_{i=0}^{n} \alpha_i t^i \ , \tag{1}$$

where n is the degree of the polynomial. To fit a polynomial to the data, the least squares method is used [9]. This method optimises the parameter, α_i, of a function $f(t, \alpha_0, \ldots, \alpha_n)$, $n + 1 < |J|$, $|J|$ is the number of elements in J, such that the function $Q(\alpha_0, \ldots, \alpha_n) = \sum_{i \in J} (f(t_i, \alpha_0, \ldots, \alpha_n) - \tilde{y}_l(t_i))^2$ becomes minimal. Therefore the equations $\partial Q / \partial \alpha_k = 0$, $k = 0 \ldots n$ have to be solved. Applying this equation to polynomials yields

$$\sum_{k=0}^{n} \alpha_k \sum_{j \in J} t_j^{k+i} = \sum_{j \in J} t_j^i \tilde{y}_l(t_j) \quad i = 0, \ldots, n \ . \tag{2}$$

These are $n + 1$ linear equations for the $n + 1$ parameters $\alpha_0, \ldots, \alpha_n$. To solve these equations an inverse matrix of the $(n + 1) \times (n + 1)$ matrix $\sum_{j \in J} t_j^{k+i}$ has to be calculated for each distinct subset J of I that occurs in the data set. To avoid large numbers in the calculation and hence a loss of precision, the time series are scaled to the time interval $[-1, 1]$.

The modelling is done using polynomials with degrees ranging from 2 to 12. Figure 1 shows examples for the degrees 4, 8 and 12. With increasing degree the models fits the data better, but also may over-fit the data.

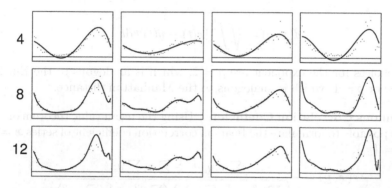

Fig. 1. Modelling of gene expression data with polynomials of different degrees

2.2 Similarity Measures

To calculate the similarity between polynomials, distance measures for functions have to be used. Usually these distance measures involve integration, which replace the sum in the equations for discrete measures. Polynomials are expandable into a Taylor-series, so that a large class of distance measures can be applied. For polynomials it is possible to calculate the anti-derivative, so that numerical integration can be avoided. Each polynomial is represented by the vector of its parameters, $(\alpha_0, \alpha_1, \ldots, \alpha_n)$. Therefore the sum of two polynomials, represented by (α_i) and (β_i), can be written as $(\alpha_0 + \beta_0, \alpha_1 + \beta_1, \ldots, \alpha_n + \beta_n)$ and the anti-derivative of (α_i) is represented by $(0, \alpha_0, 1/2\alpha_1, \ldots, 1/(n + 1)\alpha_n)$. The representations for the products and derivatives of polynomials can be found analogously. Therefore, the calculation of the integrals can be reduced to some simple algebraic operations on the $n + 1$ parameters α_i, which keeps the computational complexity for the distance measures low. The calculation of the derivative, the anti-derivative, the sum and the function value of polynomials takes $O(n)$ operations, the calculation of the product of two polynomials takes $O(n^2)$ operations. For the DC the calculation effort depends on the number of TPs T. Because the number of parameters has to be considerably smaller than the number of TPs (otherwise the models would be over-fit), the calculation of the similarity matrix takes less operations for the PMC than for the DC. Two distance measures are considered, the L_p distance and the distance based on a continuous Pearson correlation coefficient.

L_p Distance. The L_p distance is a standard distance in the space of continuous functions and is equivalent to the p-distance (Minkowski distance) for finite dimensional spaces, such as Euclidean distance,

$$d(\boldsymbol{x}, \boldsymbol{y}) = \sqrt{\sum (x_i - y_i)^2} \ , \tag{3}$$

or the Manhattan distance. Let $\boldsymbol{x} = x(t)$ and $\boldsymbol{y} = y(t)$ be continuous functions over the closed interval [a,b], then the L_p distance is given by

$$d(\boldsymbol{x}, \boldsymbol{y}) = \sqrt[p]{\int_a^b |x(t) - y(t)|^p \mathrm{d}t} \ . \tag{4}$$

Usual choices for the exponent are $p = 2$, which is analogous to the Euclidean distance or $p = 1$, which is analogous to the Manhattan distance.

Continuous Correlation Coefficient. Using the mean value theorem of calculus it is possible to formulate the Pearson correlation coefficient of series $\boldsymbol{x} = \{x_i\}$ and $\boldsymbol{y} = \{y_i\}$,

$$r(\boldsymbol{x}, \boldsymbol{y}) = \frac{\sum x_i y_i - \frac{1}{N} \sum x_i \sum y_i}{\sqrt{\left(\sum x_i^2 - \frac{1}{N}(\sum x_i)^2\right)\left(\sum y_i^2 - \frac{1}{N}(\sum y_i)^2\right)}} \ , \tag{5}$$

for integrable functions. Let $\boldsymbol{x} = x(t)$ and $\boldsymbol{y} = y(t)$ be continuous functions over the closed interval $[a, b]$ and $L = b - a$, then the correlation r can be calculated by

$$r(\boldsymbol{x}, \boldsymbol{y}) = \frac{\int_a^b xy \mathrm{d}t - \frac{1}{L}\int_a^b x \mathrm{d}t \int_a^b y \mathrm{d}t}{\sqrt{\left(\int_a^b x^2 \mathrm{d}t - \frac{1}{L}(\int_a^b x \mathrm{d}t)^2\right)\left(\int_a^b y^2 \mathrm{d}t - \frac{1}{L}(\int_a^b y \mathrm{d}t)^2\right)}} \ . \tag{6}$$

2.3 Grouping

The similarity matrices can be used with any standard clustering technique. To compare the method presented in this paper the Partitioning Around Medoids (PAM) [6] and two variations of hierarchical clustering algorithms were used, the average-linkage cluster analysis and the complete-linkage algorithm [3]. These methods are well-established and have been used for clustering microarray data with some success.

3 Data Set

The PMC is tested with a subset of the gene expression data of the malaria intraerythrocytic developmental cycle [2]. This subset was chosen, because a functional interpretation of the genes is known and can be used to assess the clusterings. It comprises 530 genes in 14 functional groups. The gene expression is measured in 48 TPs with 1 hour time differences. The data set contained 0.32% of missing data and had a low noise level, which has been verified through [2] and by visually plotting many of the functional groups.

4 Experiments

In every experiment the clustering is done with PAM, the average-linkage method and the complete-linkage method. For DC the methods were always applied to both the Euclidean and the correlation based similarity matrix. Polynomials of degrees from 2 to 12 were fitted to each variation of the data set and both the L_2 distance (4) and the correlation (6) were used for clustering. The following experiments were conducted.

1. The data set was clustered without any variations.
2. Normal distributed noise was added to the data. The standard deviation varied between 2% and 66% of the overall mean of the original gene expression values. The experiment was repeated 25 times.
3. The data set was changed by randomly deleting values. The number of missing values varied between 2% and 50%. The experiment was repeated 25 times.

To validate the clustering results, the weighted κ (WK) method [1,7] and quotient of homogeneity and separation (H/S) [4] were used. The WK is a similarity metric between clusters, with possible values between -1 and 1. The larger the WK value the better the agreement between the cluster results. For a clustering $\mathcal{C} = \{C_1, \ldots, C_K\}$ and a distance measure d H/S is given by

$$H(\mathcal{C}) = \sum_{k=1}^{K} H(C_k) = \sum_{k=1}^{K} \sum_{\boldsymbol{x} \in C_k} d(\boldsymbol{x}, \boldsymbol{r}_k)^2 \tag{7}$$

and

$$S(\mathcal{C}) = \sum_{1 \leq l < k \leq K} d(\boldsymbol{r}_j, \boldsymbol{r}_k)^2 \ , \tag{8}$$

where $\boldsymbol{r}_k = 1/n_k \sum_{\boldsymbol{x} \in C_k} \boldsymbol{x}$ are the cluster centres. A good clustering should have a low homogeneity value and a high separation value, hence a low H/S quotient. Because it is a quotient of sums of distances, H/S is always non-negative.

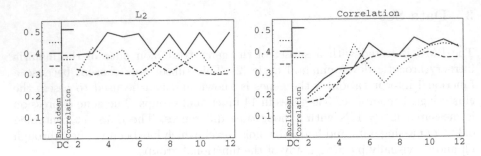

Fig. 2. Weighted κ results of DC and PMC for different similarity measures. PMC is shown depending on the degree of the model. Methods: Average Linkage – normal line, Complete Linkage – dotted line, PAM – dashed line.

5   Results

The results of the clustering of the unchanged data set for WK and H/S are shown in the Figures 2 and 3. The results of DC are marked at the left hand side of each figure, the results for PMC are shown in the main parts, one Figure for each similarity measure. The best DC result yielded the average-linkage clustering of the correlation based similarity matrix, $WK = 0.51$ and $H/S = 0.18$. The PMC showed the best results when applying the average-linkage clustering to L$_2$ similarity matrix. In particular, the models with even-numbered degrees from 4 to 10 showed constantly good values for WK (0.49) and H/S (0.16–0.19). The correlation based distance showed the best results for models with high degree, but the values for H/S were varied.

The best results in the noise experiment yielded the DC with the correlation based distance and average-linkage clustering and the PMC with L$_2$ distance and average-linkage clustering. These results are shown in Figure 4. The model of degree 4 is printed with a dotted line and the models of degree 2 and 3 are printed with dashed lines. The other lines show the models of degrees from 5 to 12.

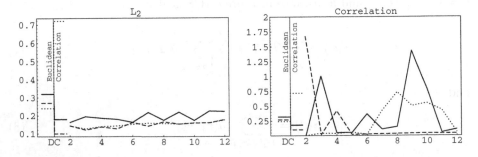

Fig. 3. Homogeneity/Separation results of DC and PMC for different similarity measures. PMC is shown depending on the polynomial degree. Methods: Average Linkage – normal line, Complete Linkage – dotted line, PAM – dashed line.

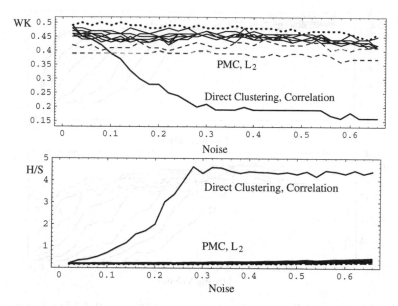

Fig. 4. Weighted κ and Homogeneity/Separation depending on the noise level (in parts of one). Cluster method: Average-linkage. Dotted line – model of degree 4, dashed lines – models of degree 2 and 3.

The PMC displayed excellent robustness, even for a high level of random noise, whereas the DC gave clearly poorer results as the noise level increased. The best choice of the model degree is 4. Models of degree 2 and 3 yielded poorer results for WK than other models, but showed consistently good H/S values. The H/S value for the model of degree 12 changed from 0.21 to 0.47, whereas the H/S value for the degree 2 model only increased from 0.17 to 0.25.

The unchanged data set has 0.32% missing values (not available values – NA). No single time series has more than 3 NAs. The best WK-results in the missing value experiment is yielded by the average linkage method with the correlation based distance for the DC and the L_2 distance for the PMC, see Figure 5 top. Good results were observed using the model with degree 4 (WK 0.49–0.4; dotted line), which worked with up to 46% NA. The DC result is printed with a bold line and aborts at 44% NA. It runs slightly below the degree 4 model (WK 0.48–0.37). The results for the polynomials of degree 2 (WK 0.39–0.38 up to 46% NA) and 3 are very stable as well (dashed lines), but showed poorer WK values. The polynomial models with higher degrees, printed with thin lines, failed sooner.

For the DC methods, the H/S could not be calculated for more than 2% of NAs in the data set. The reason is that every operation that has a NA as any operant returns a NA. For vector operations only these components are used that are not NA for every operant. As it can be seen from the definitions of the cluster centres and separation, eventually every value of a TP is used as an operant. If only one of these values is NA, than this TP is not used in the

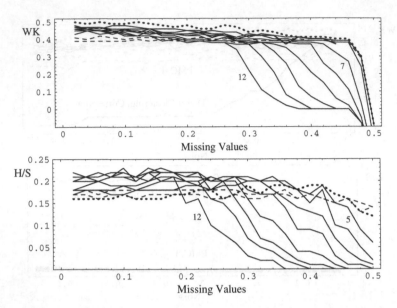

Fig. 5. Weighted κ and Homogeneity/Separation of DC and PMC (degrees 2–12) depending on the rate of missing values. Cluster method: Average Linkage. Bold line – DC, dotted line – model of degree 4, dashed lines – models of degree 2 and 3, models of degrees 5-12 – thin lines. The H/S-value could not be calculated for the DC.

calculation of the separation. In other words, if every TP has at least one NA value the separation will be NA. If p is the ratio of missing values, T the number of TPs and N the number of genes, then the probability that

- a TP contains no NA is $(1 - p)^N$,
- a TP contains at least one NA is $1 - (1 - p)^N$,
- every TP contains at least one NA is $(1 - (1 - p)^N)^T$,
- at least one TP contains no NA is $1 - (1 - (1 - p)^N)^T$.

The last expression is the probability that the separation is not NA. The likelihood to calculate a non-NA H/S value for the considered data set is 0.9999 for 0.32% NAs, 0.2 for 1% NAs and 0.001 for 2% NAs.

On the other hand, the model based clustering allows H/S to be calculated for any number of NAs in the data, see Figure 5 bottom. Moreover, the H/S value stays stable for low degree models, between 0.16 and 0.19 for the degree 4 model, and for up to 46% missing values in the data set (method: L_2, average linkage).

Summary. The DC and the PMC showed similar results for the good quality data set. Higher degree models showed better results than models of lower degree. With increasing noise the quality of DC dropped significantly, whereas PMC showed excellent robustness. The stability of DC and PMC against missing values is about the same whilst low degree models showed slightly better results.

Homogeneity and separation could not be calculated for DC. The PMC using a model of degree 4 showed the best overall result.

6 Conclusions

This paper has introduced a novel clustering method, polynomial clustering (PMC), for dealing with missing and noisy data. It has demonstrated that PMC yielded consistently good results for different levels of noise and missing values, whereas traditional clustering methods only generated good clusters when applied to high quality data. PMC showed the best robustness when using models with lower degrees, but these were too simple to reflect the features of the data set and hence yielded poorer clustering results. The models with higher degree showed better results but poorer robustness. PMC is easy to calculate. It can be used with a wide range of similarity measures and any standard clustering method.

The optimal degree of the polynomial was calculated empirically in order to balance complexity of the data with over-fitting. Future work will involve looking into ways to determine the degree automatically. The tests will be extended to other real world data sets, a wider range of distance measures and clustering techniques. It is also intended to refine the approximation technique and to analyse the modelling with other classes of functions.

References

1. Altman, D. G.: Practical Statistics for Medical Research. Chapman and Hall (1997)
2. Bozdech, Z., Llinás, M., Pulliam, B. L., Wong, E. D., Zhu, J., DeRisi, J. L.: The Transcriptome of the Intraerythrocytic Developmental Cycle of Plasmodium falciparum. PLoS Biology 1 (2003) 85–100
3. Eisen, M. B., Spellman, P. T., Brown, P. O., Botstein, D.: Cluster analysis and display of genome-wide expression patterns. Proc Natl Acad Sci USA 95 (1998), 14863–14868.
4. Hand, D. J., Mannila, H., Smyth, P.: Principles of Data Mining. MIT Press (2001)
5. Jain, A. K., Murty, M. N., Flynn, P. J.: Data Clustering: A Review. ACM Computer Surveys 32 No. 3 (1999) 264–323
6. Kaufman, L., Rousseeuw, P. J.: Clustering by means of Medoids, Y. Dodge (ed.), Statistical Data Analysis based on the L1-Norm. North-Holland, Amsterdam (1987) 405–416
7. Kellam, P., Liu, X., Martin, N., Orengo, C., Swift, S., Tucker, A.: Comparing, Contrasting and Combining Clusters in Viral Gene Expression Data. Proceedings of the IDAMAP2001 workshop, London (2001) 56–62
8. Lichtenberg, G., Faisal, S., Werner, H.: Ein Ansatz zur dynamischen Modellierung der Genexpression mit Shegalkin-Polynomen (An Approach to Dynamic Modelling of Gene Expression by Zhegalkin Polynomials). at – Automatisierungstechnik 53 12 (2005) 589–596

9. Ralston, A.: A First Course in Numerical Analysis. McGraw-Hill (1965)
10. Spellman, P. T., Sherlock, G., Zhang, M. Q., Iyer, V. R., Anders, K., Eisen, M. B., Brown, P. O., Botstein, D., Futcher, B.: Comprehensive Identification of Cell Cycle-regulated Genes of the Yeast Saccharomyces cerevisiae by Microarray Hybridization. Molecular Biology of the Cell **9**, 3273–3297, URL http://cellcycle-www.stanford.edu
11. Stekel, D.: Microarray Bioinformatics. Cambridge University Press (2003)
12. Vinciotti, V., Liu, X., Turk, R., de Meijer, E. J., 't Hoen, P. A. C.: Exploiting the full power of temporal gene expression profiling through a new statistical test: Application to the analysis of muscular dystrophy data. BMC Bioinformatics **7** 183 (2006)
13. Wit, E., McClure, J.: Statistics for Microarrays. John Wiley (2004)

A Hybrid Grid and Its Application to Orthologous Groups Clustering*,**

Tae-Kyung Kim[1], Kyung-Ran Kim[2], Sang-Keun Oh[2],
Jong-Hak Lee[3], and Wan-Sup Cho[2]

[1] Dept. of Information Industrial Engineering, Chungbuk National University, 361763
Cheongju, Chungbuk, Korea
tkkim@chungbuk.ac.kr
[2] Dept. of Management Information Systems, Chungbuk National University,
361763 Cheongju, Chungbuk, Korea
k2ran@chungbuk.ac.kr, ok@lanlinux.com, wscho@chungbuk.ac.kr
[3] Division of CICE, Catholic University of Daegu, 712702
Gyeongsan, Gyeongbuk, Korea
jhlee11@cu.ac.kr

Abstract. Orthologous groups are useful in the genome annotation, studies on
gene evolution, and comparative genomics. However, the construction of
orthologous groups is difficult to automate and takes so much time as the number
of genome sequences increases. Furthermore, it is not easy to guarantee the accuracy of the automatically constructed orthologous groups. We propose an automatic orthologous group construction system for a large number of genomes. A
hybrid grid computer system, consisting of 40 PCs, has been devised for fast
construction of the orthologous groups from large number of genome sequences.
The grid system constructs orthologous groups for 89 complete prokaryotes genomes just in a week (it takes 8 months on a single computer system). Furthermore, the system provides good extensibility for adopting new genomes in the
existing orthologous groups. In the real experiment of the orthologous group
constructions, more than 85% of the constructed orthologous groups coincide
with those of KO (KEGG Ortholog) and COGs (Clusters of Orthologous Group
of Proteins). Note that KO and COGs have been constructed manually or semi-
automatically at the sacrifice of the extensibility for newly completed genomes.

1 Introduction

Recently, complete genomes of various species are sequenced rapidly by using advanced biological tools [19]. One of the most important issues for the sequence data is
to predict the functions of the genes. Orthologous groups are the most useful concept
to estimate the function of the genes.

---

* This research was supported by the Program for the Training of Graduate Student in Regional
Innovation which was conducted by the Ministry of Commerce, Industry and Energy of Korea Government.
** This work was supported by the Regional Research Centers Program of the Ministry of Education & Human Resources Development in Korea.

M.R. Berthold, R. Glen, and I. Fischer (Eds.): CompLife 2006, LNBI 4216, pp. 11–20, 2006.
© Springer-Verlag Berlin Heidelberg 2006

An *orthologous group* is a collection of the genes in two or more species that have evolved from a common ancestor [8]. Ancestral genes have been distributed into the genome of the various organisms by the speciation during the biological evolution processes. Therefore, the distributed genes of each organism have the biologically same function and sequence pattern. Sequence comparisons are required to identify the orthologous relationships among the species. In contrast, *paralogs* are genes in different species that are originated by the duplication of a gene in a common ancient genome [8]. Paralogs evolve new functions whereas orthologs retain the same function in the course of the evolution, even if these are related to the original one.

Orthologous groups for large number of genomes are critical for reliable prediction of the function of the genes in newly sequenced genomes. However, construction of the orthologous groups are difficult to automate and takes great sequence comparison time as the number of complete genomes increases [3]. In addition, it is not easy to accommodate new genome sequences to the existing orthologous groups. Furthermore, evaluation of the functional accuracy for the constructed orthologous groups is a difficult task.

To overcome these limitations, we propose an automatic orthologous clustering system having a good performance, extensibility, and accuracy. We first devised a hybrid grid computer system with a new clustering algorithm to speed-up huge number of pair-wise sequence comparisons [1, 2, 5]. The clustering algorithm also supports good extensibility in the accommodation of new genome sequences. The algorithm exploits the cut-off value in order to raise the functional accuracy of orthologous groups.

The grid system clustered orthologous groups for 89 complete prokaryote genomes just in a week. Note that it takes more than 8 months if we use a single server. To verify the functional accuracy of the generated orthologous groups, we compared them with other well-known orthologous databases such as COGs (Clusters of Orthologous Group of Proteins) [9, 10, 11, 12, 13] and KO (KEGG Ortholog) [4, 15]. The result   shows that more than 85% of the result coincides with those of the KO and COGs. Note that KO and COGs have been constructed manually or semi-automatically at the sacrifice of the extensibility for newly completed genomes.

The paper is organized as follows. In Section 2, we introduce related work. In Section 3, we describe the method and algorithm for orthologous clustering in detail. In Section 4, we present the experimental results and comparison with other approaches. In Section 5, we conclude the paper and describe future work.

2   Related Work

In this section, we introduce conventional well-known orthologous databases such as COGs and KO. We then present necessity of grid computing system in the clustering of orthologous groups.

Two representative databases, COGs(Clusters of Orthologous Group of Proteins) by NCBI [19] and KO(KEGG Orthology) by KEGG[20], have been constructed for the orthologous groups and they provide great value in the genome annotation, studies on gene evolution, and comparative genomics.

COGs [17] released in 1999 is a second database constructed on the basis of sequence databases in NCBI. They provide orthologous groups for 66 prokaryotes and 7 eukaryotes. Although COGs have the advantage of classifying many complete sequences [9, 10, 11, 12, 13], they require manual work in the whole processes such as removing the paralogs and false positive. Note that as the number of the complete genomes increases dramatically, the limitation of the manual work makes it difficult to accommodate new genomes into the existing orthologous database.

KO [15] consists of the orthologous genes extracted from metabolic pathways and regulatory pathways. KO guarantees higher accuracy than COGs as classifying the orthologous groups by manual work with the sophisticated experiment. However, they can not accommodate new genomes easily.

In general, there are three significant issues in the construction of the orthologous database; *accuracy, construction time, and clustering algorithm with extensibility*. To overcome these challenges, we adopt INPARANOID, grid computing, and novel orthologous clustering algorithm, respectively.

- INPARANOID[8] accurately detects complex orthologous genes among protein sequences in only two species. Especially, INPARANOID is helpful to remove false-positive genes and detect functionally similar in-paralogs.
- Grid computing has been primarily adapted in scientific problems to solve large scale of data analysis such as sequence comparisons [6, 7, 14, 16] for large number of genomes. We developed a grid computer system consisting of 40 PCs spread on Internet to solve large scale data analysis.
- In the clustering of orthologous groups, there are so many tasks to process and each task takes so much time to be completed. Proposed clustering algorithm automates these complex processes into one package and provides a good extensibility in the accommodation of new genomes.

3 Clustering Orthologous Groups on a Hybrid Grid

In this section, we define notations in order to describe proposed orthologous clustering algorithm on a hybrid grid computing environment. We then discuss the orthologous clustering algorithm including sequence comparisons.

3.1 Notations

We first define notations in Table.1 to be used in the sequence comparisons of the orthologous clustering algorithm.

Table 1. Notations in the analysis

Symbols	Meaning
G	A set of target genomes in the orthologous clustering algorithm.
N	The number of species in the orthologous clustering.
G_i	The genome of the species i ($G_i \in G$).
N_{Gi}	The number of genes in the genome G_i.
$g_{i,j}$	The jth gene of the genome G_i.

Furthermore, there are three factors related to the grid performance evaluation: *SG*, *EG*, and *CG*. These factors give us the objective evaluation for our grid system.

- *SG* : Grid speedup, i.e., (total time taken by an application in a single node)/ (total time taken by an application in the grid system).
- *EG* : Grid performance efficiency, i.e., (*SG*)/(total number of grid nodes). Maximum *EG* is 1.
- *CG* : Grid maintenance cost.

We show the experimental results for these factors between our hybrid grid approach and middleware-only one.

3.2 Sequence Comparisons on a Hybrid Grid

To construct orthologous groups for multiple genomes, we should first find genes that satisfy the reciprocal best hit between two genomes. In this phase, we adopt INPARANOID to automatically remove out-paralogs. Note that out-paralogs have the similar sequence pattern but different functions. Although INPARANOID provides sophisticated results of 84% accuracy compared with phylogenetic approach, performance is still a hot issue due to the exhaustive comparisons. For two genomes G_i and G_j, *a genome-genome comparison* requires $2 \times (N_{Gi} \times N_{Gj})$ gene comparisons, each of which takes from 20 minutes to 4 hours. Thus the number of gene comparisons for *n-genomes comparisons* can be represented by Equation (1).

$$_NC_2 \times (\text{genome-genome comparison}) = {}_NC_2 \times 2 \times (N_{Gi} \times N_{Gj}) \tag{1}$$

For 89 prokaryotes genomes, it needs $_{89}C_2 = 3,916$ genome-genome comparisons. In the real experiment, it takes 8 months to complete the comparisons. To solve this performance problem, we constructed a *hybrid grid computing environment* that is a combination of middleware-only approach and hardware approach based on network booting [21]. Fig.1 shows the hybrid grid computing environment with 40 personal computers. Since each genome-genome comparison is independent task, we tried to apply such problems to the grid computing.

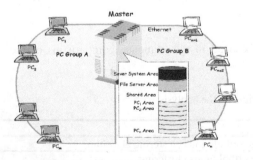 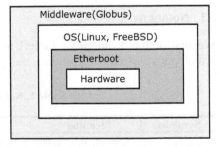

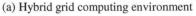

(a) Hybrid grid computing environment   (b) Node structure of hybrid grid

Fig. 1. Hybrid grid computing system

In Fig.1 (a), each PC boots LINUX from the master node and does not use his local disk while he is a member of the grid. Fig. 1 (b) shows the SW structure of the nodes. *EtherBoot* layer is responsible for the LINUX booting via the master node.

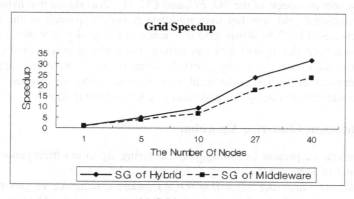

(a) Grid speedup

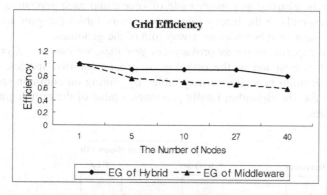

(b) Grid efficiency

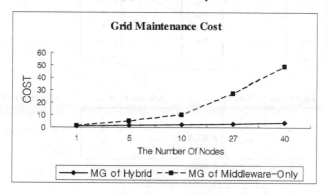

(c) Grid maintenance cost

Fig. 2. Hybrid Grid evaluation results

Since each node in the hybrid grid boots LINUX from master node before it becomes a member of the grid, we can have a unified processing environment regardless of node's heterogeneity. Therefore, it is very efficient in aspect of grid construction, management, and maintenance. Fig. 2 shows the performance evaluation results for the grid system in aspects of the *SG*, *EG*, and *CG*. Fig. 2(a) shows that hybrid grid has more proportional grid speedup than middleware-only approach as the number of nodes increases. Fig. 2(b) shows grid efficiency (*EG*): grid performance efficiency. Furthermore, note that hybrid grid can reduce maintenance cost dramatically as we only control the master node (Fig.2 (c)). In contrast, middleware-only approach has low maintenance efficiency as the number of the nodes increases because we have to install the middleware one by one and manage each node individually.

3.3 Orthologous Clustering Algorithm

In this section, we present an orthologous clustering algorithm from genome-genome comparisons in Section 3.2.

Orthologous Clustering Algorithm (OCA) makes a sequence of genome-genome comparisons to complete orthologous groups of entire genomes. The key idea is that each genome is regarded as a unary table of genes, and each *genome-genome comparison* corresponds to the binary join of two unary tables (or genomes). Similarly, *n-genomes comparison* becomes an n-way join of the genomes.

In our real experiment for 89 prokaryotes genomes, we can get $_{89}C_2=3,916$ binary tables, which correspond to the orthologous groups for two species. From binary tables, the algorithm generates ternary tables by joining on common species. By repeating this task, the algorithm finally generates a table of the orthologous groups for entire genomes.

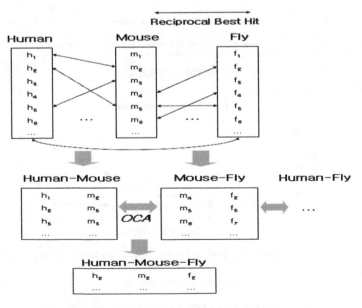

Fig. 3. Orthologous groups construction – An Example

Example 1: Orthologous groups for *human*, *mouse*, and *fly* can be constructed as follows. Let assume h_i, m_i, and f_i be the genes of *human*, *mouse*, and *fly* respectively.

(1) Orthologous groups for *human* and *mouse* can be generated by joining two unary tables **Human** and **Mouse**, and the results are stored in the **Human-Mouse** table.

(2) Similarly, we can get **Mouse-Fly** table and **Human-Fly** table.

(3) Orthologous groups for *human*, *mouse*, and *fly* can be generated by the joining of **Human-Mouse** and **Mouse-Fly** tables via the common attribute (here, Mouse).

Fig.3 shows this procedure in detail. Detailed description of the algorithm *OCA* is attached in *Appendix*.

4   Real Experiment

In this section, we compare proposed system with other well-known orthologous databases. To do this, we first select 20 prokaryote organisms, and compare the

geneid10	geneid11	geneid12	geneid13	geneid14	geneid15	geneid16	geneid17	geneid18	geneid19
15673268	13472662	NULL	NULL	15680229	16332153	15642373	15839299	11498558	15669542
15673800	13470922	15676855	15603816	15599430	16330044	15640421	15839017	NULL	15668900
15673782	13470542	15676860	15603682	15599466	16329957	15640355	15839222	11499470	15669229
15674242	13470549	15676866	15603221	15597267	16330914	15640388	15839218	11499478	15669237
15673781	13470543	15676861	15603601	15599465	16330858	15640356	15839221	11499472	15669231
15673845	13476816	15677669	15603527	15599756	16330137	15640701	15839089	11498241	15669137
15674155	13471197	15676101	15602683	15599029	16332082	15642499	15836739	11499806	15669196
15672544	13472362	15677197	15602294	15598334	16331511	15642773	15837569	NULL	NULL
NULL	13476996	15677103	15603843	15597000	16331433	15641922	15837791	11497976	15669607
15673786	13470354	15677445	15603152	15596100	16331117	15640567	15836729	11499836	15668744
15673090	13473815	15676664	15601928	15595964	16332331	15642459	15838034	NULL	NULL
15673607	13473220	15677313	15603503	15595745	16329902	15640500	15838530	NULL	15668862
15672470	13471011	15677663	15601099	15598036	16332330	15642243	15836009	NULL	NULL
15673205	13474469	15677026	15603490	15595550	16332267	15640060	15836704	11498619	15669462
15673947	13476902	15676377	15602048	15596160	16330502	15641179	15838454	11498525	15669750
30024031	13475114	15676837	15602285	15597123	NULL	15641700	15838863	NULL	15669665
15673345	13472276	15677690	15603370	15599950	16331605	15642386	15837709	11498873	15669568
15672897	13472996	15677388	15603004	15599599	16331549	15642391	15837400	NULL	NULL

Fig. 4. Clustered results in OGT for 20 species

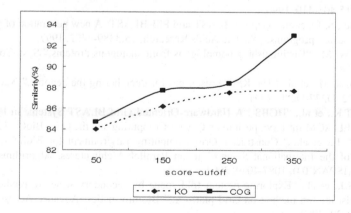

Fig. 5. Similarity comparison with KO and COG

orthologous groups constructed by our system with the well-known orthologous databases KO[18] and COGs[9,10,11,12,13]. Fig.4 shows the clustered results for the 20 species. Each row means the orthologous group of the genes that are considered functionally similar. We have 13,211 orthologous groups in the experiments.

Fig.5 shows the similarity of our orthologous database compared with KO and COGs respectively. We can get over 84% similarities when score-cutoff is greater than 50. Note that as the score-cutoff increases, we can get higher similarity; in COGs, 85% similarity when score-cutoff is 50, and 93% when score-cutoff 350, respectively. As a conclusion, proposed clustering algorithm provides an orthologous database that is similar to the KO and COGs in spite of full automation.

5 Conclusion

Construction of orthologous groups from many complete genomes is a significant task to predict the functions of the new genes. However, constructing orthologous groups has troubles in the aspects of the time, accuracy, and automation. In this paper, we proposed an automatic orthologous clustering system by using a hybrid grid technology. Proposed system provides high performance for large number of genome sequences with a reasonable accuracy in the automatic fashion. Furthermore, new genomes can be easily accommodated in the existing orthologous database. Hybrid grid computing system supports high speed analysis for huge amount of sequence data; we cut down sequence analysis time from 8 months to just one week. Real experiment for 20 genomes shows that we can get more than 85% similarity compared with well-known orthologous databases such as KO and COGs in spite of full automation. In future, we plan to provide the orthologous groups for overall complete genomes, and accommodate new genomes in a minimum overhead.

References

[1] Altschul, S. F., et al., "Basic Local Alignment Search Tool," Journal of Molecular Biology, 215:403-410, 1990.

[2] Altschul, S. F., et al., "Gapped BLAST and PSI-BLAST: A new generation of protein database search programs," Nucleic Acids Research, 25:3389-3402, 1997.

[3] Fitch, W. M., "Distinguishing homologous from analogous proteins," Syst. Zool. 19, 99-113,1970

[4] Kanehisa, M., et al., "The KEGG resources for deciphering the genome," Nucleic Acids Res., 32, D277-D280, 2004

[5] Kim, T.K., et al., "HGBS : A Hardware-Oriented Grid BLAST System,"In Proc. of the 5th IEEE/ACM Int'l Symposium on Cluster Computing and the Grid, BioGrid 2005.

[6] Kuo,Y. L., et al., " Construct a Grid Computing Environment for Bioinformatics," In Proc. of the International Symposium on Parallel Architectures, Algorithms and Networks(ISPAN'04), 1087-4089, 2004.

[7] Lee, S.J., et al., "Exploring protein fold space by secondary structure prediction using data distribution method on Grid platform," Bioinformatics, Advance Access published on July 29, 2004.

[8] Remm, M., et al., "Automatic clustering of orthologs and in-paralogs from pairwise species comparisons," J. Mol. Biol. 314: 1041-1052, 2001

[9] Roman L. Tatusov, et al., "The COG Database: A Tool for Genomic-Scale Analysis of Protein Function and Evolution," Nucleic Acids Res. 28, 33-36, 1999.

[10] Tatusov, R., et al., "A genomic perspective on protein families," Science 278, 631-637. 1997

[11] Tatusov, R., et al., "The COG database: new developments in phylogenetic classification of proteins from complete genomes," Nucleic Acids Res., 29, 22-28, 2001

[12] Tatusov, R. L., et al., "The COG database: a tool for genome-scale analysis of protein functions and evolution," Nucleic Acids Res., 28, 33-36, 2000.

[13] Tatusov RL, et al., "The COG database: an updated version includes eukaryotes," BMC Bioinformatics, 11;4:41, 2003

[14] Wang, L., et al., "Biogrid Computing Platform: Parallel computing for protein alignment analysis," HPC Asia'02, Bangalore, India, 2002.

[15] Yamanishi, Y., et al, "Extraction of Organism Groups from Whole Genome Comparisons," Genome Informatics 14, 438-439, 2003.

[16] Yong-Meng TEO, et al, "GLAD: a system for developing and deploying large-scale bioinformatics Grid," Bioinformatics, Advance Access published on September 23, 2004

[17] COGs official homepage, http://www.ncbi.nlm.nih.gov/COG/

[18] KO official homepage, http://www.genome.jp/kegg/ko.html

[19] NCBI, http://www.ncbi.nlm.nih.gov/

[20] KEGG, www.genome.ad.jp

[21] EtherBoot Project, http://etherboot.sourceforge.net/

Appendix: Detail OCA Description

The result of the *Seq_align (G$_i$, G$_j$)* for 89 prokaryotes consists of 3,916 tables, each of which has the table form $T_{ij}[g_{ip}, g_{jp}, score]$. Here, g_{ip} and g_{jp} are regarded as the functionally similar genes and score is the similarity score between two gene sequences; we apply score-cutoff 50. Score-cutoff can be used in controlling sensitivity of the orthologous groups. Based on these results of reciprocal best hit genes between two genomes, proposed algorithm performs the orthologous clustering automatically.

Orthologous Clustering Algorithm (OCA) constructs orthologous groups for the multiple species. It consists of two steps; initial clustering step for arbitrary 3 species and then adding the rest species one by one. The initial clustering can be done as shown in Fig.6.

```
Begin
    T₁₂ J(1)  T₁₃  ⇒  OGT₁₂₃ ;  OGT₁₂₃ J(2)  T₂₃  ⇒  OGT₁₂₃
    OGT₁₂₃ J(3)  T₂₃  ⇒  OGT₁₂₃ ;  T₁₂ J(2)  T₂₃  ⇒  OGT₁₂₃
    OGT₁₂₃ J(3)  T₁₃  ⇒  OGT₁₂₃ ; T₁₃ J(3)  T₂₃  ⇒  OGT₁₂₃
End
```

Fig. 6. Orthologous clustering for initial three species.

Let T_{ij} ($[g_{ip}, g_{jp}, score]$) be the result of the *Seq_align (G$_i$, G$_j$)*. Here, $J(a)$ means join operation with condition on the attribute a. OGT (Orthologous Groups Table) is composed of N attributes and contains the orthologous genes as an intermediate result of the clustering algorithm.

After joining two tables T_{12} and T_{13}, the results are inserted into OGT_{123}. The joined records of the two tables can not be contained in the other groups as checking the flag. In case of join with OGT, the new groups are inserted. After construction of the initial cluster, we add the other species by the algorithm of the following Fig.7.

```
Input : m (The number of species)

Begin                              for(i=1;i<m-2;i++)
for(i=1; i<=m-1; i++)                for(j=i+1;j<=m-1;j++)
  OGT J(i) T_im⇒ OGT;                  T_ij J(i) T_im ⇒ OGT
end for                                for(k=1;k<=m-1;k++)
                                         if(j!=k)
for(i=2; i<=m-1; i++)                      OGT J(m) T_km ⇒ OGT(insert)
  OGT J(m) T_im⇒ OGT;                    else if(j==k)
end for                                    OGT J(k) T_km ⇒ OGT(update)
                                         end if
for(i=1;i<=m-2;i++)                      end for
  for(j=i+1;j<=m-1;j++)               end for
    T_ij J(i) T_im ⇒ OGT             end for
    for(k=1;k<m-1;k++)               for(i=1;i<=m-2;i++)
    if(i!=k)                           for(j=i+1;j=m-1;j++)
      OGT J(m) T_km ⇒ OGT(insert)        T_im J(m) T_jm ⇒ OGT
    else if(i==k)                        for(k=j+1;k<=m-1;k++)
      OGT J(k) T_km ⇒ OGT(update)          OGT J(m) T_km ⇒ OGT
    end if                               end for
    end for                            end for
  end for                            end for
end for                            End
```

Fig. 7. Algorithm to add the rest species based on initial clustering of Fig.6

By applying Fig.7, all tables are joined each other, and OGT and the joined results are inserted or updated into OGT. As a result, almost 99% genes of each T_{ij} are clustered into OGT and the rests are regarded as the species-specific genes. The clustering work needs a lot of join operations that cause the performance to be delayed remarkably. To speed up the join operation, we basically applied indexing technique for condition attributes existing DBMSs support and plan to apply this algorithm on grid.

Promoter Prediction Using Physico-Chemical Properties of DNA

Philip Uren, R. Michael Cameron-Jones, and Arthur Sale

School of Computing, Faculty of Science, Engineering and Technology,
University of Tasmania, Hobart and Launceston, Tasmania, Australia
{Philip.Uren, Michael.CameronJones, Arthur.Sale}@utas.edu.au

Abstract. The ability to locate promoters within a section of DNA is known to be a very difficult and very important task in DNA analysis. We document an approach that incorporates the concept of DNA as a complex molecule using several models of its physico-chemical properties. A support vector machine is trained to recognise promoters by their distinctive physical and chemical properties. We demonstrate that by combining models, we can improve upon the classification accuracy obtained with a single model. We also show that by examining how the predictive accuracy of these properties varies over the promoter, we can reduce the number of attributes needed. Finally, we apply this method to a real-world problem. The results demonstrate that such an approach has significant merit in its own right. Furthermore, they suggest better results from a planned combined approach to promoter prediction using both physico-chemical and sequence based techniques.

Keywords: promoter prediction, support vector machine, SVM, physico-chemical, classifier, DNA, transcription.

1 Introduction

In-silico eukaryotic promoter recognition is known to be a difficult problem [1]. Objective statements about the current state of the art in terms of promoter recognition are complicated by the wide selection of metrics used for assessing performance. Moreover, an optimal trade off between sensitivity and specificity is not immediately apparent. In some applications, a high sensitivity is valued – i.e. it is best to find as many actual promoters as possible and a relatively high false positive rate is tolerable. In contrast, in other situations the specificity may be more important, particularly where it is expensive to validate predictions. As an indication of the progress to date in the field, Bajic, Tan et al. [2] report that none of the programs they tested achieved a combined sensitivity and specificity greater than 65%.

The purpose of this paper is to examine how effectively physico-chemical properties of DNA can be used to predict the location of promoters within the human genome. Previous studies have demonstrated that promoters exhibit distinct patterns in terms of these properties [3-7]. Physical properties of DNA have also been shown to be important in terms of a biological understanding of the mechanisms of transcription – for

M.R. Berthold, R. Glen, and I. Fischer (Eds.): CompLife 2006, LNBI 4216, pp. 21–31, 2006.

example [8-10]. Approaches to separating promoters from non-promoters in E-coli using physico-chemical properties have met with success [11, 12].

Ohler, Niemann et al. proposed an approach for incorporating them into their promoter recognition program, McPromoter [13]. They demonstrated that this reduced the false positive rate on a given test set by about 30%. In contrast to the approach we employ within this work, they computed the mean value for a given model within a segment of the instance. The segmentation was based upon the sequence alone.

It has also been shown in previous work that an encoding of sequence data using structural models can be more efficient than a sequence based encoding. Using a single model Baldi, Chauvin et al [14] demonstrated that similar accuracy to a sequence based approach was possible but with only about a fourth of the attributes required. We explore a different approach to reducing the size of the representation.

In their important examination of the application of physico-chemical properties to the clustering and classification of promoters, Florquin, Saeys et al. [4] examined how effectively certain properties could be used to discriminate between promoters and non-promoters. We aim to extend their exploration in several significant directions. Firstly, they examined the application of a single model at a time. We explore the application of multiple models simultaneously and assess their classification performance, demonstrating an improved accuracy over any single model. Secondly, we examine which models are of importance within which segments of the promoter. We use this information to demonstrate how comparable accuracies can be achieved with fewer attributes, reducing computational time. Finally, we explore the application of this approach within a more realistic scenario – the classification of a contiguous segment of human DNA from chromosome 21, of length approximately 10Mbps. We measure the results by means of the approaches used by Bajic, Tan et al. [2] and show that this technique has merit in its own right for promoter prediction.

To the authors' knowledge, this is the first large-scale application of a promoter prediction method that uses *only* physico-chemical properties. We demonstrate that they are actually quite effective at picking up promoters on their own. Although we are not advocating an abandonment of sequence related techniques, we present these results as further impetus to re-examine the abstraction of DNA (and biological information in general). That is, DNA is often represented in computational areas as simply a string of characters. Results such as those presented within this paper suggest that its properties as a complex molecule are not only useful, but essential for various forms of computational modelling and biological understanding.

2 Materials and Methods

2.1 Datasets and Physico-Chemical Properties

We make use of the publicly available DBTSS (which can be accessed at: http://dbtss.hgc.jp/) for the location and sequence data of human promoters [15-17]. For training the classifier, we used this dataset as the positive instances and randomly selected an equal number of negative instances from the human genome. When testing on chromosome 21, we used a modified version of this dataset for training, which excluded all data from chromosome 21.

Based upon the recommendations of Florquin, Saeys et al. [4], we selected six models for describing the physico-chemical properties of the DNA sequences. These were: A-philicity [18], DNA bending stiffness [19], DNA denaturation [20], duplex disrupt energy [21], nucleosome position preference [22] and propeller twist [23].

When training the classifier, all sequence data is already in uniformly sized instances (we use an instance size of 150bp with 100bps downstream and 50bps upstream of the TSS). From the raw sequence information, we evaluate each of the models listed above. This produces 6 sequences, of size 149 or 148 for a di- or tri-nucleotide model respectively, which we then smooth with a window of size 10 and step of 1. After smoothing, the sequences are of length 139 or 140. These 6 sequences are then concatenated and represent a single instance as presented to the classifier.

When scanning a contiguous segment of DNA for promoters, we first split the sequence into segments of 150bps in size, using a step of 1. That is, if the sequence is *n* bps in size, we convert this to *n-150* segments of 150bps in size. Put another way, each segment differs from its nearest neighbour by only 2 bps (one at each end). The processing of these segments then proceeds as described above.

When applying attribute pruning, we train the classifier on the dataset as described above and examine the weights associated with each of the attributes. Attributes with low weights are then removed from the dataset. In some cases we prune all attributes with a weight below a certain threshold. In other cases the top *n* weighted attributes are retained and the remainder are pruned. We specify which of these two approaches are being employed when we present the results.

2.2 Classifier and Post-processing

We employ a support vector machine using a linear kernel, specifically the implementation present in WEKA [24-26] using the default settings, to classify instances as either promoter or non-promoter. We also post process the output of the support vector machine. The motivation for this is explained in more detail within the results section. The approach taken here is to locate runs of positive predictions within a window of a certain size and exceeding a threshold of a given percent of positives. For all the results presented within this paper, we set this to be a window of size 400bps containing 85% positives or more. Different values for these parameters result in different balances of sensitivity and specificity. This produces a window of varying length describing the promoter region. For the purposes of comparing TSS prediction position, we examined taking the start, middle and end of this window. Given the metrics we are using, we determined that taking either the middle or end produced the same results, but taking the start was markedly worse.

Due to the fact that training times are quite long for some of the approaches we present herein (most notably the combined model with 839 attributes), we do not perform a ten-fold cross validation as is often done. Instead, we present the result of training the model on increasing segments of the data (from 1% up to 50%) and show the result of testing on the left over data (from 99% down to 50%). In each case the sets are cumulative. That is, the 5% dataset contains all of the 1% dataset; the 10% dataset contains all of the 5% and the 1%, etc.

Using the same approach as Bajic, Tan et al. [2], we count a true positive (TP) when a prediction falls within 2000bps of a TSS. We also aggregate predictions that are within 1000bp of each other. In evaluating the performance of the classifier we make use of five metrics throughout the paper. These are *Sensitivity* – abbreviated to Se (the percentage of positives that are correctly identified), *Positive Predictive Value* – abbreviated to *PPV* (the percentage of positive predictions that are correct), *Accuracy* (the percentage of predictions that are correct, be they negative or positive), *True Positive cost* (the number of FPs required to achieve a TP) and finally *specificity* – abbreviated to Sp (the percentage of negatives correctly identified). Expressed more succinctly, Se = TP/(TP+FN), PPV = TP/(TP+FP), Accuracy = (TP+TN)/(TP + FP + FN + TN), TP cost = FP / TP, and Sp. = TN/(TN+FP).

3 Results and Discussion

3.1 Single vs. Combined Model

Within this section, we examine the results of comparing the classification performance of a single model with that of a combined model, using the six models suggested as highly discriminative by Florquin, Saeys et al. [4]. We use the full promoter dataset, as described in the section materials and methods. The results are summarised in Table 1, using a simple accuracy (i.e. percentage of correct predictions) metric.

As can be seen, there is an approximately 2% increase in accuracy achieved by the combined model over the best of the single model accuracies. Although the difference is not great, the 2% improvement would account for hundreds of instances when considered from the perspective of a genome wide scan, or similar magnitude. This is clearly a significant improvement, although it comes at the cost of computational time. Using half the dataset for training (approx 8,000 instances), a single model SVM takes only approximately ten minutes to train. In contrast, the combined model,

Table 1. The accuracy of the SVM in separating promoter sequences from non-promoter sequences in a 50% positive/negative dataset. Promoter sequences are taken from DBTSS and non-promoter sequences randomly selected from the human genome.

Training set proportion (%)	1	5	10	15	20	30	40	50
A-philicity	81.4	84.8	84.7	85.0	85.4	86.1	86.5	86.6
DNA Bending Stiffness	82.9	84.8	84.8	84.8	85.1	85.6	85.9	86.2
DNA Denaturation	80.4	84.2	84.5	85.0	85.1	85.5	86.0	86.2
Duplex Disrupt Energy	85.1	86.9	87.5	87.7	87.9	88.4	88.7	88.7
Nucleosome Positional Pref.	83.2	86.5	86.5	86.7	87.0	87.2	87.3	87.7
Propeller Twist	85.1	87.0	87.3	87.5	87.6	88.0	88.3	88.4
COMBINED MODELS	81.5	87.4	88.3	88.9	89.3	90.1	90.4	90.7

using the same size training set takes approximately two hours. Putting this in perspective, however, the SVM need only be trained once and the difference in classification time between the two approaches is less of an issue due to the relatively small figures (approx. 8.15s and 0.35s respectively for slightly more than 8,000 instances).

These two approaches allow a trade off between speed and accuracy. Within the next section, we will examine a third technique that can equal the accuracies of the single model approach but uses less than half the number of attributes, resulting in a further speed improvement.

3.2 Attribute Pruning

Due to the characteristics of the learning method we are using (i.e. an SVM), we can examine the weights associated with the attributes used. As described in the section *materials and methods*, input is provided to the SVM as a series of instances. Each instance describes the result of applying the given models to the sequence data returned by a sliding window of 150bps. We will now examine these weights to determine which parts of the promoter are discriminative in terms of each model.

We graph these attribute weights in Figure 1. As can be seen, the area around the TSS is of most discriminatory power, with many models displaying clear spikes in this region (150bps are shown for each model; the TSS is at the 100th bp).

Curiously, A-philicity has the highest weighted attribute spike but when used on its own (as discussed in the previous section and shown in Table 1) it does not produce the best classification performance. This seems to indicate that A-philicity is predictive of promoter activity when combined with other models, but less so when considered independently. In contrast, Duplex Disrupt Energy clearly performs well on its own and is highly influential in a combined approach.

We applied the attribute pruning describe in the section *materials and methods* to both single models and the combined model presented in the previous section. The results are shown in Table 2. These datasets are described by only 61 attributes, as compared to the 139/140 (for a di- or tri-nucleotide model) attributes required for a single model and the 839 attributes required for the combined model. Despite the

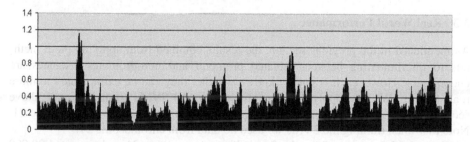

Fig. 1. The attribute weights associated with each of the 839 attributes used in the combined model approach. The weights are segmented according to model in the following order: A-philicity, DNA bending stiffness, DNA denaturation, duplex disrupt energy, nucleosome positional preference and propeller twist.

Table 2. The accuracy of the models after attribute pruning. The combined model was pruned to include only those attributes weighted higher than 0.6 – 61 attributes. The single models were pruned to include the 61 attributes with the highest weights.

Training set proportion (%)	1	5	10	15	20	30	40	50
A-philicity	82.6	84.9	85.2	86.0	85.7	86.3	86.5	86.6
DNA Bending Stiffness	83.5	84.9	84.8	84.9	85.3	85.6	85.9	86.1
DNA Denaturation	82.2	84.3	85.1	85.1	85.6	85.7	85.9	86.0
Duplex Disrupt Energy	85.8	87.1	87.7	87.9	88.1	88.5	88.7	88.8
Nucleosome Positional Pref.	84.5	86.7	86.8	86.8	87.1	87.3	87.5	87.7
Propeller Twist	85.8	87.5	87.5	87.7	87.8	88.1	88.2	88.5
COMBINED MODELS	83.0	86.0	86.3	86.6	86.8	87.1	87.3	87.5
DDE + PROP. TWIST	86.7	88.1	88.3	88.4	88.6	89.0	89.2	89.3

significantly smaller size, one can see that performance is comparable to evaluating each model over the entirety of the instance. The combined model loses enough accuracy from the pruning to no longer be the best choice under the simple accuracy metric used here. We also present an alternate approach to pruning, listed as DDE + prop. twist. This method first prunes the duplex disrupt energy and propeller twist models to 30 attributes and then combines them to make a model of 60 attributes. As can be seen, this approach produces the highest accuracy. The training time required for these pruned models is roughly half that of the full models, at approximately 5 minutes. Most importantly though, the classification time on approximately 8000 instances is only 0.17s. With such a small number, classification time is now insignificant in comparison to the time required for operations such as file loading. On large datasets, this fast classification time is invaluable.

To briefly summarise, we have demonstrated that a combined approach using multiple physico-chemical properties improves accuracy over any single model considered on its own. We have also shown that if speed is more important than accuracy, these models can be evaluated over a smaller portion of the input data for an almost negligible lose of accuracy but a significant improvement in execution time.

3.3 Real World Performance

As mentioned in the previous section, the results obtained from applying these methods to discriminating between isolated promoter and non-promoter instances are encouraging. However, it is not yet known whether they can be extrapolated to more realistic uses. We now apply the two combined model SVMs produced in the above section to contiguous sections of sequence data, taken from human chromosome 21. Note that when training the SVMs we used no data from chromosome 21.

We took 25 promoters from the first half of chromosome 21 and extracted 20,000 bps around each one producing 25 testing sets. We found that both SVMs on average labelled 20% of the dataset (or approx 4000 instances) as being promoters.

Although this is a poor result, we observed that the distribution of predictions was correlated with the location of actual positives. This is best illustrated with a diagram

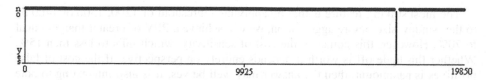

Fig. 2. The classification diagram produced by the SVM. The X-axis is the position within the input sequence data. The top line represents the locations of negative predictions. Notice the clear gap in negative predictions centred on the TSS position (indicated by the vertical line).

and an example is presented in Figure 2. As can be seen, the actual location of the TSS coincides with a gap in the prediction of negatives by the classifier. We used this observation as the basis for a second level to the classifier, which searches for these gaps in the negative output. The results obtained using the 25 segments from the first-half of chromosome 21 were used to select a size for the gap and the percentage of instances within the gap that must be positive. We also determined that taking the middle or the end of a gap as the predicted TSS was equally accurate, but taking the beginning produced significantly worse results.

3.4 Combined Classifier Performance

We now examine the performance of the combined classifiers, using a section of DNA from the second half of chromosome 21 of about 10Mbps. Interestingly, we found that the SVM produced from taking a 0.6 cut-off to the full combined model produced fewer false positives than the most highly weighted attributes from the duplex disrupt energy and propeller twist models. This in turn leads to a worse PPV for the DDE + Prop Twist model. It does however have one advantage: producing the highest sensitivity (at 66%) of the two models examined, although this comes at the cost of a very low PPV (approx. 8%). It is not clear whether the other combined model might also produce sensitivity within this range if more relaxed parameters allowed a lower PPV. Because of space restrictions, we choose to omit a detailed comparison and focus our attention on the model labeled "combined model" above. We feel that both the higher PPV values and the ability to provide a better balance between the PPV and Se. make this model more practical in a realistic environment.

Within the region we have chosen for testing there are 65 promoters and the classifier correctly identifies 36 of them (a sensitivity of 55.4%). In total, 224 positive predictions are made, equating to a positive predictive value (PPV) of 16.1%. The sensitivity is comparable to many other current approaches for promoter prediction – of the 14 programs examined by Bajic, Tan et al [2] 9 of them score sensitivity of approximately 55% or less. However, we are aware that the low PPV value is a drawback to the proposed method and now suggest an approach for improving it.

As we have mentioned previously, the output from the upper level of the classifier is a window of predicted promoter activity. We now consider the size of the window as an indication of confidence in respect to the prediction. This leads to the ability to discard low confidence predictions. Because we are aggregating windows that are within 1000bp of each other, this allows for two possible approaches in respect to applying a window-size threshold; before aggregation and after. We present the results obtained from both approaches graphically in Figure 3.

The most striking feature is that by applying a threshold of 1200, 1300 or 1400 bp to the window size before aggregation, we can achieve a PPV of greater than or equal to 70%. However, this comes at the cost of sensitivity, which falls to less than 15%. Whether this trade-off is worth it depends entirely on perspective: if the cost of false positives is paramount, then the answer may well be yes. It is also interesting to note that by applying a threshold of about 300bp, we can achieve an improvement in PPV for an almost negligible loss of sensitivity. A reasonable trade off is also apparent with a threshold of about 500bp, where PPV and Se of approximately 40% is achieved. It is also apparent that thresholding before aggregation favours PPV whereas after aggregation favours sensitivity. Intuitively, this is as one would expect. Because of the nature of the aggregation, it is possible that, for example, two small windows which are 900bps from each other may be aggregated into one large window of more than 1000bps. Hence, thresholding before aggregation favours large uninterrupted windows, which are more likely to represent actual promoters, while discarding the interrupted windows which are less likely.

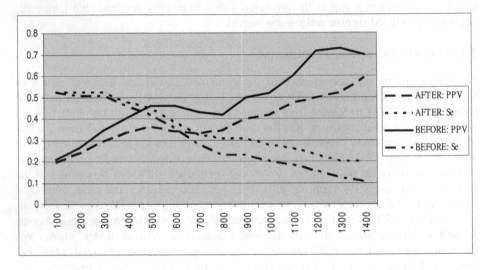

Fig. 3. The performance of the classifier when applying a post-processing step of discarding small windows. The window threshold is shown on the x-axis. We graph both sensitivity and positive predictive value for threshold application both before and after aggregation.

4 Conclusions

We have set out within this paper to address two distinct points. Firstly, we intended to demonstrate that a combined model of physico-chemical properties could be used for promoter prediction, improving the accuracy of using a single model. We have demonstrated this fact, showing an increase of approximately 2% over the best of the single models. We stress once again the importance of such a result in considering large datasets, or within applications where sensitivity is of importance.

We have also explored the relative importance of the different properties in different segments of the promoter in terms of predictive power by examining the attribute weighting of the support vector machine produced. Using these results, we produced a reduced model, which is faster but loses little in terms of accuracy. These two results combined demonstrate that this approach can be tailored for sensitivity by using the full combined model or for speed by using attribute pruned models.

The second component of this paper has looked at how these approaches perform when confronted with large, real world problems. We showed that although the raw results of the SVM do not immediately appear encouraging, by post-processing the output, it is possible to predict promoter locations with a sensitivity of 55.4% and a PPV of 16.1%. By using a threshold approach to the window size, we also demonstrated that a balanced 40% PPV and Se. was possible. Further to this, by modifying the threshold size, the approach can be tailored for either sensitivity or specificity. These results demonstrate that a promoter prediction technique based only on physico-chemical properties is possible and capable of performance that is competitive with established approaches. The use of the physico-chemical properties of DNA for promoter prediction is a promising direction for further research, both on its own and in conjunction with sequence based methods.

5 Further Work

There are several distinct directions to pursue as a result of this work. Firstly, Florquin, Saeys et al. [4] suggest that promoters be treated as distinctly separate groups. Within this work, we have not implemented this idea, treating all promoters as being the same. By learning separate classifiers for distinct types of promoters, improvements in sensitivity and PPV may be possible. Alternately, a simpler approach of retraining the classifier using the instances that were not correctly classified on the first pass may produce equivalent performance. There is also the question of how to combine these multiple classifiers.

Bajic, Tan et al. [2] also suggest that the masking of repeats may improve the performance of promoter prediction programs. They also present evidence to suggest that experiments run on single chromosomes may not be representative of results produced on the whole genome. Future work on this approach would require the exploration of both these ideas: the application of repeat masker and larger-scale tests.

Furthermore, we have presented proof that combined models can outperform single models, but have only examined a few possible combinations. Our results indicate that the combination of models may not be as simple as originally thought and a more thorough exploration is called for.

Finally, we have shown that the application of physico-chemical properties to the problem of promoter prediction can produce competitive results on their own, but we are not advocating the abandonment of other approaches. A combined approach, employing these techniques and other complementary (possibly sequence related) approaches may produce better results. In particular, this approach is currently not strand specific. That is, predictions are equally likely to be on either strand. By employing sequence based techniques, this could be overcome, allowing a strand specific prediction to be made.

Acknowledgements

This work was supported by an Australian Postgraduate Award.

References

1. Fickett, J. W., and A. G. Hatzigeorgiou, 1997, Eukaryotic Promoter Recognition: Genome Research, v. 7, p. 861-878.
2. Bajic, V. B., S. L. Tan, Y. Suzuki, and S. Sugano, 2004, Promoter prediction analysis on the whole human genome: Nature Biotechnology, v. 22, p. 1467 - 1473.
3. Pedersen, A. G., P. Baldi, Y. Chauvin, and S. Brunak, 1998, DNA Structure in Human RNA Polymerase II Promoters: J. Mol. Biol., v. 281, p. 663-673.
4. Florquin, K., Y. Saeys, S. Degroeve, P. Rouze, and Y. Van de Peer, 2005, Large-scale structural analysis of the core promoter in mammalian and plant genomes: Nucl. Acids Res., v. 33, p. 4255-4264.
5. Fukue, Y., N. Sumida, J.-i. Nishikawa, and T. Ohyama, 2004, Core promoter elements of eukaryotic genes have a highly distinctive mechanical property: Nuc. Acids Res., v. 32, p. 5834-5840.
6. Fukue, Y., N. Sumida, J.-i. Tanase, and T. Ohyama, 2005, A highly distinctive mechanical property found in the majority of human promoters and its transcriptional relevance: Nuc. Acids Res., v. 33, p. 3821-3827.
7. Kanhere, A., and M. Bansal, 2005, Structural properties of promoters: similarities and differences between prokaryotes and eukaryotes: Nucleic Acids Research, v. 33, p. 3165-3175.
8. Choi, C. H., G. Kalosakas, K. Rasmussen, M. Hiromura, A. R. Bishop, and A. Usheva, 2004, DNA dynamically directs its own transcription initiation.: Nucleic Acids Res., v. 32, p. 1584-90.
9. Tsai, L., L. Luo, and Z. Sun, 2002, Sequence-dependent flexibility in promoter sequences.: J. Biomol. Struct. Dyn., v. 20, p. 127-34.
10. Gabrielian, A., D. Landsman, and A. Bolshoy, 1999-2000, Curved DNA in promoter sequences: In Silico Biol., v. 1, p. 183-96.
11. Lisser, S., and H. Margalit, 1994, Determination of common structural features in Escherichia coli promoters by computer analysis: Eur J Biochem, v. 223, p. 823-830.
12. Wang, H., M. Noordeweier, and C. J. Benham, 2004, Stress-Induced DNA Duplex Destabilization (SIDD) in the E. coli Genome: SIDD Sites Are Closely Associated With Promoters: Genome Research, v. 14, p. 1575-1584.
13. Ohler, U., H. Niemann, G. Liao, and G. Rubin, 2001, Joint modeling of DNA sequence and physical properties to improve eukaryotic promoter recognition: Bioinformatics, v. 17, p. S199-206.
14. Baldi, P., Y. Chauvin, S. Brunak, J. G. Anders, and G. Pedersen, 1998, Computational Applications of DNA Structural Scales: Int. Conf. Intell. Syst. Mol. Biol., p. 35-42.
15. Ota, T., Y. Suzuki, T. Nishikawa, T. Otsuki, T. Sugiyama, R. Irie, A. Wakamatsu, K. Hayashi, H. Sato, K. Nagai, K. Kimura, H. Makita, M. Sekine, M. Obayashi, T. Nishi, and T. Shibahara, 2004, Complete sequencing and characterization of 21,243 full-length human cDNAs: Nat Genet., v. 36, p. 40-5.
16. Suzuki, Y., and S. Sugano, 2003, Construction of a full-length enriched and a 5'-end enriched cDNA library using the oligo-capping method: Methods Mol. Biol., v. 221, p. 73-91.

17. Suzuki, Y., R. Yamashita, S. Sugano, and K. Nakai, 2004, DBTSS, DataBase of Transcriptional Start Sites: progress report 2004.: Nucleic Acids Res, v. 32(Database issue), p. D78-81.
18. Ivanov, V. I., and L. E. Minchenkova, 1994, The A-form of DNA: in search of the biological role: Mol Biol (Mosk), v. 28, p. 1258-71.
19. Sivolob, A. V., and S. N. Khrapunov, 1995, Translational positioning of nucleosomes on DNA: the role of sequence-dependent isotropic DNA bending stiffness: J Mol Biol., v. 247, p. 918-31.
20. Blake, R. D., and S. G. Delcourt, 1998, Thermal stability of DNA: Nucleic Acids Res., v. 26, p. 3323-32.
21. Breslauer, K., R. Frank, H. Blocker, and L. Marky, 1986, Predicting DNA duplex stability from the base sequence: Proc. Natl. Acad. Sci. USA, v. 83, p. 3746-50.
22. Satchwell, S. C., H. R. Drew, and A. A. Travers, 1986, Sequence periodicities in chicken nucleosome core DNA: J. Mol. Biol., v. 191, p. 659-75.
23. el Hassan, M., and C. Calladine, 1996, Propeller-twisting of base-pairs and the conformational mobility of dinucleotide steps in DNA: J Mol Biol., v. 259, p. 95-103.
24. Witten, I. H., and E. Frank, 2005, Data Mining: Practical machine learning tools and techniques: San Francisco, Morgan Kaufmann.
25. Platt, J., 1998, Fast Training of Support Vector Machines using Sequential Minimal Optimization, in B. Schoelkopf, C. Burges, and A. Smola, eds., Advances in Kernel Methods - Support Vector Learning, MIT Press.
26. Keerthi, S. S., S. K. Shevade, C. Bhattacharyya, and K. R. K. Murthy, 2001, Improvements to Platt's SMO Algorithm for SVM Classifier Design: Neural Computation, v. 13, p. 637-649.

Parametric Spectral Analysis of Malaria Gene Expression Time Series Data

Liping Du[1,2], Shuanhu Wu[1,3], Alan Wee-Chung Liew[4],
David Keith Smith[5], and Hong Yan[1,6]

[1] Department of Electronic Engineering, City University of Hong Kong,
Tat Chee Avenue, Kowloon, Hong Kong
[2] Information Engineering School, University of Science and Technology Beijing,
Beijing 100083, China
[3] School of Computer Science and Technology, Yantai University,
Shandong, Yantai 264005, China
{duliping, itwush, h.yan}@cityu.edu.hk
[4] Department of Computer Science and Engineering,
Chinese University of Hong Kong, Shatin, Hong Kong
wcliew@cse.cuhk.edu.hk
[5] Department of Biochemistry, University of Hong Kong, Pok Fu Lam, Hong Kong
dsmith@hku.hk
[6] School of Electronic and Information Engineering,
University of Sydney, NSW2006, Australia

Abstract. Spectral analysis of DNA microarray gene expressions time series data is important for understanding the regulation of gene expression and gene function of the *Plasmodium falciparum* in the intraerythrocytic developmental cycle. In this paper, we propose a new strategy to analyze the cell cycle regulation of gene expression profiles based on the combination of singular spectrum analysis (SSA) and autoregressive (AR) spectral estimation. Using the SSA, we extract the dominant trend of data and reduce the effect of noise. Based on the AR analysis, high resolution spectra can be produced. Experiment results show that our method can extract more genes and the information can be useful for new drug design.

Keywords: Singular Spectrum Analysis, Autoregressive Model, Spectral Estimation, Microarray Time Series Analysis, *Plasmodium Falciparum*.

1 Introduction

One of the four Plasmodium species that cause human malaria is the protozoan parasite Plasmodium falciparum, for which there are no effective vaccines currently. Massive efforts to eradicate the disease have payed off, but drug resistance in the parasite is now still widespread and few antimalarial chemotherapeutics are available as both prophylaxis and treatment. In 2002, a complete genome sequence of *P. falciparum* was reported [1]. The database provides valuable information for

M.R. Berthold, R. Glen, and I. Fischer (Eds.): CompLife 2006, LNBI 4216, pp. 32–41, 2006.
© Springer-Verlag Berlin Heidelberg 2006

researchers to find potentially unique or at least substantially different genes in *P. falciparum* compared with other species. These genes may be useful in designing drugs that can cause less risk of negative side effects.

DNA microarray gene expressions time series data are useful for gene discovery, disease diagnosis, treatment, and drug design. From the fact that certain genes are regulated only at specific stages of the cell cycle, these genes consequently exhibit a periodic pattern of expression. Therefore, the identification of periodically expressed genes is important for understanding the cell control over structure and function of *P. falciparum*. Many studies have aimed at identifying the cell cycle-regulated (or periodically expressed) genes [2-5]. Among these works, spectral analysis is one common method due to its unique characteristic to periodic data. In 2003, Bozdech et al [2] applied the Fourier transform to analyze the periodicity of transcriptome of the intraerythrocytic developmental cycle (IDC) and offered a detailed description of the four major morphological stages, namely, ring/early trophozoite, trophozoite/early schizont, schizont and early ring. The paper revealed that 70% of expression profiles of transcriptome of the IDC showed a high periodicity. As discussed below, however, there are only 46 points with missing values in gene expression time series for the IDC of *P. falciparum*. This kind of time series would be considered extremely short and in the field of signal processing. The commonly used Fourier transform is unfit to analyse this kind time series due to the so-called windowing effect and missing samples. Furthermore, microarray data contain a high level of noise, which can degrade the performance of data analysis algorithms. Thus, effective methods are needed to process microarray gene expression time series data.

In this paper, we propose a new scheme for analyzing the periodicity of the trancriptome of the IDC by combining singular spectrum analysis (SSA) and autoregressive (AR) modeling. By using the SSA, the dominative component trend can be extracted from the noise contaminated expression profiles and the effect of noise can be reduced effectively. Then utilizing the advantage of AR model in spectral analysis, we are able to analyze the periodicities of the data more accurately. The combination of SSA and AR methods can identify about 90% of genes in *P. falciparum* and thus gain an advantage over the Fourier analysis for gene identification.

2 Materials and Methods

2.1 Dataset and Data Preprocessing

The microarray dataset used in this paper is downloaded from http://malaria.ucsf.edu. It contains the expression profiles of 5080 oligonucleotides measured at 46 time points spanning 48 hours during the IDC with one hour time resolution for the HB3 strain. A logarithm transform was applied to the expression ratio of channel Cy5 to channel Cy3. The mean of each expression profile was subtracted from whole profiles so that the average $\log_2(Cy5/Cy3)$ value over the time span was equal to zero.

2.2 Trend Estimation of Expression Profiles Using the SSA

SSA has been proven to be a powerful tool for processing many types of time series in geophysics, economics, biology, medicine and other sciences [6,7], which may have nonlinear characteristics. It is also a useful technique for gene expression data processing. Let each expression profiles be a time series $\{s_1, s_2, ..., s_n, ..., s_N\}$. The SSA can be performed as follows,

(1) Construct the trajectory matrix $X_{M,K} = (x_{ij} = s_{i+j-1})$ from the original series by sliding a window of length M $(M \leq N/2)$, $K = N - M + 1$.

(2) Decompose the trajectory matrix into a series of components X_i, where $X_i = \sqrt{\lambda_i} U_i V_i^T$, $(i = 1, 2, ..., M)$, λ_i is the eigenvalues of the matrix $R = XX^T$, U_i and V_i are the left and right singular vectors of the matrix X, respectively.

(3) Group a specified number of leading eigenvalues λ_i and sum the corresponding components X_i, then the resultant matrix is $X'_{M,K}$. The number of eigenvalues to use is discussed below.

(4) Reconstruction the data series $\{s'_1, s'_2, ..., s'_n, ..., s'_N\}$ by averaging the elements of matrix X' over the 'diagonals' $i + j = n + 1$.

The SSA decomposition can be three parts: low frequencies corresponding to the trend, the higher frequencies describing the oscillatory component, and the residual as noise. The Fourier analysis demonstrates that the power spectra of the expression profiles are dominated by low frequency components. Therefore, SSA could be used to extracts the trend curve that represents the dominative component in order to reduce the effect of noise.

In the process of SSA, one of the important steps is how to group the eigenvalues to reconstruct the expression profiles. If we plot the eigenvalues, the diagram contains an initial steep slope, representing the signal, and a "flat floor", representing the noise level [8]. The ratio of the leading eigenvalues and the total eigenvalues is defined as follows to evaluate the quality of reconstruction:

$$E = \frac{\sum_{i=1}^{d} \lambda_i}{\sum_{i=1}^{M} \lambda_i} . \tag{1}$$

where d is the number of the leading eigenvalues.

For the majority of transcriptome of IDC, the leading two eigenvalues of expression profiles contain most energy and correspond to the signal (see Fig.1). Therefore, we group the leading two eigenvalues to reconstruct the expression profiles [9]. An example of trend extraction using the SSA is shown in Fig.2.

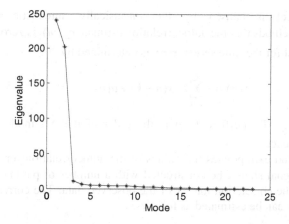

Fig. 1. The eigenvalue spectra of Dihydrofolate Reductase-Thymidylate Synthase (DHFS-TS). The spectra contains a steep slope. The leading two eigenvalues which contain most energy represent the signal, and a "flat floor", representing the noise level. Then the signal can be reconstructed from the expression profiles by grouping the leading two eigenvalues. The window length is $M = 23$.

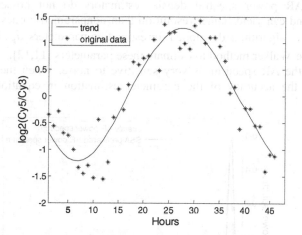

Fig. 2. Signal extraction of DHFS-TS using the SSA. The leading two eigentriples which represent 90% energy are chosen for the signal reconstruction in order to reduce the effect of noise. The asterisk indicates the original data of expression profiles and the curve is the dominative component trend extracted from the original data.

2.3 AR Model Based Spectral Estimation

Periodicity in genome-wide gene expression datasets has been widely used to identify cell-cycle-regulated genes. If the data are highly periodic, the calculated power spectrum would show sharp peaks at the corresponding frequency points. Let the discrete data sequence be $x(n)$, $0 < n < N - 1$, the nonparametric power spectrum estimation method can be computed by fast Fourier transform (FFT). The well-known windowing effect is caused by the intrinsically unrealistic assumption of nonparametric methods that the autocorrelation estimate $r_{xx}(m)$ is zero for $m > N$.

The AR model for the time series $x(n)$ is determined by

$$x(n) = -\sum_{k=1}^{p} a_k x(n-k) + u(n) \ .$$ (2)

where a_k are the AR coefficients, p is the order of the AR model, and $u(n)$ is a white noise sequence.

The AR method extrapolates the values of the autocorrelation for $m > N$. Then a model for the signal should be constructed with a number of parameters that can be estimated from the correlation matrix of the observed data. The corresponding power spectrum density can be estimated as follows,

$$P_{xx}(\omega) = T \sum_{m=-\infty}^{\infty} r_{xx}(m) e^{-jm\omega T} \ .$$ (3)

where $Pxx(\omega)$ represents the power spectral density of the signal x at frequency ω. Equation (3) shows that the support of autocorrelation r_{xx} in it is $(-\infty, \infty)$. For this reason, the AR power spectral density estimators do not possess the sidelobe phenomena and can yield higher resolution than nonparametric ones [10-12]. There are a number of algorithms developed for estimating parameters a_k. In this work, we adopt the Yule-Walker method to estimate these parameters [11,12].

However, the AR spectrum is very sensitive to noise. When the signal to noise ratio is low, the accuracy of the parameter estimation in equation (2) would be

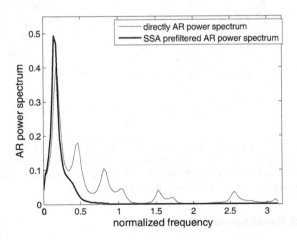

Fig. 3. The AR spectra of the expression profile of DHFR-TS with and without SSA filtering. Due to the removal of the noise using the SSA, spurious peaks are eliminated from the spectrum. The dash-dot line represents the AR power spectrum without SSA filtering the noise and the solid line is the AR power spectrum with SSA prefiltered the noise.

reduced substantially. A higher order AR model has to be used to improve the frequency resolution, but the usage of higher order would induce the appearance of spurious peaks. According to Keppenne and Penland [13,14], SSA can be used to be a data-adaptive filter. Then by applying AR power spectrum estimation to the SSA pre-filtered data series, noise-free or nosie-reduced frequency spectrum could be obtained. As shown in Fig.3, due to the removal of the noise using the SSA, spurious peaks are eliminated from the spectrum.

2.4 The Periodicity Detection Using the AR Model

Assuming that the AR spectrum reaches its peak point at the frequency f_m and considering the frequency band [f_{m-1} , f_{m+1}] as f_m's influencing region. We take the ratio of the power in f_m's influencing region to the total power of the signal to qualify the periodicity of the expression profile of each gene:

$$S = \frac{power_m}{power_{total}} . \tag{4}$$

where $power_m$ and $power_{total}$ represent the power over f_m's influencing region and the total power of the signal, respectively.

The process of periodicity detection is: Firstly, each expression data is reconstructed using SSA. Only the expression profiles with the eigenvalue ratios in Equation (1) ($d = 2$) greater than 0.6 are considered to be reconstructed. Secondly, the AR spectrum is calculated for each reconstructed expression profiles. Then the frequency f_m at peak value point and the ratio of the power in f_m's influencing region to the total power are calculated according to Equation (4). Thirdly, all the profiles were filtered according to following rule: if the power ratio calculated previously is larger than 0.7 (which is the same value as that used in [2]), the corresponding profile would be selected for further processing, otherwise, we consider it lacking of periodicity and discard it.

3   Results

3.1   Statistical Significance

We have tested our algorithm on the expression data of the IDC of *P. falciparum*. There are total 5080 expression profiles in the dataset. We used the proposed periodicity detection process and have selected those profiles by utilizing the power ratio to filter the reconstructed expression profiles. The histogram of the power ratio of the express profiles of the IDC data is shown in Fig.4. By using our proposed periodicity detection method, 4496 periodic profiles are found from the original 5080 expression profiles. Compared with the Bozdech et al's result, additional 777 oligonucleotides are detected using our algorithm.

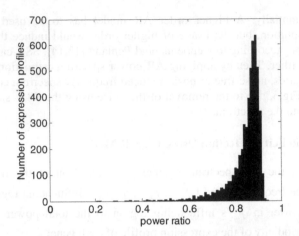

Fig. 4. The histogram of the power ratios of the express profiles of the *P. falciparum* IDC transcriptome. There are total 5080 expression profiles in the dataset. The power of the signal qualify the periodicity of the expression profile of each gene. The histogram indicates the majority of *P. falciparum* genes are periodical signal. Compared with the Bozdech et al's result, additional 777 oligonucleotides are detected using our algorithm.

3.2 Analysis of Classification Results

It is generally accepted that the function of a gene is related to the initial phase of its expression profile [2,15]. So for the convenience of gene function classification, we have ordered the expression profiles of the extracted 4496 oligonucleotides according to their peak time points of expression profiles. Figure 5 shows a continuous cascade of gene expressi+ons, which correspond to the developmental stages throughout the IDC, that is, ring, trophozoite and schizont stages. According to the sharp transitions of ring-to-trophozoite (at the 17h time point), trophozoite-to-schizont (at the 29h time point) and schizont-to-rings stages (at the 45h time point) [2], 4496 selected genes could be categorized into four stages on the basis of the peak time points of expression profiles in Fig.5. The comparison of the classification results of the oligonucleotides assigned to these stages by our method with those in [2] is shown in Table 1. In Table 2, we provide the list of functional genes identified by our algorithm that was not found by the method in [2].

Genes expressed during the mid to late schizont and early-ring stage encode proteins predominantly involved in highly parasite-specific functions facilitating various steps of host cell invasion. The highly parasite-specific functions implied that they should serve as good targets for both drug discovery and vaccine-based antimalarial strategies [2]. The additional identified genes are distributed in 777 additional oligonucleotides found by our method and they would be useful in the future identification of novel targets for anti-malarial therapies. There are often two gene targets to be used as implications for new drug discovery: apicoplast-targeted genes and proteases. In Table 3, we provide a comparison of the number of the two gene targets detected by our method with that by Bozdech et al [2].

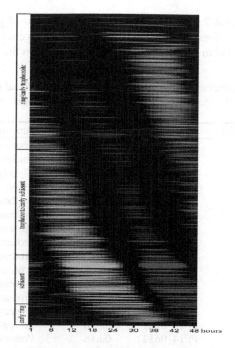

Fig. 5. The phaseogram of the transcriptome of the IDC of P. falciparum. 4496 genes which expressed periodically are ordered along the y axis in the order of the time of their peak expression for the convenience of gene function classification. The different developmental stages throughout the IDC are shown on the left. The phaseogram reveals a continuous cascade of gene expression.

Merozoite invasion is one of the most promising target areas for antimalarial vaccine developpment. Among these invasion proteins are seven of the well known malaria vaccine candidates, including Apical Merozite Antigen-1 (AMA1), Merozoite Surface Protein-1 (MSP1), Merozoite Surface Protein-3 (MSP3), Merozoite Surface Protein-5 (MSP5), Erythrocyte Binding Antigen-175 (EBA175), Rhoptry-Associated Protein-1 (RAP1) and Ring-infected Erythrocyte Surface Antigen-1 (RESA1). It is now widely accepted that genes with the same or similar function are likely to have similar expression profiles. Therefore, we have used the similarity to identify genes with possible involvement in the merozoite invasion process. The similarity of 4835 expression profiles to seven known vaccine candidates is evaluated using the Euclidian

Table 1. The total number of the classification results of oligonucleotides assigned to different stages by our method and those in [?] More genes can be identified using our method in each stage.

Stages	Our method	Method in [2]
Ring/early Trophozoite	1970	1563
Trophozoite/early Schizont	1524	1296
Schizont	709	625
Early ring	293	235

distance. There are total 267 genes constituting the top 5% of genes in the IDC with minimum distance to these seven genes. Additional 5 genes are detected by our method, which are listed in Table 4. They are identified from the late schizont stage.

Table 2. Fifteen genes detected by our method for different functional groups of *P. Falciparum* with high power ratio after the preprocess of SSA, that was not found by the method in [2].

Oligonucleotide	Gene ID	Power ratio	Functional groups
b471	PFB0715w	0.904	Transcription machinery
opff72413	MAL6P1.189	0.798	Glycolytic pathway
j22_5	PF10_0086	0.794	Ribonucleotide synthesis
c345	PFC0520w	0.844	Proteasome
F20448_1	MAL8P1.142	0.899	Proteasome
I14413_3	PFI0630w	0.911	Proteasome
j110_4	PF10_0174	0.838	Proteasome
j64_2	PF10_0298	0.914	Proteasome
j73_12	PF10_0081	0.905	Proteasome
kn3744_1	PF10_0174	0.867	Proteasome
ks101_10	NULL	0.776	Proteasome
M21665_2	PF13_0063	0.898	Proteasome
n135_24	PF14_0632	0.818	Proteasome
n137_50	PF14_0676	0.795	Proteasome
n155_11	PF14_0025	0.853	Proteasome

Table 3. The total number of the two types of potential gene targets to be used as implications for drug discovery detected by our method and those in [2]

Potential genes targets	Our method	The method in [2]
Apicoplast-targeted	409	358
proteases	115	88

Table 4. Additional 5 genes detected by our method out of those genes in the IDC with minimum distance to these seven genes the for antimalarial vaccine. All of them are from the late schizont stage.

Oligonucleotides	Genes	Description
n139_5	PF14_0757	Hypothetical protein
n143_54	PF14_0183	RNA helicase
n168_2	PF14_0530	Ferlin, putative
opfblob0035	PFD0430c	Hypothetical protein
opfl0111	PFL2110c	Hypothetical protein

4 Conclusion

Spectral analysis of the transcriptome of the asexual intraerythrocytic development cycle (IDC) of *P. falciparum* offers an effective tool for the studies of gene expression and future drug and vaccine targets. Based on the SSA and the AR power spectral analysis, we have proposed a new scheme for detecting genes with periodicity in the IDC of *P. falciparum*. The result shows our method can not only detect more genes but also can find genes that could be useful for potential drug design.

Acknowledgments. This work is supported by a CityU interdisciplinary research grant (project 9010003) and a grant from the Hong Kong Research Grant Council (project CityU122005).

References

1. Gardner, M.J., Hall, N., Fung, E., White, O., Berriman, M., et al.: Genome sequence of the human malaria parasite *Plasmodium falciparum*. Nature 419 (2002) 498-511
2. Bozdech, Z., Llinas, M., Pulliam, B.L., Wong, E.D., Zhu, J.C., DeRisi, J.L.: The transcriptome of the intraerythrocytic developmental cycle of *Plasmodium falciparum*. Plos Biology 1 (2003) 1-16
3. Spellman, P.T., Sherlock, G., Zhang, M.Q., Iyer, V.R., Anders, K., et al.: Comprehensive identification of cell cycle-regulated genes of the yeast saccharomyces serevisiae by microarray hybridization. Mol. Biol. Cell. 9 (1998) 3273-3297
4. Le Roch, K.G., Zhou, Y.Y., Blair, P.L., Grainger, M., Moch, J. K., et al.: Discovery of gene function by expression profiling of the malaria parasite life cycle. Science 301 (2003) 1503-1508.
5. Rajarajeswari, B., Eyke, H., Nils, W., Jörg, K.: Clustering of gene expression data using a local shape-based similarity measure. Bioinformatics 21 (2005) 1069-1077
6. Vautard, R., Yiou, P., Ghil, M.: Singular-spectrum analysis: A toolkit for short, noisy chaotic signals. Physica D 58 (1992) 95-126
7. Golyandina, N., Nekrutkin, V., Zhigljavsky, A.: Analysis of time series structure: SSA and related techniques. Chapman & Hall/CRC, Boca Raton Florida (2001)
8. Vautard, R., Ghil, M.: Singular spectrum analysis in nonlinear dynamics, with applications to paleoclimatic time series. Physica D 35 (1989) 395-424
9. Liu, L, Hawkins, D.M., Ghosh, S., Yong, S.S.: Robust singular value decomposition analysis of microarray data. PNAS 100 (2003) 13167-13172
10. Marple, S.L.: Digital spectral analysis: with applications. Prentice-Hall, Englewood Cliffs New Jersey. (1987)
11. Yan, H. (ed.): Signal Processing for Magnetic Resonance Imaging and Spectroscopy. Marcel Dekker, New York (2002)
12. Yeung, L.K., Szeto, L.K., Liew, W.C., Yan, H.: Dominant spectral component analysis for transcriptional regulations using microarray time series data. Bioinformatics 20 (2004) 742-749
13. Keppenne, C.L., Ghil, M.: Adaptive filtering and prediction of the Southern Oscillation index. J. Geophys. Res. 97 (1992) 20449-20454
14. Penland, C., Ghil, M., Weickmann, K.M.: Adaptive filtering and maximum entropy spectra with application to changes in atmospheric angular momentum. J. Geophys. Res. 96 (1991) 22659-22671
15. Rustici, G., Mata, J., Kivinen, K., Lió, P., Penkett, C.J., et al.: Periodic gene expression program of the fission yeast cell cycle. Nature Genetics 36 (2004) 809-817

An Efficient Algorithm
for Finding
Long Conserved Regions Between Genes

Tak-Man Ma[1], Yuh-Dauh Lyuu[2,*], and Yen-Wu Ti[2]

[1] Dept. of Computer and Information Science,
University of Pennsylvania, Philadelphia, USA
mata3@seas.upenn.edu
[2] Dept. of Computer Science and Information Engineering,
National Taiwan University, Taipei, Taiwan
{lyuu, d91010}@csie.ntu.edu.tw

Abstract. We study the problem of approximate non-tandem repeat (conserved regions) extraction among strings (genes). Basically, given a string S and thresholds L and D over a finite alphabet, extracting approximate repeats is to find pairs (β, β') of substrings of S under some constraints such that β and β' have edit-distance at most D and their respective lengths are at least L. Previous works mainly focus on the case that D is small, so they are not appropriate for extracting approximate repeats with relatively large D. In contrast, this paper focuses on extracting long approximate repeats with large D and it is more efficient than previous works. We also show that our algorithm is optimal in time when D is a constant.

In this paper, given an input string S and thresholds L and D, we would like to extract all (D, L)-supermaximal approximate repeats (β, β') of S. One useful application of extracting all (D, L)-supermaximal approximate repeats (β, β') is to find all longest possible substrings β of S such that there exist some other substring β' of S where β and β' have edit-distance at most D and their respective lengths are at least L. This algorithm can be easily applied to the case where there are multiple input strings $S_1, S_2, \ldots, S_n$ if we first concatenate the input strings into one long subject string S with a special symbol "$\sharp$" for separation: $S_1 \sharp S_2 \sharp \ldots \sharp S_n$. The running time complexity of our algorithm is $O(DN^2)$ where $N = |S_1| + |S_2| + \cdots + |S_n|$.

1 Introduction

1.1 Related Work

The repetitive structure of genomic DNA holds many secrets that await discovery. Conserved regions between genes have important meanings in biology and play an important role in the identification of novel functional units [4]. For example, the gene sequences of a gene family may share conserved regions that

---

* Corresponding author.

M.R. Berthold, R. Glen, and I. Fischer (Eds.): CompLife 2006, LNBI 4216, pp. 42–51, 2006.
© Springer-Verlag Berlin Heidelberg 2006

correspond to meaningful domains in their protein structures. A useful approach for identifying meaningful conserved regions is to find non-tandem approximate repeating patterns that occur more frequently than expected by chance. The problem of exact repeat extraction seems settled since it already has an optimal linear-time algorithm by suffix tree [4]. However, as mutations often render DNA copies imperfect, it is much more important to study approximate repeats.

The first paper that dealt with model-based recognition of degenerate repeats solved the problem of finding the highest-scored pair of (possibly overlapping) substrings in a string in $O(N^2)$ time and space, where N is the length of input string [3]. Kannan and Myers [5] and Benson [2] restrict the outputs to pairs of non-overlapping substrings. Both algorithms run in $O(N^2 \log^2 N)$ time. The space usage is $O(N^2 \log N)$ [5], which was later improved to $O(N^2)$ [2]. A related question is to find all approximate repeats of a string which are within a fixed edit-distance threshold, which is the main focus of this paper. A detailed survey of previous work on finding approximate repeats can be found in [6].

A popular idea for finding maximal approximate repeats (as defined later in Section 1.2) which are within a fixed edit-distance threshold is to first search for small exact repeats as seed strings and then form maximal approximate repeats by extension. Kurtz et al. [6] have presented one such algorithm for finding maximal approximate repeats of length L which runs in $O(N + D^3 z)$ by constructing a suffix tree where N is the length of the input string, z is the number of seeds and D is the maximum edit-distance expected in the resulting repeats. The expected number of seeds is $E(z) = O(N^2 / |\Sigma|^{\lfloor L/D+1 \rfloor})$. This is suitable for finding maximal approximate repeats with small edit-distances. For long repeats and large edit-distances (that is, $L = c_1 N$ and $D = c_2 L$ where $c_1, c_2 < 1$), the expected running time of this algorithm is $O(N^5)$. This algorithm, therefore, is inefficient for extracting repeats for queries like "Find all maximal approximate repeats which are 10% different in edit-distance." Moreover, this algorithm would suffer from the poor locality behavior of the suffix tree [7], which results in many cache misses and dominates the running time.

Adebiyi, Jiang, and Kaufmann [1] designed algorithms for finding approximate non-tandem repeats by an idea similar to "seed expansion." The expected and practical running times of their algorithm are $O(DN^{1.7} \log N)$ and $O(N^{1.9} \log N)$, respectively, for DNA and protein sequences. However, it can only find short repeats with length $O(\log N)$. The main use of their algorithm is therefore for extracting short motifs only.

Instead of finding short approximate repeats of sequences as mentioned above, our paper finds the long conserved regions with relatively large edit-distance between multiple input sequences (genes). Moreover, our algorithm outputs only "significant" approximate repeats for diminishing the size of output such that the later analysis of the long conserved regions can be more efficient.

1.2 Definitions

Let S be a string of length $|S| = N$ over an alphabet Σ. $S[i]$ denotes the i-th character of S, for $i \in [1, N]$. For $i \leq j$, $S[i, j]$ denotes the substring of

S starting with the i-th and ending with the j-th character. Substring $S[i,j]$ can also be denoted by the pair of positions (i,j) sometimes for brevity and sometimes because the contents of the substrings are irrelevant. The length of the substring (i,j) is $\ell(i,j) = j - i + 1$.

There are three kinds of edit operations: deletions, insertions, and mismatches of single characters. The edit-distance of (i_1, j_1) and (i_2, j_2), denoted by $d((i_1, j_1), (i_2, j_2))$, is the minimum number of edit operations needed to transform $S[i_2, j_2]$ to $S[i_1, j_1]$.

A pair of positions (i_1, j_1), $i_1 \leq j_1$, covers a pair (i_2, j_2), $i_2 \leq j_2$, if and only if, $i_1 \leq i_2$ and $j_2 \leq j_1$. A pair $((i_1, j_1), (i_2, j_2))$ of pairs of string positions covers another pair $((i_3, j_3), (i_4, j_4))$ with respect to the first position if and only if (i_1, j_1) covers (i_3, j_3) and $\ell(i_1, j_1) \geq \ell(i_3, j_3) + 1$. A pair $((i_1, j_1), (i_2, j_2))$ of pairs of string positions completely covers another pair $((i_3, j_3), (i_4, j_4))$ if and only if (i_1, j_1) covers (i_3, j_3) and (i_2, j_2) covers (i_4, j_4). The length of $((i_1, j_1), (i_2, j_2))$ is defined to be $\ell((i_1, j_1), (i_2, j_2)) = \min\{j_1 - i_1 + 1, j_2 - i_2 + 1\}$.

A pair $((i_1, j_1), (i_2, j_2))$ of substrings is a (D, L)-approximate repeat if and only if $(i_1, j_1) \neq (i_2, j_2)$, $d((i_1, j_1), (i_2, j_2)) \leq D$ and $\ell((i_1, j_1), (i_2, j_2)) \geq L$. A (D, L)-approximate repeat is maximal if and only if it is not completely covered by any other (D, L)-approximate repeats. A (D, L)-maximal approximate repeat is supermaximal if and only if it is not covered by any other (D, L)-maximal approximate repeats with respect to the first position.

Two distinct (D, L)-maximal approximate repeats $((i_1, j_1), (i_2, j_2))$ and $((i_3, j_3), (i_4, j_4))$ are closely positioned if $i_1 = i_3$, $i_2 = i_4$. Note that, although $((i_1, j_1), (i_2, j_2))$ and $((i_3, j_3), (i_4, j_4))$ do not completely cover each other, (i_1, j_1) highly overlaps with (i_3, j_3), and (i_2, j_2) highly overlaps with (i_4, j_4). Figure 1 shows one such example. Worse, there are $O(D)$ such closely positioned (D, L)-maximal approximate repeats $((i_1, j_1), (i_2, j_2))$ given $i_1, i_2 \in [1, N]$ [6]. In the case that D is large, not all closely positioned (D, L)-maximal approximate repeats are interesting. Instead, if we only output (D, L)-supermaximal approximate repeats, then, given $i_1, i_2 \in [1, N]$, only one (D, L)-maximal approximate repeat $((i_1, j_1), (i_2, j_2))$ is output whose (i_1, j_1) is not covered by any other (i_3, j_3) for all (D, L)-maximal approximate repeats $((i_3, j_3), (i_4, j_4))$. This approach can efficiently diminish the size of output for the later analysis of the long conserved regions. Obviously, one useful application of extracting all (D, L)-supermaximal approximate repeats (β, β') of S is to find all longest possible substrings β of S such that there exists some other substring β' of S where $d(\beta, \beta') \leq D$ and $\ell(\beta, \beta') \geq L$.

This paper presents an algorithm that extracts all (D, L)-supermaximal approximate repeats of S given an input string S and thresholds L and D. This algorithm can be easily applied to the case where there are multiple input strings $S_1, S_2, \ldots, S_n$ by first concatenating the input strings into one long subject string S with a special symbol "$\sharp$" for separation: $S_1 \sharp S_2 \sharp \cdots \sharp S_n$.

The rest of this paper is organized as follows. Section 2 gives an efficient algorithm for a simplified problem. Section 3 gives an efficient algorithm for extracting all (D, L)-supermaximal approximate repeats based on the solution

Assume $D = 3$, $S[x, x + 7]=$ GAATCGGT and $S[y, y + 6] =$ CTAGGCT. There are two (3,4)-maximal approximate repeats $((i_1, j_1), (i_2, j_2))$ and $((i_3, j_3), (i_4, j_4))$ where $i_1 = i_3 = x + 1$, $i_2 = i_4 = y + 1$, $j_1 = x + 6$, $j_2 = y + 4$, $j_3 = x + 4$ and $j_4 = y + 5$.

The first repeat $((i_1, j_1), (i_2, j_2))$:

$(i_1, j_1) = S[x + 1, x + 6]=$AATCGG
$(i_2, j_2) = S[y + 1, y + 4]=$TAGG

Alignment for $((i_1, j_1), (i_2, j_2))$:

(i_1, j_1) GAATCGGT
(i_2, j_2) CTA..GGC

The second repeat $((i_3, j_3), (i_4, j_4))$:

$(i_3, j_3) = S[x + 1, x + 4]=$AATC
$(i_4, j_4) = S[y + 1, y + 5]=$TAGGC

Alignment for $((i_3, j_3), (i_4, j_4))$:

(i_3, j_3) G.AATCG
(i_4, j_4) CTAGGCT

Note that, although $((i_1, j_1), (i_2, j_2))$ and $((i_3, j_3), (i_4, j_4))$ do not completely cover each other, (i_1, j_1) highly overlaps with (i_3, j_3), and (i_2, j_2) highly overlaps with (i_4, j_4). Outputting both of these maximal approximate repeats may not be interesting.

Fig. 1. An example of two closely positioned (3,4)-maximal approximate repeats

to the simplified problem. Section 4 proves the correctness of the algorithm and analyzes its time and space complexity. Moreover, Section 4 also shows that our algorithm is optimal in time when D is a constant.

2   The Simplified Problem

Before attacking the problem of finding all (D, L)-supermaximal approximate repeats, we first solve the problem of finding (D, L)-approximate repeats of the input string $S[1, N]$. Formally, we find, for each i, j, at least a pair of indices r and m such that $d((i, r), (j, m)) \leq D$, where $\ell(i, r) \geq L$ and $\ell(j, m) \geq L$ — if such r and m exist.

The idea is to find alignments with edit-distance k for every pair of S's substrings starting at i and j, respectively, where $1 \leq i, j \leq N$ and $1 \leq k \leq D$. More precisely, our algorithm computes $f_1(i, j, D)$ and $f_2(i, j, D)$ as the index $w + 1$ of $S[i, w]$ and the index $h + 1$ of $S[j, h]$, respectively, where $d((i, w), (j, h)) \leq D$ for the maximum possible index w. The complete algorithm appears in Figure 2.

2.1   Analysis of FIND-SIM-REPEAT

Step 1 initializes the boundary values. For $S[i] \neq S[j]$, step 8 considers the possible edit-operations to align substrings starting from index i and j: insertion, deletion, or substitution. It is obvious that $d((i, f_1(i, j, D) - 1), (j, f_2(i, j, D) - 1)) \leq D$ for $i > j$. It is also obvious that steps 9–18 assign $f_1(i, j, D)$ as the maximum value such that $d((i, f_1(i, j, D) - 1), (j, m)) \leq D$ for some m. If more than one of $f_1(i, j + 1, k - 1)$, $f_1(i + 1, j + 1, k - 1)$ and $f_1(i + 1, j, k - 1)$ are maximum, then FIND-SIM-REPEAT always chooses $f_1(x, y, k - 1)$ such that $f_2(x, y, k - 1)$ is the maximum of $f_2(i, j + 1, k - 1)$, $f_2(i + 1, j + 1, k - 1)$ and $f_2(i + 1, j, k - 1)$. Steps 21–26 prevent overlapping repeats to be output.

2.2   Time and Space Complexity

We analyze the time and space complexity. The running time is obviously $O(DN^2)$. Instead of using $O(DN^2)$ space as traditional dynamic programming does, we can reduce it to $O(DN)$. This is based on the observation that the only values needed to compute $f_1(i, j, k)$ are $f_1(i, j + 1, k - 1)$, $f_1(i + 1, j + 1, k - 1)$, $f_1(i + 1, j, k - 1)$, and $f_1(i + 1, j + 1, k)$. Similarly, $f_2(i, j, k)$ depends only on $f_2(i, j + 1, k - 1)$, $f_2(i + 1, j + 1, k - 1)$, $f_2(i + 1, j, k - 1)$, and $f_2(i + 1, j + 1, k)$. Note that when we compute the values $f_1(i, j, k)$, $j < i$ and $k = 1, \ldots, D$, those previously computed $f_1(i', j, k)$ with $i' > i + 1$ are no longer needed. Only the "wavefront" is needed for the computation. This reduces the space complexity to $O(DN)$.

3   Extracting Supermaximal Approximate Repeats

The number of (D, L)-approximate repeats output by FIND-SIM-REPEAT can be very large, so this fact makes difficult the later analysis of long conserved regions. In order to extract only "significant" repeats, one idea is to output only (D, L)-maximal approximate repeats as in [6]. However, as discussed in Section 1.2, the number of (D, L)-maximal approximate repeats can still be large, especially when D is large. Therefore, we would like to output (D, L)-supermaximal approximate repeats only.

We present a two-phase procedure for extracting all (D, L)-supermaximal approximate repeats of the input string S. First, a modified version of FIND-SIM-REPEAT outputs substrings β for all (D, L)-supermaximal approximate repeats (β, β'). Second, according to those β output by the first phase, we apply another modified version of FIND-SIM-REPEAT to extract all (D, L)-supermaximal approximate repeats (β, β') and output them. We call the resulting algorithm FIND-SUPER-REPEAT.

1: $f_1(N, j, k) = N, j = 1, \ldots, N, k = 0, \ldots, D;$
$\quad f_2(N, j, k) = j, j = 1, \ldots, N, k = 0, \ldots, D;$
$\quad f_1(j, j, k) = N, j = 1, \ldots, N, k = 0, \ldots, D;$
$\quad f_2(j, j, k) = N, j = 1, \ldots, N, k = 0, \ldots, D;$
2: **for** $i = N - 1, \ldots, 1, j = i - 1, \ldots, 1$ **do**
3: $\quad$ **if** $S[i] = S[j]$ **then**
4: $\quad\quad$ $f_1(i, j, k) = f_1(i + 1, j + 1, k), k = 0, \ldots, D;$
$\quad\quad$ $f_2(i, j, k) = f_1(i + 1, j + 1, k), k = 0, \ldots, D;$
5: $\quad$ **else**
6: $\quad\quad$ $f_1(i, j, 0) = i;$
$\quad\quad$ $f_2(i, j, 0) = j;$
7: $\quad\quad$ **for** $k = 1, \ldots, D$ **do**
8: $\quad\quad\quad$ Compare the values of $f_1(i, j + 1, k - 1), f_1(i + 1, j + 1, k - 1)$ and $f_1(i + 1, j, k - 1);$
9: $\quad\quad\quad$ **if** $f_1(i, j + 1, k - 1)$ is the maximum **then**
10: $\quad\quad\quad\quad$ $f_1(i, j, k) = f_1(i, j + 1, k - 1);$
11: $\quad\quad\quad\quad$ $f_2(i, j, k) = f_2(i, j + 1, k - 1);$
12: $\quad\quad\quad$ **else if** $f_1(i + 1, j + 1, k - 1)$ is the maximum **then**
13: $\quad\quad\quad\quad$ $f_1(i, j, k) = f_1(i + 1, j + 1, k - 1);$
14: $\quad\quad\quad\quad$ $f_2(i, j, k) = f_2(i + 1, j + 1, k - 1);$
15: $\quad\quad\quad$ **else**
16: $\quad\quad\quad\quad$ $f_1(i, j, k) = f_1(i + 1, j, k - 1);$
17: $\quad\quad\quad\quad$ $f_2(i, j, k) = f_2(i + 1, j, k - 1);$
18: $\quad\quad\quad$ **end if**
19: $\quad\quad$ **end for**
20: $\quad$ **end if**
21: $\quad$ **if** $f_2(i, j, D) \geq i$ **then**
22: $\quad\quad$ $f_2(i, j, D) = i;$
$\quad\quad$ $k = 1;$
23: $\quad\quad$ **while** $f_2(i, j, k) < i$ **do**
24: $\quad\quad\quad$ $k + +;$
25: $\quad\quad$ **end while**
$\quad\quad$ $f_1(i, j, D) = f_1(i, j, k - 1) + (i - f_2(i, j, k - 1));$
26: $\quad$ **end if**
27: $\quad$ **if** $f_1(i, j, D) - i \geq L$ and $f_2(i, j, D) - j \geq L$ **then**
28: $\quad\quad$ Output $((i, f_1(i, j, D) - 1), (j, f_2(i, j, D) - 1))$
29: $\quad$ **end if**
30: **end for**

Fig. 2. FIND-SIM-REPEAT

Lemma 1. *The set of (D, L)-approximate repeats of a string S output by FIND-SIM-REPEAT includes all (D, L)-supermaximal approximate repeats of S.*

Proof. Assume there is a (D, L)-supermaximal approximate repeat $((i, q), (j, m))$ which is not output by FIND-SIM-REPEAT. Then, this situation immediately results in a contradiction because, for the indices i and j, FIND-SIM-REPEAT always outputs the (D, L)-approximate repeat $((i, w), (j, h))$ such that $d((i, w), (j, h)) \leq D$ for the maximum possible index w. $\qquad\square$

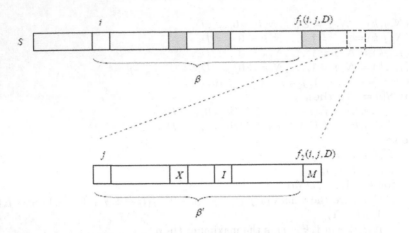

M: Mismatch   X: Deletion   I: Insertion

Fig. 3. An example of a $(2, L)$-approximate repeat (β, β') output by FIND-SIM-REPEAT where $D = 2$, $\beta = (i, f_1(i, j, D) - 1)$, $\beta' = (j, f_2(i, j, D) - 1)$ and the lengths of β and β' are $\ell(i, f_1(i, j, D) - 1) = f_1(i, j, D) - i \geq L$ and $\ell(j, f_2(i, j, D) - 1) = f_2(i, j, D) - j \geq L$, respectively.

Let U be the set of (D, L)-approximate repeats of S output by FIND-SIM-REPEAT. By Lemma 1, the first phase of FIND-SUPER-REPEAT finds β_1 for all (D, L)-supermaximal approximate repeats (β_1, β'_1) of S in U. Moreover, it only needs to focus on the substring β_2 of each (D, L)-approximate repeat (β_2, β'_2) in U for finding all of those β_1, so we can reduce the remaining work for the first phase of FIND-SUPER-REPEAT to maximal substrings problem defined below.

Finding Maximal Substrings. Given a string S and a set Q of substrings of S where $\mid S \mid = N$ and $\mid Q \mid = M$. We would like to find a subset W of Q such that none of the strings in W are covered by any other strings in Q. The time and space complexity are $O(M)$ and $O(N)$, respectively.

Let $P[1 \ldots N]$ be an array where $P[i] = -1$ for $i = 1, \ldots, N$ initially. The complete algorithm FIND-MAX-SUBSTRING appears in Figure 4. Recall that substring $S[x, y]$ is denoted by the pair of positions (x, y).

Lemma 2. *A substring* (x, y) *of* S *is output by FIND-MAX-SUBSTRING if and only if* (x, y) *is not covered by any other substrings of* S *in* Q.

Proof. If (x, y) is not covered by any other substrings of S in Q, then when (x, y) is picked up from Q in step 2, the "if condition" in step 3 must be satisfied. So, $P[x] = y$ before step 8 is executed. Moreover, $P[x]$ is always larger than maxTo in the x-th iteration of step 9. Otherwise, there must exist another substring (i, q) where $i < x$, $q \geq y$ and $P[x] = q$ such that (i, q) covers (x, y). This situation contradicts our assumption of (x, y). So, (x, y) is output.

```
 1: while Q ≠ ∅ do
 2:     Pick a substring (x, y) from Q arbitrarily;
 3:     if y > P[x] then
 4:         P[x] = y;
 5:     end if
 6: end while
 7: Let maxTo= −1;
 8: for i = 1, ..., N do
 9:     if P[i] > maxTo then
10:         Output (i, P[i]);
            maxTo= P[i];
11:     end if
12: end for
```

Fig. 4. FIND-MAX-SUBSTRING

On the other hand, assume (x, y) is covered by some other substrings (i, q) of S in Q. If $i = x$ and $q > y$, then $P[x] = q$ before step 8 is executed and (x, y) will not be output. If $i < x$, $q \geq y$ and $P[x] = y$ before step 8 is executed, then the value of maxTo is at least q in the x-th iteration of step 9. Since maxTo $\geq q \geq y$, the "if condition" in step 9 will not be satisfied. (x, y) will not be output.   □

Similar to FIND-MAX-SUBSTRING, in order to output β of all (D, L)-supermaximal approximate repeats (β, β'), we need to modify steps 27–29 and add some steps at the end of FIND-SIM-REPEAT. Figures 5 and 6 show this approach. The first phase of FIND-SUPER-REPEAT is this modified version of FIND-SIM-REPEAT. The functionality of REPEAT$[1 \ldots N]$ is similar to that of $P[1 \ldots N]$ in FIND-MAX-SUBSTRING and REPEAT$[i] = -1$ for $i = 1, \ldots, N$ initially.

```
if f₁(i, j, D) − i ≥ L and f₂(i, j, D) − j ≥ L and f₁(i, j, D) − 1 > REPEAT[i] then
    REPEAT[i]= f₁(i, j, D) − 1;
end if
```

Fig. 5. Modification of steps 27–29 of FIND-SIM-REPEAT for outputting β for all (D, L)-supermaximal approximate repeats (β, β')

```
maxTo = −1;
for i = 1, ..., N do
    if REPEAT[i] > maxTo then
        Output (i,REPEAT[i]);
        maxTo = REPEAT[i];
    end if
end for
```

Fig. 6. Additional steps at the end of FIND-SIM-REPEAT for outputting β for all (D, L)-supermaximal approximate repeats (β, β')

The second phase of FIND-SUPER-REPEAT is trivial. It simply follows FIND-SIM-REPEAT and outputs only (D, L)-supermaximal approximate repeats (β, β') according to those β output by the first phase.

4   Correctness and Analysis

4.1   Correctness

Theorem 1. *A (D, L)-approximate repeat of a string S is a (D, L)-supermaximal approximate repeat if and only if it is output by FIND-SUPER-REPEAT.*

Proof. Lemma 1 and 2 imply that the first phase of FIND-SUPER-REPEAT outputs β for all (D, L)-supermaximal approximate repeats (β, β'). Moreover, because Lemma 1 shows that the output of FIND-SIM-REPEAT must include all (D, L)-supermaximal approximate repeats of S, therefore, according to those β output by the first phase, the second phase of FIND-SUPER-REPEAT is able to extract all (D, L)-supermaximal approximate repeats (β, β'). As a result, a (D, L)-approximate repeat of S is a (D, L)-supermaximal approximate repeat if and only if it is output by FIND-SUPER-REPEAT. $\qquad\square$

4.2   Time and Space Complexity

Recall that the time and space complexity of FIND-SIM-REPEAT are $O(DN^2)$ and $O(DN)$, respectively. The first phase of FIND-SUPER-REPEAT follows FIND-SIM-REPEAT with modifications as shown in Figures 5 and 6. Obviously, the steps of FIND-SUPER-REPEAT in Figure 5 do not increase the time complexity of FIND-SIM-REPEAT. Also, the time complexity of the steps of FIND-SUPER-REPEAT in Figure 6 is $O(N)$. On the other hand, the second phase of FIND-SUPER-REPEAT, in fact, is just to run FIND-SIM-REPEAT once. Therefore, the overall time complexity of FIND-SUPER-REPEAT is $O(DN^2)$.

Besides, the space complexity of FIND-SUPER-REPEAT is $O(DN)$. It is because FIND-SUPER-REPEAT only needs one more array REPEAT$[1 \ldots N]$ than FIND-SIM-REPEAT does and the space complexity of REPEAT$[1 \ldots N]$ is clearly $O(N)$.

4.3   Optimality of FIND-SUPER-REPEAT

Finally, we would like to show FIND-SUPER-REPEAT is optimal in time complexity when the input D is a constant. In addition, the space complexity of our algorithm is linear in N.

From Theorem 1, it is obvious that the time and space complexity of FIND-SUPER-REPEAT are $O(N^2)$ and $O(N)$, respectively, when D is a constant. Then, for each $e \in \Sigma^*$, we define the powers of e by $e^1 = e$, $e^2 = ee$, $e^3 = ee^2, \ldots, e^{n+1} = ee^n$ for any positive integer n. Let $\Sigma = \{A, T, G\}$, strings $S_1 = (A^{c_1}T^{c_2})^{N/2}$, and $S_2 = (G^{c_1}T^{c_2})^{N/2}$, where c_1, c_2 are positive constants. We concatenate S_1, S_2 into one string $S = S_1 S_2$ to supply as the input of

FIND-SUPER-REPEAT. Then, given $D < c_1$, $L = c_2$, the number of all (D, L)-supermaximal approximate repeats of S is $\Omega(N^2)$. It is because there are $\Theta(N)$ of S's substrings $\beta = A^D T^{c_2}$ which can be aligned with $\Theta(N)$ of S's substrings $\beta' = G^D T^{c_2}$ such that the combinations of all (β, β') are (D, L)-supermaximal approximate repeats. As the time for outputting all (D, L)-supermaximal approximate repeats in this tight example is $\Omega(N^2)$, FIND-SUPER-REPEAT is optimal in time complexity when D is a constant.

References

1. E. F. Adebiyi, T. Jiang and M. Kaufmann (2001). "An Efficient Algorithm for Finding Short Approximate Non-tandem Repeats." *Bioinformatics*, 17(90001), pp. S5–S12.
2. G. Benson (1994). "A Space Efficient Algorithm for Finding the Best Nonoverlapping Alignment Score." In M. Crochemore and D. Gusfield, eds., *Proceedings of the 5th Annual Symposium on Combinatorial Pattern Matching (CPM94)*, volume 807 of *LNCS*, pp. 1–14. Berlin: Springer Verlag.
3. W. Fitch, T. Smith and J. Breslow (1986). "Detecting Internally Repeated Sequences and Inferring the History of Duplication." In J. P. Segrest and J. J. Albers, eds., *Plasma Proteins. Part A: Preparation, Structure, and Molecular Biology*, volume 128 of *Methods in Enzymology*, pp. 773–788. San Diego, CA: Academic Press.
4. D. Gusfield (1997). *Algorithms on Strings, Trees and Sequences: Computer Science and Computational Biology*. New York: Cambridge University Press.
5. S. K. Kannan and E. W. Myers (1996). "An Algorithm for Locating Nonoverlapping Regions of Maximum Alignment Score." *SIAM J. Computing*, 25(3), pp. 648–662.
6. S. Kurtz, E. Ohlebusch, et al. (2000). "Computation and Visualization of Degenerate Repeats in Complete Genomes." In *Proceedings of the Eighth International Conference on Intelligent Systems for Molecular Biology (ISMB2000)*, pp. 228–238.
7. S. Kurtz (1999). "Reducing the Space Requirement of Suffix Trees." *Software Practice and Experience*, 29(13), pp. 1149–1171.

The Reversal Median Problem, Common Intervals, and Mitochondrial Gene Orders

Matthias Bernt, Daniel Merkle, and Martin Middendorf

Parallel Computing and Complex Systems Group
Department of Computer Science
University of Leipzig, Germany
{bernt, merkle, middendorf}@informatik.uni-leipzig.de

Abstract. An important problem for phylogenetic investigations that
are based on gene orders is to find for three given gene orders a fourth
gene order that has a minimum sum of reversal distances to the three given
gene orders. This problem is called Reversal Median problem (RMP). The
RMP is studied here under the constraint that common (combinatorial)
structures are preserved which are modeled as common intervals. An existing branch-and-bound algorithm for RMP is extended here so that it
can solve the RMP with common intervals optimally. This algorithm is
applied to mitochondrial gene order data for different animal taxa. It is
shown that common intervals occur often for most taxa and that many
common intervals are destroyed when the RMP is solved optimally with
standard methods that do not consider common intervals.

1 Introduction

The order of genes in the genomes of many species has changed during their
evolution. The differences between the gene orders of species can be used to
infer information on their evolutionary relation. A common method to compare
the gene order of two species is to compute a sequence of genome rearrangement
operations that transfers one order into the other. Often most parsimonious rearrangement scenarios are computed which use a minimal number of rearrangement operations. Recently, several works appeared that consider rearrangement
scenarios where certain common combinatorial structures within the gene order
are not destroyed ([1,2,6]). Such combinatorial structures can model groups of
genes that have a common control. Often such a control requires that the corresponding genes occur neighboured along the gene strand. Common intervals
([3]) and conserved intervals [12] are examples of such combinatorial structures
that describe groups of neighboured genes.

Formally, gene orders can be described as signed permutations over a set of
integers where each integer denotes a gene and its sign denotes the orientation.
The most often considered rearrangement operations are reversal operations (reversals) which are permutations that reverse the order of a subsequence of neighboured genes and change the sign of each reversed gene. The minimal number of
reversals that are required to convert one given signed permutation into another

M.R. Berthold, R. Glen, and I. Fischer (Eds.): CompLife 2006, LNBI 4216, pp. 52–63, 2006.

given signed permutation is called reversal distance ([4]) and can be used to measure the phylogenetic distance between the species.

The Reversal Median problem (RMP) is to find for three given signed permutations a fourth signed permutation which has a minimum sum of reversal distances to the given signed permutations. RMP has been used by several authors as a starting point for finding a phylogenetic tree that describes a parsimonious rearrangement scenario for more than three given signed permutations (Multiple Genome Rearrangement problem). It is known that RMP is NP-complete ([5]).

GRAPPA ([7]) and MGR ([8]) are two well known algorithms for solving the Multiple Genome Rearrangement problem. GRAPPA is a significantly extended version of the BPAnalysis algorithm ([9]) that uses the reversal distance measure. A key element of GRAPPA and MGR is an algorithm to solve the RMP. GRAPPA has integrated two different branch-and-bound algorithms for RMP — Siepel's median solver ([10]) and Caprara's median solver ([5]). MGR uses a heuristic method to solve the RMP. Another heuristic for RMP is rEvoluzer [11].

In this paper we study RMP under the additional constraint that certain common combinatorial structures which can describe groups of genes of the given permutations are preserved. The biological motivation is that reversals which destroy gene groups can often be lethal and therefore are unlikely events during phylogeny. Two such structures for gene groups that have been studied in the literature are conserved intervals [12] and common intervals [3]. RMP with the aim not to destroy conserved intervals has been studied studied in [11] and is realized in the heuristic rEvoluzer. Here we study RMP with the constraint that common intervals are not destroyed, i.e., the problem is to find for three given signed permutations π^1, π^2, π^3 a fourth signed permutation π^4 such that $\{\pi^1, \pi^2, \pi^3\}$ and $\{\pi^1, \pi^2, \pi^3, \pi^4\}$ have the same common intervals and π^4 has a minimum sum of reversal distances where only reversals that do not destroy common intervals are allowed. We call this problem Common Interval Preserving Reversal Median problem (CIP-RMP). Whereas rEvoluzer is a heuristic that has some similarities with the heuristic MGR we follow a different approach in this paper. The idea is to change Caprara's median solver in GRAPPA so that the changed version can solve CIP-RMP optimally.

The following example taken from [12] illustrates the idea of finding a median without destroying common intervals. Table 1 shows the condensed mitochondrial gene orders of three Arthropods. As noted in [12] the gene orders are very similar. E.g. the element set $\{4, 5, 6\}$ is a common interval of the input permutations (for a formal introduction of common intervals see Section 2). The same holds for the set $\{13, 14, 15\}$. The RMP median destroys these common intervals. In the example 44 of the 74 common intervals are destroyed by the given RMP median. Table 1 also gives an CIP-RMP median which preserves all common intervals of the input permutations.

In the first part of this paper we describe the changes that have been applied to Caprara's median solver in GRAPPA. Two new versions of Caprara's median solver are proposed. The first version called CIP makes slight changes in the

Mosquito	1 2 3 4	5	6	8	7	9	-10	11	12	13	14	15	16	17
Silkworm	1 2 3 4	5	6	7	8	9	10	11	12	14	13	15	16	17
Centipede	1 3 4 5	6	7	8	9	10	11	-2	12	16	13	14	15	17
RMP median	1 2 3 4	-14	-13	-12	-11	-10	-9	-8	-7	-6	-5	15	16	17
CIP-RMP median	1 2 3 4	5	6	7	8	9	10	11	12	13	14	15	16	17

Fig. 1. Condensed mitochondrial gene orders of three Arthropods

branch-and-bound algorithm of Caprara's median solver with the effect that median solutions which destroy common intervals are not accepted. The modified distance measure is computed based on the algorithm described in [13] (see also [14]). One problem with this distance measure is that the corresponding decision problem is NP-complete. It is not hard to see, that CIP-RMP is also NP-complete. In order to increase the speed of the modified algorithm we propose a second version called Extended-CIP (E-CIP) which avoids computing gene orders that destroy common intervals. It is shown experimentally that E-CIP obtains a significant speedup over CIP. It should be noted that our versions can also handle circular genomes. In the second part of this paper we apply the new versions of Caprara's median solver to RMP instances with biological data. RMP and CIP-RMP are analysed for mitochondrial gene orders for various taxa of animals. It is shown how important common intervals are in the different taxa. Moreover, it is analysed how many common intervals are destroyed when standard GRAPPA for RMP is used.

The rest of the paper is organized as follows. In Section 2 we give basic definitions. Section 3 presents Caprara's median solver. Section 4 describes the modifications to Caprara's median solver. Experimental results are discussed in Section 5, and Conclusions are drawn in Section 6.

2   Basic Definitions

A *permutation of size* n is a permutation of the elements $\{1, 2, \ldots, n\}$. A *signed permutation of size* n is a permutation of size n where every element has an additional sign ("+" or "−") that defines its orientation (the "+" sign is usually omitted). A *reversal* $\rho(i, j)$, $1 \leq i \leq j \leq n$ applied to a signed permutation π of size n transforms it into $\pi \circ \rho = (\pi_1, \ldots, \pi_{i-1}, -\pi_j, \ldots, -\pi_i, \pi_{j+1}, \ldots, \pi_n)$. The *reversal distance* $d(\pi, \sigma)$ between two signed permutations π and σ is the minimum number of reversals that is necessary to transform π into σ.

An interval of a permutation π is a set of consecutive elements of the permutation π. Let Π be a set of signed permutations of size n. A *common interval* [1] of Π is a subset of $\{1, 2, \ldots, n\}$ of size ≥ 2 that is an interval in all $\pi \in \Pi$. Let $C(\Pi)$ be the set of all common intervals of Π. Two common intervals c and c' overlap if $c \cap c' \neq \emptyset$, $c \not\subset c'$, and $c' \not\subset c$. The overlap graph of C is the graph $G(C) = (C, E)$ with node set C where there is an edge between nodes c and c' if c and c' overlap. A sequence $c_1, \ldots, c_k$ of common intervals is a *chain* if every two successive intervals overlap. A common interval c is called *reducible* if there exists a chain $c_1, \ldots, c_k$ of length at least two such that $c = \cup_{i=1}^{k} c_i$ and otherwise it is called *irreducible*.

3   Caprara's Median Solver

Caprara's median solver ([5]) is a branch-and-bound algorithm that solves RMP optimally. As coded in GRAPPA it uses the multiple breakpoint graph (MB graph) data structure. Let $\pi^1, \ldots, \pi^q$ be signed permutations of size n. The corresponding MB graph is a graph on the node set $V = \{0, 1, \ldots, 2n + 1\}$. The edge set is defined using the fact that there is a correspondence between signed permutations and perfect matchings M of V (i.e., sets of pairs of different elements of V such that each element of V occurs in exactly one pair) with the property that $M \cup \{(2i - 1, 2i) \mid i = 1, \ldots, n\}$ form a Hamiltonian cycle. Such perfect matchings are called *permutation matchings*. Formally, the multiple edge set of the MB graph is $M(\pi^1) \cup \ldots \cup M(\pi^q)$ where for a signed permutation π the set $M(\pi)$ is the so called permutation matching defined by $M(\pi) = \{(2|\pi_i| - \nu(\pi_i), 2|\pi_{i+1}| - 1 + \nu(\pi_{i+1})) : i \in \{0, \ldots, n\}\}$, where $\nu(\pi_i) = 0$ if $\pi_i \geq 0$ and $\nu(\pi_i) = 1$ if $\pi_i < 0$. Let $c(M(\pi^1), M(\pi^2))$ (resp. $c(\pi^1, \pi^2)$) be the number of cycles in the MB graph defined by two arbitrary permutation matchings. The heart of Caprara's algorithm is to solve the Cycle Median problem (CMP), which is to find a permutation matching T such that $q(n+1) - \sum_{i=1}^{q} c(T, M(\pi^i))$ is minimised. This is inspired by the well known (see, e.g., [15]) inequality $d(\pi^1, \pi^2) \geq n+1-c(M(\pi^1), M(\pi^2))$ that relates the reversal distance between two permutations to the number of cycles in $M(\pi^1) \cup M(\pi^2)$. From this inequality follows that $\sum_{i=1}^{q} d(\sigma, \pi^i) \geq q(n + 1) - \sum c(\sigma, \pi^i)$. This implies a lower bound on the sum of reversal distances for a candidate solution of the RMP that is used in the branch-and-bound algorithm.

The branch-and-bound algorithm that is used to solve the CMP (for details see [5]) has a branching operation that corresponds to the so called edge contraction operation. This operation removes an edge together with its adjacent vertices from the MB graph and modifies the affected permutation matchings in an appropriate way. Contracting an edge in the MB graph, means fixing the edge in the solution permutation matching. The algorithm enumerates in a depth-first way all combinations of edges which are permutation matchings and performs a lower bound test after each fixing. The current best known solution is always updated when a complete permutation matching has been constructed. It was shown in [5] that the quality of the optimal solutions often equals the lower bound that can be computed for the input permutations. Therefore the algorithm uses a target value t for the quality of a solution. This target value is initialised with the lower bound. The algorithm then searches for a CMP solution with value t. If such a solution is found, it is optimal and the algorithm stops. Otherwise no solution with this value exists and the algorithm is started again with t increased by one. With slight modifications the branch-and-bound algorithm finds also optimal solutions for RMP.

4   Computing Optimal Structure Preserving Medians

In this section we explain how Caprara's Median Solver can be adapted to find structure preserving medians which are optimal under the corresponding structure preserving reversal distance measure. In particular, we consider the case of common intervals.

4.1   Algorithm CIP

An easy way to adapt Caprara's Median Solver is to check always when a complete permutation matching, i.e., a complete permutation, has been constructed, if this permutation destroys any intervals of the input permutations. If this is the case it is not accepted. Clearly, this check has to be done before updating the best known solution. Additionally the reversal distance measure that is used in the median solver has to be replaced by the corresponding structure preserving reversal distance.

4.2   Algorithm E-CIP

A more efficient method to adapt Caprara's Median Solver is to constrain the branch-and-bound search. The main idea is that a branching step of the algorithm, i.e. the addition of an edge to the partial permutation matching, can also be seen as adding a new element to the partial permutation by fixing an adjacency (by definition there is a one-to-one relationship between edges of a permutation matching and adjacencies of the corresponding permutation). Thus the algorithm can be adapted to produce only structure preserving permutations by forbidding certain branching steps, i.e., by excluding adjacencies of the permutation which are impossible when preserving the given structures. Again the reversal distance measure has to be replaced by the corresponding structure preserving reversal distance measure. In the remainder of this section we describe the constraints that can be used for common intervals. We distinguish static and dynamic constraints. In the following let Π be a set of signed permutations of size n and $C = C(\Pi)$ the set of common intervals of Π.

The static constraints are impossible adjacencies which are independent of the decisions that are made during the assembly of a candidate median gene order. The static constraints can be described by an $n \times n$ zero-one-matrix $M = (m_{ij})$ which can be computed before the run of the algorithm. For each adjacency between i and j that is allowed the corresponding matrix element is one, i.e., $m_{ij} = 1$.

It can be observed that a single common interval does not imply static constraints because every element can be moved freely within this interval, and thus any adjacency with an element from inside or outside is possible. But, if there are overlapping common intervals their elements can not be moved freely within both intervals. To described the static constraints that can be derived from overlapping intervals consider a connected component of $G(C)$ of size at least two and let C' be the nodes of this component. Clearly the set of integers that appear in at least one common interval in C' is a subinterval $[r : s]$ of $[1 : n]$. Then there exist integers $r = i_1, \ldots, i_k \leq s$, $1 \leq i_1 < \ldots < i_k < i_{k+1} = s + 1$, $k > 1$ such that for each subinterval $I_j := [i_j : i_{j+1} - 1]$, $j \in \{1, \ldots, k\}$ and each common interval c in C' either $I_j \subset c$ or $I_j \cap c = \emptyset$ and the subintervals are maximal with this property. It is not hard to show that two integers $i \in I_h$ and $j \in I_l$, $1 \leq h \leq l \leq k$ can only be adjacent when $|h - l| \leq 1$. Moreover, an integer $i \in I_1 \cup I_k$ can be adjacent to elements in $[1 : i_1 - 1] \cup [i_{k+1} : n]$. Clearly, all these constraints are static constraints.

Unfortunately this is not enough to describe all possible constraints when common intervals have to be regarded. There are also dynamic constraints because choosing an element for an interval implies that all elements of this interval have to be chosen before elements from outside. Recall that the branch-and-bound algorithm constructs a candidate permutation by successively fixing adjacencies during the branching steps. This can be seen as appending one element to the partial permutation that has been constructed so far. If the element that has been chosen in a branching step is from one of the subintervals of the form I_h, $h \in [1 : k]$ (as described in the previous paragraph) it is clear that the next elements that can be chosen must be chosen from the same interval until all elements in I_h have been chosen. This holds analogously also for intervals $[i : j]$ that correspond to common intervals which are isolated nodes in the graph $G(C)$. Clearly, all these constraints are not static because in principle an element from an interval I_h, $h \in [2 : k - 1]$ can also be neighboured to an element in I_{h-1} or I_{h+1} (similar arguments hold for the cases $h = 1$ and $h = k$). All these dynamic constraints are therefore computed during the run of our version of Caprara's median solver used for E-CIP.

The sketch of Algorithm E-CIP and the modifications in GRAPPA are given in Algorithms 1, 2, and 3. Algorithm 1 describes the preprocessing phase of E-CIP for computing the static and dynamic constraints. Algorithm 2 presents the modifications done in GRAPPA to restrict the branching step. Note, that the number of remaining elements in the subintervals used for the dynamic constraints are recomputed when GRAPPA changes its recursion level. Algorithm 3 is used for checking if branching is allowed with respect to the static and dynamic constraints.

4.3   Extensions and Improvements

The described method also applies to circular chromosomes, with some slight modifications. The main idea is that each common interval of a circular genome has a complementary common interval, and if one of the complementary pair is preserved the other is preserved, too. Circular genomes are usually represented as linear genomes by cutting them at an arbitrary position (in all genomes the same). This has to be done with more care to ensure that the above arguments still hold. The cut has to be done between two adjacent subintervals I_k of an arbitrary connected component $G(C)$. It should be noted that it is enough to consider only irreducible common intervals in order to compute the static and dynamic constraints. This is advantageous because the number of irreducible

---

Algorithm 1. Preprocessing of algorithm E-CIP: computation of the constraints

1: **INPUT:**
    permutations π^1, π^2, π^3
    OUTPUT:
    m - matrix describing the static constraints
    $\forall c, j : l_c[j]$ - length of all subintervals I_j of a component c
    $\forall c, e : R_c[e]$ - index of the subinterval of component c where element e is included
2: initialise $m_{ij} = 1$ $\forall i, j \in \{1, \ldots, n\}$
3: C = common intervals of $\{\pi^1, \pi^2, \pi^3\}$
4: find the connected components in the overlap graph $G(C)$
5: **for all** connected components c **do**
6:     determine the i_j $(1 \le j \le k + 1)$
7:     $i_0 \leftarrow 1$ and $i_{k+2} \leftarrow n + 1$
8:     $I_j := [i_j : i_{j+1} - 1]$ $\forall j \in [0 : k + 1]$
9:     **for all** I_h and I_l with $|h - l| > 1$ **do**
10:         $m_{ij} \leftarrow 0$ $\forall i \in I_h, \forall j \in I_l$
11:     **end for**
12:     **for all** $e \in \{1, \ldots, n\}$ **do**
13:         let j be the index of I_j for which $e \in I_j$
14:         $R_c[e] \leftarrow j$
15:     **end for**
16:     **for all** $j \in \{0, \ldots, k + 1\}$ **do**
17:         $l_c[j] \leftarrow i_{j+1} - i_j$
18:     **end for**
19: **end for**

---

common intervals is always smaller n whereas the number of common intervals is only smaller than $n \cdot (n + 1)/2$ in general.

5 Results

In this section we describe our results for applying the new versions of Caprara's median solver — CIP and E-CIP — to biological data. The frequency of common intervals in biological data is studied and the new GRAPPA versions were used to compare solutions for the RMP and CIP-RMP problem. Moreover, we measured the speedup of E-CIP over CIP. As test instances we choose mitochondrial gene order data from animal genomes of different taxa.

The test instances were chosen as follows. All mitochondrial genomes that were marked as complete were taken from the Mitochondrial Database (see [16]). From these mitochondrial gene orders we selected all those gene orders that contain all the 37 standard genes that can be found in animal mitochondrial genomes. These standard genes are: i) 13 protein coding genes cox1, cox2, cox3, nad1, nad2, nad3, nad4, nad4L, nad5, nad6, atp6, atp8, cob, ii) 2 rRNA genes rrnL and rrnS and iii) 22 tRNA genes L(nag), Y, A, V, E, D, G, F, I, H, K, M, Q, P, R, W, S(nct), N, L(yaa), C, S(nga), T. The species of the resulting mitochondrial gene orders were grouped into taxa in order to get subsets of reasonable size.

Algorithm 2. Algorithm E-CIP: Modification of branching in GRAPPA

1: **if** branching allowed for (i, j) **then**
2:    **for all** connected components c **do**
3:       $l_c[R_c[j]] \leftarrow l_c[R_c[j]] - 1$
4:    **end for**
5:    branch
6:    **for all** connected components c **do**
7:       $l_c[R_c[j]] \leftarrow l_c[R_c[j]] + 1$
8:    **end for**
9: **end if**

Table 1. Total number (#) of mitochondrial gene orders and number of different gene orders ($\#_\delta$) for different taxa; average reversal distance (d); percentage of test instances where at least one common interval is destroyed by the optimal solution of RMP computed with standard GRAPPA (%); in brackets abbreviations used in the result section

Taxon	#	$\#_\delta$	d	%
Annelida (Ann)	3	3	9.67	100
Brachiopoda (Bra)	3	3	25.66	100
Arthropoda (Art)	81	41	15.04	29
Arthropoda Crustacea (Cru)	23	18	15.90	18
Arthropoda Hexapoda (Hex)	38	15	12.50	35
Mollusca (Mol)	10	10	25.02	24
Chordata (Cho)	452	45	6.97	25
Chordata Actinopterygii (Act)	200	23	5.81	8
Chordata Sarcopterygii (Sar)	234	22	4.73	19
Echinodermata (Ech)	8	5	10.60	0

This was done with the help of taxonomy information in the Mitochondrial Database and manual inspection of the phylogenetic tree given from the NCBI taxonomy browser. Table 1 shows the absolute number of gene orders grouped in the different taxa and the number of different gene orders in each taxon. As test instances we used for each taxon in the table all possible triples of different mitochondrial gene orders. Table 1 contains also the abbreviations used for the different taxa in this section. For all tests we used the versions of GRAPPA, CIP, and E-CIP that handle the gene orders as circular orders.

The number of common intervals for the test instances in the different taxa is shown in Figure 2. The figure shows that the number of common intervals in the test sets is very different between the taxa. For Brachiopoda (only one test instance) and Mollusca the average number of common intervals is zero or nearly zero and it is nearly 1000 for the Chordata (and in both subtaxa Actinopterygii and Sarcopterygii). Note, that the number of common intervals is higher for Deuterostomia as for Protostomia, which is due to the fact, that the pairwise reversal distance is smaller in Deuterostomia (see Table 1).

Furthermore Figure 2 depicts the fraction of the common intervals that are destroyed by an optimal solution of the RMP obtained with standard GRAPPA,

---

Algorithm 3. Algorithm E-CIP: Branching restriction in GRAPPA

1: **INPUT:**
(i, j) - a potential adjacency
m, R, l - see algorithm 1
OUTPUT:
true if branching is allowed, false otherwise
2:
3: // check static constraints
4: **if** $m_{ij} = 0$ **then**
5:    return false
6: **end if**
7:
8: // check dynamic constraints
9: **for all** connected components c **do**
10:    **if** $R_c[i] = 0 \lor R_c[i] = k + 1$ **then**
11:       continue
12:    **end if**
13:    **if** $(l_c[R_c[i]] > 0)$ **then**
14:       // forbid branching, as there are still elements in subinterval
15:       return false
16:    **end if**
17: **end for**
18: return true

---

for those test instances where at least one common interval is destroyed. It can be seen that for most taxa the fraction of destroyed intervals is less 5%. An exception are the Chordata where the fraction is nearly 20%. It can also be seen that in most taxa there exist test instances where approximately 50% of the common intervals are destroyed. The percentage of test instances where at least one common interval is destroyed by that optimal solution of RMP is shown in Table 1. It can be seen that in most taxa the percentage is nearly 20% or more. Only for the Actinopterygii and the Echinodermata the percentage is less than 10%. Note also, that the fraction is not significant different for Deuterostomia and Protostomia. It can be concluded that for nearly all investigated taxa there exist a significant fraction of test instances where many of the existing common intervals are destroyed by an optimal solution of RMP obtained with standard GRAPPA.

Table 2 shows the average run times for CIP and E-CIP for (maximally) 100 randomly chosen test instances per each different taxa. Note, that the run times of CIP and E-CIP can be several 1000 times larger than the run time of standard GRAPPA (one of the Arthropod instances could not be solved in reasonable time). For example the Arthropoda Crustacea data set was solved in 0.19 seconds per instance on average by GRAPPA. The main reason is that the distance sums for CIP-RMP are at least as big as for RMP and therefore the modified algorithms need more iterations of increasing the target value t. For the 100 Arthropoda Crustacea instances the largest difference between the lower bound used for initialization of the target value t and the optimal solution

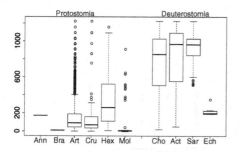 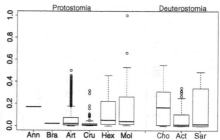

Fig. 2. Left: number of common intervals for sets of three mitochondrial genomes: right: fraction of common intervals that are destroyed by the optimal solution of RMP computed with standard GRAPPA where the average is computed over all test instances where at least one common interval is destroyed

Table 2. Average run time of CIP, and E-CIP; for each taxon 100 randomly chosen test instances were used (all test instances were used if less than 100 test instances are available)

Taxon	#	CIP	E-CIP
Ann	1	0.18	0.11
Bra	1	61259.6	47208.8
Art	100(99)	1324.14	873.81
Cru	100	2716.86	1507.07
Hex	100	218.21	63.50
Mol	100	3383.88	2259.89
Cho	100	6.09	1.54
Act	100	0.17	0.13
Sar	100	0.08	0.016
Ech	10	3.01	1.60

found was 4 when GRAPPA was used (1.62 on average). This difference was 9 for CIP and E-CIP, when common intervals are considered to find the optimal solution (2.67 on average for both algorithms). Since each iteration needs much longer than the previous, the algorithms need more time. On the other hand as CIP and E-CIP use a more complicated distance measure it is clear that the run time will be much slower than GRAPPA. For the 100 Arthropoda Crustacea instances GRAPPA had to compute 268.9 times the reversal distance between two sequences on average (recall, that this distance can be calculated in linear time). CIP had to compute the Common Interval Preserving distance 166828.9 times per instance on average, E-CIP had to compute this distance 166787.9 times per instance on average (recall, that computing the Common Interval Preserving distance is NP-complete). Nevertheless our results show that both new versions CIP and E-CIP can be used with reasonable run times for test instances that are important for phylogenetic studies in biology. The table shows also that E-CIP obtains a significant speedup over CIP. A more detailed analysis as shown in Figure 3 clearly shows, that the speedup can be up to 100.

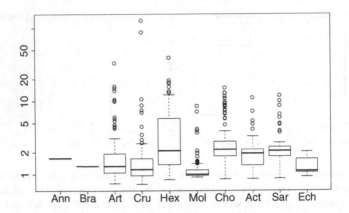

Fig. 3. Speedup of E-CIP over CIP; for each taxon 100 randomly chosen test instances were used (all test instances were used if less than 100 test instances were available)

6   Conclusion

Two algorithms to solve the Reversal Median problem with the additional constraint that common intervals are not destroyed (CIP-RMP) have been proposed. Both algorithms are based on Caprara's median solver. They use a method of [3] to compute common intervals and a method of [13] to compute the common interval preserving reversal distances. Our algorithms have been applied to mitochondrial gene order data for different animal taxa. It has been shown that common intervals occur often within most taxa and that many common intervals are destroyed by standard algorithms which solve RMP without considering common intervals. The new algorithms can solve the CIP-RMP for mitochondrial gene order data for various animal taxa in reasonable time. It was shown that common intervals occur often for most taxa and that many common intervals are destroyed when solving the RMP without considering common intervals. Currently we extend our algorithms so that they can preserve other combinatorial structures. We are currently extending our algorithms with techniques that have been proposed recently in [14] to speedup the computation of common interval preserving reversal distances.

Acknowledgements

This work was supported by the German Research Foundation (DFG) through the project "Deep Metazoan Phylogeny" within SPP 1174.

References

1. Bérard, S., Bergeron, A., Chauve, C.: Conservation of Combinatorial Structures in Evolution Scenarios. Proc. 2nd RECOMB Workshop. LNBS 3388, (2004) 1–14.
2. Bergeron, A., Blanchette, M., Chateau, A., Chauve, C.: Reconstructing Ancestral Gene Orders Using Conserved Intervals. Proc. WABI. LNCS 3240, (2004) 14–45.

3. Heber, S., Stoye, J.: Algorithms for Finding Gene Clusters. Proc. WABI, LNCS 2149, (2001) 252–263.
4. Sankoff, D.: Edit distance for genome comparison based on non-local operations. Proc. CPM, LNCS 644, (1992) 121–135.
5. Caprara, A.: The Reversal Median Problem. INFORMS Journal on Computing 15(1), (2003) 93–113.
6. Landau, G.M., Parida L., Weimann, O.: Using PQ Trees for Comparative Genomics. Combinatorial Pattern Matching, 16th Annual Symposium, CPM 2005. LNCS 3537, (2005) 128–143.
7. Moret, B.M.E., Tang, J., Warnow, T.: Reconstructing phylogenies from gene-content and gene-order data. Mathematics of Evolution and Phylogeny, Olivier Gascuel editor. Oxford University Press (2004) 321–352.
8. Bourque, G., Pevzner, P.A.: Genome-Scale Evolution: Reconstructing Gene Orders in the Ancestral Species. Genome Res. (2002) 12(1):26–36,.
9. Blanchette, M., Bourque, G., Sankoff, D.: Breakpoint phylogenies. Genome Informatics (1997) 25–34.
10. Moret, B., Siepel, A.: Finding an Optimal Inversion Median: Experimental Results. Proc. WABI , LNCS 2149, (2001) 189–203.
11. Bernt, M., Merkle, D., Middendorf, M.: Genome Rearrangement Based on Reversals that Preserve Conserved Intervals. IEEE/ACM Transactions on Computational Biology and Bioinformatics, to appear.
12. Bergeron, A., Stoye, J.: On the Similarity of Sets of Permutations and Its Applications to Genome Comparison. Proc. COCOON, LNCS, (2003) 68–79.
13. Figeac, M., Varré, J.: Sorting by Reversals with Common Intervals. Proc. WABI. LNBI 3240, (2004) 26–37.
14. Bérard, S., Bergeron, A., Chauve, C., Paul, C.: Perfect sorting by reversals is not always difficult. Proc. WABI, LNCS 3692, (2005) 228–238.
15. Hannenhalli, S., Pevzner, P.: Transforming cabbage into turnip: polynomial algorithm for sorting signed permutations by reversals. In: Proc. 27 Ann. ACM Symp. on Theory of Comput., (1995) 178–189.
16. Boore, J.L.: Mitochondrial database. http://evogen.jgi.doe.gov/ (2005)

Building Structure-Property Predictive Models Using Data Assimilation

Hamse Y. Mussa[1], David J. Lary[2], and Robert C. Glen[1]

[1] Unilever Centre for Molecular Sciences Informatics,
Department of Chemistry, University of Cambridge,
Lensfield Road, Cambridge CB2 1EW, U.K
[2] NASA, Goddard Space Flight Centre, Greenbelt,MD 20071, USA

Abstract. In Chemometrics it is often the norm to develop regression methods for analysing non-linear multivariate data by using the observations (measurements) as the sole constraint. This is the case regardless of the nature of the regression method (parametric or non-parametric)[1]. In this article we present the development of a regression model using *data assimilation*[2] - A technique that takes into account additional available information about the "system" which the model is to represent. The new approach shows substantial improvement over the "conventional" methods[3] against which it has been compared.

1 Introduction

To study a process manifested in the real world, a mathematical abstraction of the process (model) is created to capture the essence of the process. In order to predict the output of the process, one generally uses governing model equations (GME). However, in many systems the GMEs are usually imperfect due to lack of appropriate knowledge of the system. Generally GME solutions are therefore only approximations. One way to correct for errors in the numerical model solutions is to take advantage of observed data of the process.

In chemometrics and other parameter estimation areas however, models are developed for analysing accurately linear and non-linear multivariate data. In these areas data are the sole constraint[1] for fitting models which are supposed to give a mathematical representation of the underlining process(es) of the observation. In other words, the functional forms of the models are approximated solely from the data. This means that all other available information, such as *prior* knowledge (and its errors), errors in the chosen model, and errors in the data, all of which can tremendously improve fitting of the models is ignored.

For both scenarios there is a mathematical procedure which takes into account all available information about the dynamics and data, then combines them to yield accurate estimates of the system state variables. This procedure is known as data assimilation(DA)[2].

Nowadays DA is used routinely in engineering, geophysics and in meteorology[4]. It is also taking on an increasing role in the validation of observations, especially those observations made by satellites.

M.R. Berthold, R. Glen, and I. Fischer (Eds.): CompLife 2006, LNBI 4216, pp. 64–73, 2006.

In this article we present a predictive model that is built by employing a DA technique in order to explore the possible usefulness of DA in modelling molecular data.

The paper is organised as follows: Section 2 gives a brief introduction to DA. Section 3 describes our predictive model. In Section 4 we present a test case and the performance of the algorithm against other more conventional methods. In the final section we give discussions and our concluding remarks.

2   Data Assimilation Overview

Data assimilation is a method for improving the use of observations and mathematical models. The basic theory and mathematics of DA have been developed in estimation and control theories[5]. It is basically an inverse method[4] for solving the following equation for $\mathbf{x}$:

$$\mathbf{A}(\mathbf{x}) = \mathbf{y} \tag{1}$$

where $\mathbf{A}$ is

$$\mathbf{x}_k = \mathbf{f}(\mathbf{x}_{k-1}, \mathbf{u}_{k-1}) + \mathbf{d}_k \tag{2}$$

$$\mathbf{y}_k = \mathbf{h}(\mathbf{x}_{k-1}, \mathbf{u}_{k-1}) + \nu_k \tag{3}$$

Eq. 2 is a model equation describing the evolution of the state variables from $k-1$ step to k step; Eq. 3 is a data (measurement) equation relating the observations to model state variables. The two equations together are usually known as the dynamical equation; k is time (or iteration) step index, $\mathbf{d}_k$, and ν_k are the model and measurement noises respectively. $\mathbf{x}_k$ is a vector which represents the unknown state variables that describe the internal behaviour of the system; $\mathbf{u}_k$ is a vector of system inputs; $\mathbf{f}_k$ the state transition function; $\mathbf{y}_k$ is another vector which contains the observation; $\mathbf{h}_k$ is the system impulse response function which gives the output of the process in $\mathbf{x}$ space.

There are different types of data assimilation techniques proposed and applied [4]. All the various data assimilation schemes are alternative ways of looking for the optimum solution of Eq.1 for $\mathbf{x}$[4]. They are classified in different ways[6,7] (deterministic or probabilistic, global or sequential, etc). For more details see [2,6,7]. Variational data assimilation is an example of global and deterministic methods, whereas the Kalman-filter[8] is an example of sequential and probabilistic/deterministic methods. The latter scheme was employed for developing the regression model presented in this article.

Generally Eq.1 is ill-posed, which makes it quite difficult to solve. To overcome this, DA employs estimate of the error covariance $\mathbf{P}$ of the estimate of $\mathbf{x}_k^t$, which can be given by

$$\mathbf{P}_k = < (\mathbf{x}_k - \mathbf{x}_k^t)^T (\mathbf{x}_k - \mathbf{x}_k^t) > \tag{4}$$

Where the $\mathbf{x}_k^t$ vector is the true values of the state variables at step k.

With the Kalman filter, the solution is obtained by evolving forward *a priori* estimate, $\mathbf{x}_0$, to step k where the observations are available. The predicted state variables of the system at k are known as the forecast or background state variables and are denoted by $\mathbf{x}_k^f$. The difference between the predicted observation vector (Eq.3) given by the forecast state variables and the vector of the measured observations at k known as the innovation vector $\left[\mathbf{h}(\mathbf{x}_k^f, \mathbf{u}_k) - \mathbf{y}_k\right]$ is weighted. The weighted vector is then used to make a correction to $\mathbf{x}_k^f$ in order to yield an improved estimate of $\mathbf{x}_k^t$, $\mathbf{x}_k^a$, which is known as the analysis state variables. The transition equation is evolved forward again from the analysis state variables to the next step, $k+1$, where an observation is available. This time the error covariance of the state variables is also projected forward. The process is repeated until $\mathbf{x}_k^a$ converges to $\mathbf{x}_k^t$.

If $\mathbf{f}$ and $\mathbf{h}$ in the dynamical equation (Eqs. 1 and 2) are linear, and in matrix form[1], which can be written as

$$\mathbf{x}_k^f = \mathbf{F}_k \mathbf{x}_{k-1}^a + \mathbf{G}_k \mathbf{u}_{k-1} + \mathbf{d}_k \tag{5}$$

$$\mathbf{y}_k = \mathbf{H}_k(\mathbf{x}_k^f) + \nu_k \tag{6}$$

then mathematically the above procedure of finding the optimum solution of Eq. 1, $\mathbf{A}(\mathbf{x}_k^f) = \mathbf{y}$, using a Kalman-filter based data assimilation technique may be written as

$$\mathbf{x}_k^a = \mathbf{x}_k^f + \mathbf{K}_k \left[\mathbf{y}_k - \mathbf{H}_k(\mathbf{x}_k^f)\right] \tag{7}$$

Where $\mathbf{x}_k^a$, $\mathbf{x}_k^t$ and $\mathbf{x}_k^f$ are as described above; $\mathbf{K}_k$ (the Kalman gain matrix) weighs the innovation vector and it is given by

$$\mathbf{K}_k = \mathbf{P}_k^f \mathbf{H}_k \left[\mathbf{H}_k^T \mathbf{P}_k^f \mathbf{H}_k + \mathbf{R}_k\right]^{-1} \tag{8}$$

Where, $\mathbf{P}_k^f$ is the forecast error covariance; $\mathbf{R}_k$ is the observation error covariance, $E[\nu_k \nu_k^T]$.

The analysis error covariance, $\mathbf{P}_k^a$, is given

$$\mathbf{P}_k^a = \mathbf{P}_k^f - \mathbf{P}_k^f \mathbf{H}_k \left[\mathbf{H}_k^T \mathbf{P}_k^f \mathbf{H} + \mathbf{R}_k\right]^{-1} \mathbf{H}_k^T \mathbf{P}_k^f + \mathbf{Q}_k \tag{9}$$

$\mathbf{Q}_k$ is the covariance of the process noise, $[\mathbf{d}_k \mathbf{d}_k^T]$.

If the dynamical equation (Eqs. 1 and 2) is non-linear, it is linearised. This means that $\mathbf{F}$ and $\mathbf{H}$ in Eqs. 5 and 6 respectively are the derivative matrices of $\mathbf{f}$ and $\mathbf{h}$ with respect to the state variable vector - Jacobian matrices whole elements are

$$H_k^{ij} = \frac{\partial h(i)_k}{\partial x^f(j)_k} \tag{10}$$

$$F_k^{ij} = \frac{\partial f(i)_k}{\partial x^f(j)_k} \tag{11}$$

---

[1] $\mathbf{F}$ is transition matrix; $\mathbf{G}$ is input matrix and $\mathbf{H}$ is observation matrix.

It is obvious from equations 7 to 9 that the Kalman Filter method is basically a set of equations that implements a predictor-corrector type estimator. For linear systems, the estimator is optimal as it optimally minimizes the error covariance of the estimated states. Due to the linearization, however, the state variable estimator is sub-optimal for non-linear systems.

In summary, to perform a data assimilation these are the main elements:

1. Transition equation (numerical mathematical model)
2. Transition process (model) uncertainties.
3. Observations and a measurements equation.
4. Observation and representativeness uncertainties.
5. Using priori knowledge.
6. Assimilation scheme.

3 Parameter Estimation (Model Fitting)

Having given a brief introduction of data assimilation, in this section we present how to use data assimilation for model fitting.

Like the dynamic process, in model fitting it is required to solve Eq. 1, $\mathbf{A}(\mathbf{x}) = \mathbf{y}$, for $\mathbf{x}$ where $\mathbf{x}$ is the parameters of the model which is being fitted.

Transition Equation:

In fitting, parameters are generally time independent. So the transition equation can be seen as representing a stationary process, $\mathbf{F}_k = \mathbf{I}$, corrupted by noise $\mathbf{d}_k$. Hence Eq. 2 becomes

$$\mathbf{x}_k^f = \mathbf{x}_{k-1}^f + \mathbf{d}_k \tag{12}$$

Measurement Equation and Observations :

The known data contain both $\mathbf{y}$, $\mathbf{u}$ and probably ν. Generally the relationship between $\mathbf{y}$, and $\mathbf{u}$ is non-linear which requires the use of a non-linear impulse function

$$\mathbf{y}_k = \mathbf{h}_k(\mathbf{x}_k^f, \mathbf{u}_{k-1}) + \nu_k \tag{13}$$

As mention before, the transition process (model) and observation uncertainties are assumed to be characterised as white noise and it is also assumed that $\mathbf{d}_k$ and ν_k are not correlated:

$$E[\mathbf{d}(i)\mathbf{d}(j)^T] = \delta_{ij}\mathbf{Q}(j), \ E[\nu(i)\nu(j)^T] = \delta_{ij}\mathbf{R}(j), \ E[\mathbf{d}(i)\nu(j)^T] = 0, \ \forall_{i,j}.$$

To solve Eq. 1 for the parameters, employing a Kalman-filter scheme, the measurement impulse function $\mathbf{h}$ should be linearised with respect to the parameters. The linearization gives the observation matrix, $\mathbf{H}_k$ (Eq. 10). Note that $\mathbf{F}_k = \mathbf{I}$ here.

In the following section the computational realisation of the new approach is described.

3.1 Computational Realisation

For model selection, three way data splits were employed. The data set was randomly divided into a training set and test set [13]. The training set was further portioned into two disjoint subsets: an estimation subset was used in conjunction with the dynamical equation for selecting the model; a validating set was used to guide the tuning of the model. Note that the validating data set was only implicitly involved in the model selection. However, there could be the distinct possibility that the model with the "best" parameter values might overfit the validating set. To guard against this, the test data set was employed for testing the generalisibility of the selected model[13]. Three way data splits were also used for cross-validating the fitted model.

$\mathbf{h}(\mathbf{x}_k^f, \mathbf{u}_k)$ in Eq. 13 can be expressed in a number of different forms. Using a neural network representation is one of them. So a neural network with one input layer, a hidden layer, and one input node, expressed the impulse function. All the nodes had a *tanh* activation function.

Following the neural network notation:

$\mathbf{x}_k$ is a *[m(n+1)+ m+1]* vector.
$\mathbf{u}_k$ is a *(n+1)* vector.
$\mathbf{P}_k$ and $\mathbf{Q}_k$ are *[m(n+1)+ m+1]×[m(n+1)+ m+1]* matrices.
$\mathbf{R}_k$ and $\mathbf{d}_k$ are *1×1* vectors (scalars).
$\mathbf{K}_k$ is a *[m(n+1)+ m+1]* vector.
$\mathbf{H}_k$ is a *[m(n+1)+ m+1]×[m(n+1)+ m+1]* matrix.

m is the number of hidden nodes; n is the number of inputs excluding biases.

For the fitting algorithm to work the values of $\mathbf{x}_0^f$, $\mathbf{R}_k$, $\mathbf{Q}_0$, and $\mathbf{P}_0^f$ are required. $\mathbf{x}_0^f$ and $\mathbf{P}_0^f$ are initialised by setting them to random values. As mentioned in Section 2, $\mathbf{R}_k$ is just the error covariance of the observation and it is easy to calculate or estimate, but $\mathbf{Q}_0$ is slightly more difficult. As equations 8 – 9 show, $\mathbf{R}_k$ and $\mathbf{Q}_k$ play crucial roles in the fitting process. So choosing right values for them is important. For a slowly varying dynamic process whose observation errors are not known, forgetting factors are often used for $\mathbf{R}_k$ [15]; $\mathbf{Q}_0$ is set to a scaled identity matrix, $\xi\mathbf{I}$. (ξ is usually between 0 and 0.9).

In this work the observations errors are not known, thus an iteration-varying forgetting factor was employed as $\mathbf{R}_k$(see below). As the role of $\mathbf{Q}_k$ (see Eqs. 8 – 9) is the opposite of that of $\mathbf{R}_k$, ξ was weighed by the inverse of the iteration-varying forgetting factor. This allows us to dynamically estimate $\mathbf{R}_k$ and $\mathbf{Q}_k$.

Below we give the algorithm for fitting a model (solving Eq. 1) using a DA scheme:

1. Give pre-determined accuracy value, ς
2. Invoke three-way-data-splits
3. Initializations
   - Choose m.
   - Choose random values for $\mathbf{x}_0^f$.
   - Set the off-sets to nonzero constants.
   - Initialise $\mathbf{P}_0^f$ and $\mathbf{Q}_0$ to small nonzero numbers.
4. **For** : Choose input elements, $\mathbf{u}_{k-1}$, which is propagated through the dynamical equation to yield an output.

5. $\mathbf{R}_k$
   - If the errors of the input pattern are known, calculate $\mathbf{R}(i)_k = \delta_{i,j} E[\nu(i)_k \nu(j)_l^T]$.
   - If not, use iteration-varying factor, $\mathbf{R}(i)_k = \mathbf{I}\lambda_k$. Where λ_k is a forgetting factor given by [15]:

$$\lambda_k = \lambda^0 \lambda_{k-1} + (1 - \lambda^0)$$

   λ^0 and λ_0 are tunable parameters [15].
6. Compute $\mathbf{H}_k$, Eq. 10
7. Calculate $\mathbf{K}_k$, Eq. 8
8. Update
   - $\hat{\mathbf{x}}_k^a$, the best estimated solution so far, using the Kalman matrix and the innovation vector, Eq. 7
   - $\mathbf{P}_k^a$, Eq. 9
   - $\mathbf{Q}_k \leftarrow \mathbf{Q}_{k-1}\lambda_k$
9. **End** *of the loop.*
10. Evaluate the model using the validation set.
11. If the stopping criterion is met, exit. Otherwise go back to step 4.
12. Select the best architecture.
13. Test the model, using the test data set.
14. Cross-validate the model by repeating all the above steps except step 1 and 3 - The ς and initialised parameters are left unchanged. This final step is performed L times.

4   Case Study: TLC Behaviour Using Theoretically Derived Molecular Properties

It is generally accepted that there is a strong relationship between the structure of a compound and its chromatographic properties. The structural features which are important in defining the relationship between a compound and its retention ratio in High Performance Liquid Chromatography (HPLC) or its R_f values in prediction of Thin Layer Chromatography (TLC), are, still the subject of research [9,10,11].

Glen *et al*[3] presented some preliminary studies on the use of quantitative structure-chromatography relationships for the prediction of the R_f values of a series of benzoic acids in six solvents. They employed molecular mechanics and computational chemistry calculations to obtain a set of physicochemical descriptors for the molecules. Then they performed statistical analysis to relate these properties to experimentally determined chromatographic retention ration (R_f) and to identify which factors were important in determining chromatographic behaviour. A neural network (NN) using a back propagation method was also used in their work. Their results implied that the neural network performed better than multiple linear regression.

Using the same data, we present a comparison between our model based on a DA scheme and their NN algorithm. But first the data set is briefly described.

4.1 Data

Molecular Properties Calculations. Molecular structures for 22 benzoic acid derivatives were constructed using the Sybyl molecular modeling system[12] on a Silicon Graphics Irix. Geometries were optimised using molecular mechanics. Then 17 molecular properties were calculated for each compound. The full details of the programs used for the calculations, the list of the 22 compounds and that of the 17 properties can be found in Ref[3].

Experiment. TLC was performed on 10×20 cm glass-backed plates pre-coated with C-18 bonded silica gel[3]. The compounds were dissolved in methanol and applied to the TLC plates. Ascending chromatography was performed in glass TLC tanks containing mobile phase which comprised mixtures of different proportions of methanol or acetonitrile. Following chromatography, the compounds were examined under UV illumination and the R_f values determined. The process was repeated for the 22 compounds in six different mixtures of the two solvents with water. Thus 132 different values of R_f were measured. For full details of the solvents and mixtures is given in Table 1 of Ref.[3]. Glen *et al* analysed the calculated properties with respect to the chromatographic measured values, employing NN approaches. The authors used the physicochemical parameters (5 molecular properties and 2 solvent properties) as input variables and the related measured R_m values as response variables. Note that R_m is given by $Log_{10} \frac{1.0 - R_f}{R_f}$. They range scaled their variables from -0.5 to + 0.5 prior to analysis. 90% of the data set was employed for training and the rest for testing the predictive power of their model.

We followed a similar procedure, only the NN algorithm was replaced with the data assimilation scheme.

4.2 Calculations and Data Analysis

In this work:

- 6 physical properties of the benzonic acids and 2 physical properties of the solvents were used as $\mathbf{u}_{k-1}$, *i.e*, $n=8$.
- The observation was the measured retention value R_f.
- $\mathbf{x}_0^f$ was initialised randomly.
- $\mathbf{P}_0 = 10.0$
- $\mathbf{Q}_k = 0.5$
- $\lambda_0 = 0.99$, $\lambda^0 = 0.95$
- m = 6
- $\hat{y}_k = \mathbf{h}(\mathbf{x}_k, \mathbf{u}_k) = \tanh[\mathbf{x}_k(m(n+1)+1 : m+1)^T \mathbf{z}_k(1 : m+1)]$, where $\mathbf{z}_k(2 : m+1) = \tanh[\mathbf{x}_k(1 : n+1)^T \mathbf{x}(1 : n+1)]$ and $\mathbf{u}_k(1)$ and $\mathbf{z}_k(1)$ are the offsets in the input and hidden layers respectively.
- 122 elements of the data set (132 in total) were used as a training set. The optimal value ratio [14] which is given by

$$r_{opt} = 1 - \frac{\sqrt{2[size(\mathbf{x})] - 1} - 1}{2[size(\mathbf{x}) - 1]}$$

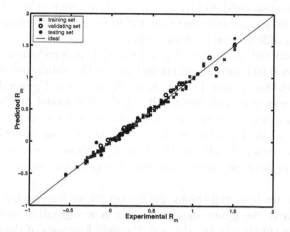

Fig. 1. The performance of the model which was generated by employing the dataassimilation technique

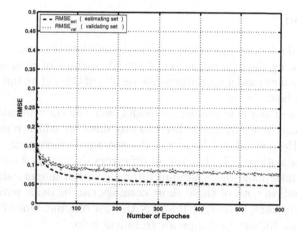

Fig. 2. The root mean square curves for the new algorithm

has been employed to determine how many of the set should be used as a training set[2]. For $n=8$, $r_{opt} \approx 0.92$ which means about 92% of the total data set should be used as a training set, ≈ 122 patterns.

- Prior to the start of parameter fitting, $\mathbf{u}_{k-1}$ and y_k were scaled between -0.9 and 0.9; $\hat{y}_k$ and y_k were re-scaled back when testing the generalisibility of the model.
- $L = 12$
- $\varsigma = 0.01$

---

[2] The derivation of r_{opt} can be found in Amari *et al*[16].

The model was cross-validated by 12 three way random splits. To make sure that the same model was not created twice, the index generating random function was seeded with a different value at each time.

The model predicted 10 unseen R_m values with a correlation coefficient of 0.980. It also gives 0.991 correlation coefficient for the validating set and 0.995 for the training set. See Figure 1. The model shows a good generalisibilty. The algorithm reproduces very well the results of the NN model of Glen *et al* (see their Figure 2). In this specific case study, the data assimilation predictive model is also fast taking about 5.0 minutes on a 1.33GHz x86 Family 6 Model 4 PC.

To quantify the generalisibilty of the model, a number of diagnostics, such as skill forecast, percentage errors, and root mean square error analysis, were performed.

Figure 2 illustrates how well the model converges with respect to the number of iterations. Clearly the figure shows that the new algorithm meets the third criterion of the rule of thumb. In addition, the skill forecasts of the model which was obtained by calculating the biases showed that the prediction of 82% of the unseen data by the L models lie within ± 0.01 of the exact values.

5   Conclusion

We have developed a fitting algorithm which is based data assimilation methods, which can be seen as the generalisation of the algorithms of Singhal *et al* and others [19,20]. The new model reproduces the results of Glen *et al*, but generalizes better and is also fast.

The advantages of the new predictive model over the conventional NN algorithms is that it can be easily generalised to any data set which represent state space processes. However, for a large $\mathbf{x}_k$, the storage requirement can become prohibitive. It was noticed [17] that round off errors due to poor computer precision can sometimes make the Kalmnan filter algorithm numerically unstable. Fortunately, nowadays both of these short comings can be dealt with [18,15].

We would like to thank UNILEVER and NASA for funding. One of the authors would like to thank Robert Easthope for technical help.

References

1. T. B. Brown, and S .D. Brown, *Journal of Chemometrics*, **1994**, 8, 391.
2. R. Daley, *Atmospheric Data Analysis. Cambridge Atmospheric and Space Science Series*, Cambridge University Press (**1991**).
3. R. C. Glen, V. S. Rose, J. C. Lindon, R. J. Ruane, I. D. Wilson, and J. K. Nicholson,*J. Pl. Chrom.*, **1991**, 4, 432.
4. S. Kim, *Korean Journal of Remote Sensing*, **2001**, 17, 345
5. A. Gleb, *Applied Optimal Filtering*, MIT Press (1974).
6. I. Fukumori,'Data assimilation by models', in *Satellite Altimetry and Earth Sciences*, L. -L. Fu and A. Cazenavo, Eds., Academic,**2001**, 237.
7. O. Talagrand, and P. Courtier,*Quart. J. Roy. Meteor. Soc.*, **1987**, 113, 1311.
8. R. E. Kalman, *Transaction of the ASME–Journal of Basic Engineering*,**1960**, 35.

9. K. Petritis ,L. J. Kangas, P. L. Ferguson, G. A. Anderson, L. Pasa-Tolic, M. S. Lipton, K. J. Auberry, E. F. Strittmatter, Y. Shen, R. Zhao, and R. D. Smith,*Anal Chem*, **2003**, 75, 1039-1048.
10. M. H. Abraham, A. Ibrahim and A. M. Zissimos,*Journal of Chromatography A*, **2004**, 1037, 29-47.
11. T. Hanai,*Journal of Chromatography A*, **2004**, 1027, 279-2871.
12. Sybyl, Tripos Associates, St.Louis, Mo., USA.
13. S. Haykin,*Neural Networks: A Comprehensive Foundation*, Second Edition, Prentice Hall(**1999**).
14. A. Verikas, and M. Bacauskiene, *Chemometrics and Intelligent Laboratory Systems*,**2003**, 67, 187.
15. Y. M. Zhang and X. R. Li, *IEEE Transactions on Neural Networks*, **1999**, 10, 930.
16. S. Amari, N. Munta, K. R. Muller,M. Finke,H. Yang,'Statistical theory of overfitting–is cross-validation asymptotically effective?' in *Advances in Neural Information Processing Systems*; S. Touretzky, M. C. Mozer, M. Hasselmo,Ed.;MIT Press: Massachusetts, **1996**, 8, pp. 176–178.
17. G. Bierman, *Factorization Methods for Discrete Sequential Estimation*, Academic Press (**1977**).
18. D. J. Lary and H. Y. Mussa, *IEEE Transactions on Neural Networks* (Submitted).
19. S. Singhal and L. Wu, 'Training Feedforward Neural Networks', in *Proc. Int. Conf. ASSP*, **1989**, pp. 1187.
20. G. V. Puskorius and L. A. Feldkalmp, *IEEE Transactions on Neural Networks*, **1994**, 5, 279.

Set-Oriented Dimension Reduction: Localizing Principal Component Analysis Via Hidden Markov Models*

Illia Horenko[1], Johannes Schmidt-Ehrenberg[2], and Christof Schütte[1]

[1] Freie Universität Berlin, Department of Mathematics and Informatics,
Arnimallee 6, D-14195 Berlin, Germany
{horenko, schuette}@mi.fu-berlin.de
[2] Zuse Institute Berlin (ZIB), Takustr. 7, D-14195 Berlin
schmidt-ehrenberg@zib.de

Abstract. We present a method for simultaneous dimension reduction and metastability analysis of high dimensional time series. The approach is based on the combination of hidden Markov models (HMMs) and principal component analysis. We derive optimal estimators for the log-likelihood functional and employ the Expectation Maximization algorithm for its numerical optimization. We demonstrate the performance of the method on a generic 102-dimensional example, apply the new HMM-PCA algorithm to a molecular dynamics simulation of 12–alanine in water and interpret the results.

Introduction

Let us assume that the observation of the physical process under consideration (e.g. conformational dynamics of some biological molecule) is given in the form of a high dimensional time series in some molecular degrees of freedom (e.g. torsion angles or distances between some important groups of atoms in the molecule). The general task which arises in many practical applications is to find the few important or *essential* degrees of freedom that can explain most of the observed process and thus can help to understand the physical mechanism [1,2,3,4].

The increasing amount of "raw" simulation data and growing dimensionality of these simulations have led to a persistent demand for modeling approaches which allow to extract physically interpretable information out of the data. What is needed is *automatized* generation of low–dimensional physical models based on (noisy) data, i.e., interesting approaches should provide *data–based dimension reduction*. This should be carefully distinguished from analytical approaches like, e.g., the Zwanzig-Mori approach, the Karhunen-Loève expansion, or averaging techniques. The latter approaches allow to reduce the dimension of a given physical model, but the problem of finding essential coordinates must be solved

* Supported in part by the DFG Research Center MATHEON, Berlin, and Microsoft Research Ltd., Cambridge, UK (Contract No. 2005-042).

M.R. Berthold, R. Glen, and I. Fischer (Eds.): CompLife 2006, LNBI 4216, pp. 74–85, 2006.
© Springer-Verlag Berlin Heidelberg 2006

previously and may be data–driven as well. See the textbook [5], or the excellent review article [6] for an overview. Compare also [7] for a related approach.

The problem of dimension reduction becomes crucial when dealing with data-bases of molecular dynamics trajectories [8,1]. Recent works show that even such simple linear dimension reduction strategies as principal component analysis (PCA) or independent component analysis (ICA) [9] allow for a significant compression of the time–series information (factor 10 in [10]). However, such a linear technique as PCA applied to in general nonlinear phenomena as, e.g., transitions between the metastable conformations of biological molecules can be misleading and produce difficulties in the interpretation [1,11,8]. One way of trying to circumvent these problems is non–linear extension of PCA (NLPCA) [12]. However, this non-linear strategy is numerically expensive and not robust enough, thus resulting in restricted applicability of the technique [13]. Another possibility to extend the linear dimension reduction techniques is contained in the theory of the indexing of high dimensional data-bases, where the problem was partially solved by combining correlation analysis with clustering techniques (possibly in low-dimensional projections of high-dimensional data-sets like in *projected clustering methods*) [14,15,16]. But due to the fact that the proposed methods rely on *geometrical clustering* of possibly high dimensional data–spaces, the resulting algorithms rely on some sort of distance-metric and scale polynomially wrt. the length of the data-set. Alternatively, for the time series analysis of molecular dynamics trajectories, due to additional information encapsulated in the time component, it is possible to employ *dynamical clustering* techniques like hidden Markov models (HMMs) which scale linear wrt. the length of the time series [17,18,19,20,21,22].

In this paper we present a novel method for simultaneous dimension reduction *and* clustering of the time series into metastable states. The approach is based on the combination of the HMM with PCA. The problem of simultaneous dimension reduction and metastability analysis is solved by the optimization of an appropriate log–likelihood functional by means of the Expectation Maximization algorithm (EM) [19]. The performance of the resulting HMM–PCA algorithm is demonstrated by application to some model examples and to a microsecond simulation of 12–alanine protein in water. The same likelihood based strategy can similarly be applied to combine HMM with ICA. This topic will be the subject of ongoing investigations.

1   Principal Component Analysis (PCA)

The simplest form of the dimension reduction is known in statistics as principle component analysis (PCA). Let the data be given in form of a sequence $\{x_t\}_{t=1,...,T}$ of states. The idea of the method consists in identification of m *principal* directions with highest variance in n-dimensional observed data $x_t : \mathbb{R}^1 \to \mathbb{R}^n$ ($m << n$). These directions are defined with the help of linear projectors $\mathbf{T} \in \mathbb{R}^{n \times m}$, i.e., $\mathbf{T}$ is understood to project onto the subspace spanned by the principal directions. Mathematically the problem of identifying $\mathbf{T}$ can be stated as minimization of the residuum–functional (which describes the least–squares difference between the original observation and its reconstruction by means of the m-dimensional projection):

$$\mathbf{L}(x_t, \mathbf{T}, \mu) = \sum_{t=1}^{T} \left\| (x_t - \mu) - \mathbf{T}\mathbf{T}^{\mathsf{T}}(t)\,(x_t - \mu) \right\|_2^2. \tag{1}$$

The functional $\mathbf{L}$ depends on the projector matrices $\mathbf{T}$ and *center vectors* $\mu \in \mathbb{R}^n$. Moreover, the projectors $\mathbf{T}$ are subjected to the orthogonality condition:

$$\mathbf{T}^{\mathsf{T}}\mathbf{T} = Id^{m \times m}, \tag{2}$$

The functional (1) can equivalently be written as

$$\mathbf{L} = \sum_{t=1}^{T} (x_t - \mu)^{\mathsf{T}} \left(Id^{n \times n} - \mathbf{T}\mathbf{T}^{\mathsf{T}} \right) (x_t - \mu). \tag{3}$$

This functional can be minimized analytically subject to the orthogonality conditions resulting in the expressions for optimal parameters μ and $\mathbf{T}$:

$$\mu = \frac{1}{T} \sum_{t=1}^{T} x_t, \tag{4}$$

$$C\mathbf{T} = \mathbf{T}S, \qquad C = \sum_{t=1}^{T} (x_t - \mu)(x_t - \mu)^{\mathsf{T}} \tag{5}$$

where $\mathbf{T}$ is the matrix of m dominant eigenvectors and $S = diag(\lambda_1, .., \lambda_m)$ contains m corresponding largest eigenvalues of the covariance matrix C. This result means that the optimal value of the parameter μ is given simply by the expectation value of the data and the corresponding optimal projector $\mathbf{T}$ is defined by the dominant eigenvectors of the data covariance–matrix. It is important to mention that nowhere in the derivation of the optimal estimator the assumption about the form of the x_t distribution is needed.

However, in many interesting cases the standard PCA–approach does not result in a meaningful dimension reduction. Let us assume, for example, that the time series given results from a realization of the process governed by a two–dimensional double-well Langevin–dynamics of the form

$$\ddot{x}(t) = -\operatorname{grad} V\left(x(t)\right) - \gamma \dot{x}(t) - \sigma \dot{W}(t), \tag{6}$$

with friction matrix $\gamma = \begin{pmatrix} 0.25 & 0.125 \\ 0.125 & 0.25 \end{pmatrix}$, noise matrix $\sigma = \begin{pmatrix} 0.6 & 0 \\ 0 & 0.6 \end{pmatrix}$ and potential energy defined as the sum of two Gaussian wells orthogonal to each other added to a harmonic potential:

$$V(x) = \sum_{l=1}^{2} a_l \exp\left(-(x - \mu_l^{sys})^{\mathsf{T}} D_l^{sys}(x - \mu_l^{sys})\right)$$

$$+6(x - 0.5(\mu_1^{sys} + \mu_2^{sys}))^{\mathsf{T}}(D_1^{sys} + D_2^{sys})(x - 0.5(\mu_1^{sys} + \mu_2^{sys})), \quad (7)$$

$$D_1^{sys} = \begin{pmatrix} 20 & 0 \\ 0 & 0.5 \end{pmatrix}, \quad D_2^{sys} = \begin{pmatrix} 0.5 & 0 \\ 0 & 20 \end{pmatrix}, \quad \mu_1^{sys} = \begin{pmatrix} 0 \\ 1.5 \end{pmatrix}, \quad \mu_2^{sys} = \begin{pmatrix} 1.5 \\ 0 \end{pmatrix}$$

The Langevin dynamics in this case produces two clusters of states each associated with the corresponding *metastable* well. The application of PCA via (4)-(5) with $m = 1$ to this time series results in an inadequately reconstructed dynamics. If we first cluster the time series into two clusters and then apply (4)-(5) to each of the clusters separately, we can reduce the value of the residuum–functional (1) from 191.1 in a "global" PCA case to 46.3 in a "local" one (by "local" PCA we understand the PCA for each of the clusters, the value of the residuum-functional is then given by the sum of the "local" functionals). This also results in a much better quality of the data-reconstruction.

This leads us to a simple idea: if we want to enhance the performance of PCA–based dimension reduction we should exploit the internal structure of the data, i.e., we should decompose the time series of the observed process into metastable aggregates and then make the "local" dimension reduction by means of PCA. Furthermore, we can state more ambitious question: Is it probably possible to use the local principle dimensions as tokens in the clustering of the time series itself. If it is possible this will allow to combine clustering of data and dimension reduction in one algorithmic step hopefully leading to synergetic effects and allowing both clustering of the time series in metastable sets and the dimension reduction.

1.1   Hidden Markov Models (HMM)

A hidden Markov model (HMM) is a stochastic process with hidden and observable states. The hidden process consists of a sequence $X_1, X_2, X_3, \ldots$ of random variables taking values in some "state space", the value of X_t being "the state of the system at time t". In applications these states are not observable, and therefore called *hidden*. Each state causes a specific output that might be either discrete or continuous. This output is distributed according to a certain conditional distribution (conditioned to the hidden state). Thus, realizations of HMM are concerned with two sequences, an observation sequence and a sequence of hidden states.

The dynamics under consideration is assumed to be a Markov process, that is, the state sequence has the Markov property which means that the conditional distribution of the "future" X_{n+1} given the "past", $X_1, \ldots, X_n$, depends on the past only through X_n. Since the HMM state space in general is finite, we thus are concerned with a Markov chain, which is characterized by the so-called transition matrix, whose entries correspond to the probabilities of switching from one state to another. The sum of all coefficients in one row is the probability of taking any state, therefore being one, which means that the transition matrix is a row-stochastic matrix.

An HMM is designed to describe the situation in which part of the information of the system is unknown (or hidden) and another part is observed. In molecular dynamics the information initially is hidden in which metastable subset (conformation) the molecular system is at a certain instant in time, while the information on the state of the selected torsion or backbone angles is completely known. An HMM then consists of a Markov chain model for the hidden (metastable) states that encodes with which probability one switches from one hidden state to another, and a *conditional probability* of observation of specific torsion angles *if* one is in a certain hidden state. To describe the whole system, we need to know the number of hidden states, the transition matrix between them, an initial distribution, and for each state a certain probability distribution for the observation. Therefore, an HMM is formally defined as a tuple $\lambda = (S, V, A, B, \pi)$ where

- $S = \{s_1, s_2, ..., s_L\}$ is a set of a finite number L of states,
- $V \subset \mathbb{R}^k$ is the observation space,
- $A = (a_{ij})$ is the transition matrix, where $a_{ij} = P(X_{t+1} = s_j | X_t = s_i)$ describes the the transition probability from state s_i to state s_j,
- $B_k, k = 1, ..., N$ are probability density functions in the observation space,
- $\pi = \pi_i$ is a stochastic vector, that describes the initial state distribution, $\pi_i = P(X_1 = s_i)$.

Often, the short notation $\lambda = (A, B, \pi)$ is used since S and V are implicitly included. HMMs can be set up for discrete or continuous observations. For continuous observations the most popular choice is to use (multivariate) normal distributions for the output distributions B_k.

1.2   HMM-PCA

The fitting of the parameters can be performed with the help of the *maximum likelihood principle*. The likelihood function is $L(\lambda) = P(x_t, X_t | \lambda)$, i.e., we consider the observation sequence as being given and ask for the variation of the probability in terms of the parameters. The maximum likelihood principle then simply states, that the optimal parameters are given by the absolute maximum of L. Thus, similarly to the PCA dimension reduction, the maximum likelihood principle is an optimization problem in parameter space.

In order to combine both approaches, we first make two assumptions on the observation process: (i) the observed data in the hidden states are distributed according to a multivariate Gaussian distributions ρ_B, (ii) the hidden process switching between the metastable states is Markovian, i.e., the probability of the conformational change depends on the current conformation only. The first assumption is approximately valid for a large class of observables in molecular dynamics (e.g. for the torsion angles or chemical bond lengths in the molecule).

The second assumption is connected to the characteristic timescale at which the memory kernel of the molecular system is decaying and is also satisfied for a wide class of applications.

These assumptions allow to design a statistical model for the observed data and to construct the likelihood function for a reduced system. In analogy to the residuum–functional (3) we have

$$P(x_t, X_t|\lambda) = \pi_{X_0} e^{-\frac{1}{2}(x_0 - \mu_{X_0})^{\mathsf{T}} \mathbf{T}_{X_0} S_{X_0} \mathbf{T}_{X_0}^{\mathsf{T}} (x_0 - \mu_{X_0})} \prod_{k=1}^{T-1} \frac{A(X_k, X_{k+1})}{\sqrt{(2\pi)^m \det(S_{X_{k+1}})}}$$

$$\times e^{-\frac{1}{2}(x_{k+1} - \mu_{X_{k+1}})^{\mathsf{T}} \mathbf{T}_{X_{k+1}} S_{X_{k+1}} \mathbf{T}_{X_{k+1}}^{\mathsf{T}} (x_{k+1} - \mu_{X_{k+1}})} \tag{8}$$

where $B_i = (\mu_i, \mathbf{T}_i, S_i)$ is a set of multivariate Gaussian distribution parameters where $\mu_i \in \mathbb{R}^n$ are the centers of the clusters, $\mathbf{T}_i \in \mathbb{R}^{n \times m}$ the corresponding optimal projectors, and $S_i \in \mathbb{R}^{m \times m}$ a diagonal matrix of dominant variances. Functional (8) should be additionally subjected to constraints: (i) the projector orthogonality condition (2),(ii) and the condition for stochasticity of the transition matrix A (i.e., the row sums of the matrix should be 1.0).

For numerical reasons it is much more convenient to take the logarithm of the likelihood functional and optimize the resulting log-likelihood functional. Writing the log-likelihood together with both constraints in Lagrange–form, taking the derivatives wrt. to the model parameters and setting them to zero we get:

$$\mu_i = \frac{1}{T} \sum_{t=1}^{T} \alpha_i(t)\beta_i(t)x_t, \tag{9}$$

$$C_i \mathbf{T}_i = \mathbf{T}_i S_i \qquad C_i = \sum_{t=1}^{T} \alpha_i(t)\beta_i(t)(x_t - \mu)(x_t - \mu)^{\mathsf{T}} \tag{10}$$

where $\alpha_i(t), \beta_i(t)$ are forward and backward variables (as usually defined in the context of HMMs, see [22]). They are related to the Markov process (A, π) and describe the probabilities to observe the hidden process X_t in the state i in the time t. We observe direct correspondence between the estimator formulas (9)-(10) and those given by standard PCA (4)-(5). That is, in the case of a single hidden state the minimization of the HMM-PCA functional (8) is equivalent to optimization of the residuum–functional (3). Only the dominant eigenvectors are needed for the construction of matrix $\mathbf{T}$. One can compute them efficiently with some iterative subspace method (e.g. Lanczos, cf. [23]). In the case of several hidden states we suggest to use the standard Expectation-Maximization algorithm [24], often also called the Baum-Welch algorithm [17,18]. The Expectation-Maximization (EM) algorithm is a maximum likelihood approach that improves iteratively an initial parameter set, and converges to a local maximum of the likelihood function. Its two steps, the E- and M-steps, are iteratively repeated until the improvement of the likelihood becomes smaller than a given limit. In all other details the EM algorithm used herein follows standard procedures.

To apply the EM algorithm to a given observation sequence, we have to set up an HMM $\lambda = \lambda(A, B, \pi)$ by assuming a finite number L of hidden states, a

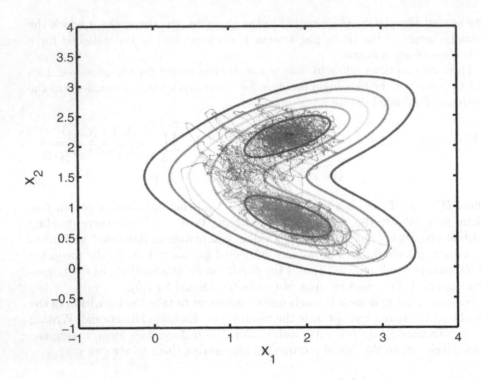

Fig. 1. 2D projection of a 102–dimensional Langevin Dynamics in a two–hole potential. Continuous curves are isolines of the projected potential function. The respective potential values are color–coded ranging from blue (low) to brown (high). The dotted curve denotes the projected dynamics and its subdivision into two dynamic clusters (red/green) corresponding to the two holes of the potential. The clustering of the time series results from the application of the Viterbi-algorithm [25] using the previously fitted HMM-PCA model.

distribution function for the output of each state, and an initial values for all remaining parameters.

2   Numerical Examples

2.1   Langevin Dynamics in 102 Dimensions

As a first example we consider realizations of the Langevin equation (6) with $x = (q, y) \in \mathbb{R}^2 \times \mathbb{R}^N$ and the perturbed two–hole potential

$$V(x) = \sum_{l=1}^{2} a_l \exp\left(-(q - \mu_l^{sys})^\mathsf{T} D_l^{sys}(q - \mu_l^{sys})\right) + \frac{1}{2} y^\mathsf{T} D_{bath} y \qquad (11)$$

$$+ \delta_0\Big(\cos(2\pi k(x_1 + x_2)) + \cos(2\pi k(x_1 - x_2))\Big), \qquad (12)$$

where $\delta_0 \ll 1$ is a small perturbation parameter. The N harmonic bath variables are denoted by y, whereas x labels the two "metastable" dimensions that live in the plane of the double well potential. We have chosen the following parameter values: $\mu_1^{sys} = (1.8, 2.2)^{\mathsf{T}}$, $\mu_2^{sys} = (1.8, 0.8)^{\mathsf{T}}$, $D_1^{sys} = \begin{pmatrix} 1 & -1 \\ -1 & 3 \end{pmatrix}$, $D_2^{sys} = \begin{pmatrix} 1 & 1 \\ 1 & 3 \end{pmatrix}$, $a_1 = -6$, $a_2 = -6$ such that we get two contiguously placed skew wells and make identification of the metastable sets more challenging compared to a well–separated situation. The parameter matrices D_{bath} and γ have been chosen to be symmetric, positive definite, and tri-diagonal, with 10.0 on the main diagonal and 5.0 on secondary diagonals for D_{bath} (5.0 and 2.5 respectively for γ). The noise parameter σ was taken as a diagonal matrix with 4.0 on the diagonal. The system is metastable because the barrier is sufficiently larger than the average kinetic energy in the system.

Simulation of the model has been realized with the Euler–Maruyama integrator (discretization time step $\Delta t_{Euler} = 0.0002$) and total time length 500. Each hundredth instance of the resulting time series has been taken for a subsequent parameter estimation (resulting in observation time step $\tau = 0.02$) such that $T = 25.000$.

Furthermore, in order to make our model system more realistic and mimic the features inherent in biological systems, we rotate the resulting time series in the (N+2) dimensional space. We do it in such a way, that the metastability of the system becomes *hidden* in all the dimensions of the system in such a way that it is impossible to identify the metastable sets using projected clustering methods [14]. Application of the HMM–PCA method indicates the presence of two metastable states in the time series. In order to interpret the quality of the resulting model, we rotate the time series back, color the elements according to the corresponding metastable state and plot them atop of the original potential surface in (x_1, x_2). As we can see in Fig. 1, the local Langevin models are correctly situated at the wells of the double–well potential in the metastable dimensions and the elements of the time series are assigned in a proper way.

2.2   Analysis of the Long–Time Behavior of 12-Alanine in Water

As a second application we analyzed a molecular dynamics trajectory of 12-Alanine for conformational changes. The molecule was simulated using CHARMM with an implicit water environment. The data were kindly provided by Jeremy Smith and Frank Noé (IWR Heidelberg). We analyzed a $1\mu s$ long simulation with $2fs$ time steps. The analysis was performed on the basis of the 33 backbone torsion angles. We identified 3 metastable states that were analyzed with $m = 3$ local PCA modes. Fig. 2 illustrates the shape variability of the three conformations using a technique from [26]. The geometries of all time steps are are aligned with respect to some atoms in the middle of the backbone and accumulated into a pseudo density that is visualized using direct volume rendering. In Fig. 3 the mean backbone shapes of the three states are depicted together with arrows that indicate the dominant PCA-dimensions inside the conformations.

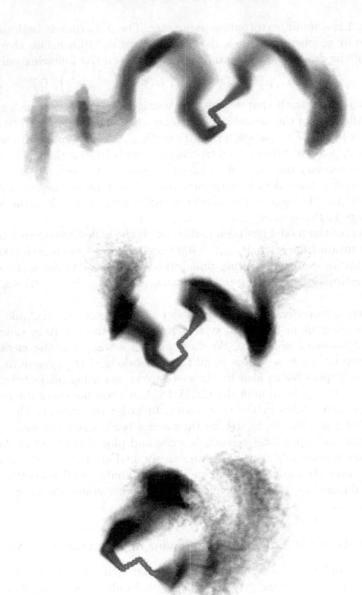

Fig. 2. Three dynamic conformations of 12-Alanine. Direct volume renderings of pseudo densities corresponding to the conformations in the investigated trajectory.

The identified metastable sets are the conformations of the molecule [27], which have life times of 10 ns. As we can see, the dominant dimensions can be interpreted as principal movements of the backbone which are characteristic for

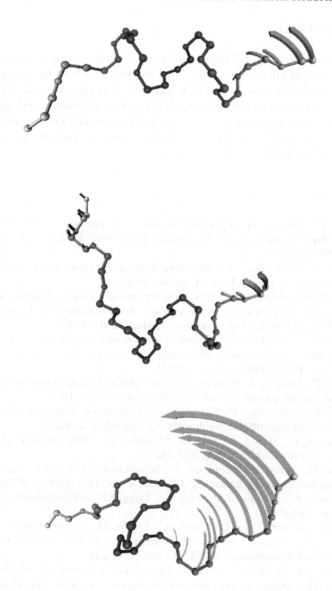

Fig. 3. Three dynamic conformations of 12-Alanine. Mean forms of the conformations in the investigated trajectory. Arrows indicate the flexibilities of the corresponding HMM-PCA-modes with highest variance. Fig. 2 shows flexibility at both ends of the backbone, while here a concentration on one end is visible. The flexibility of the other end is covered by the respective second PCA-modes, which are not shown here.

the corresponding conformations. The analysis shows that in terms of HMM-PCA these characteristic movements can be used to distinguish between the metastable sets and can be helpful in the clustering of the time series.

3   Conclusion

We presented a novel HMM based method for simultaneous dimension reduction and clustering of time series data. The method is based on a combination of an HMM approach and local PCA analysis. Incorporation of the local PCA analysis helps to map the clustering problem into low dimensional space. We have demonstrated the application of the method for a model system and a molecular dynamics trajectory of 12-Alanine. The numerical examples demonstrate the usefulness of the HMM-PCA approach.

References

1. Ichiye, T., Karplus, M.: Collective motions in proteins – a covariance analysis of atomic fluctuations in molecular dynamics and normal mode simulations. Proteins **11** (1991) 205–217
2. Frenkel, D., Smit, B.: Understanding Molecular Dynamics: From Algorithms to Applications. Academic Press, London (2002)
3. E, W., Vanden-Eijnden, E.: Metastability, conformation dynamics, and transition pathways in complex systems. In Attinger, S., Koumoutsakos, P., eds.: Multiscale, Modelling, and Simulation, Springer, Berlin (2004) 35–68
4. Deuflhard, P., Schütte, C.: Molecular conformation dynamics and computational drug design. In: Applied Mathematics Entering the 21st Century: Invited Talks from the ICIAM 2003 Congress. (2004)
5. Holmes, P., Lumley, J., Berkooz, G.: Turbulence, Coherent Structures, Dynamical Systems and Symmetry. Cambridge University Press (1996)
6. Givon, D., Kupferman, R., Stuart, A.: Extracting macroscopic dynamics: Model problems and algorithms. Nonlinearity **17** (2004) R55–R127
7. Kupferman, R., Stuart, A.: Fitting sde models to nonlinear kac-zwanzig heat bath models. Physica D **199** (2004) 279–316
8. Balsera, M., Wriggers, W., Oono, Y., Schulten, K.: Pricipal Component Analysis and long time protein dynamics. J. Chem. Phys. **100** (1996) 2567–2572
9. Hyvarinen, A., Karhunen, J., Oja, E.: Independent Component Analysis. John Wiley & Sons (2001)
10. Meyer, T., Ferrer-Costa, C., Perez, A., Rueda, M., Bidon-Chanal, A., Luque, F., Laughton, C., Orozco, M.: Essential dynamics: a tool for efficient trajectory compression and management. JCTC **2** (2006) 251–258
11. Hünenberger, P., Mark, A., van Gunsteren, W.: Fluctuation and cross-correlation analysis of protein motions observed in nanosecond molecular dynamics simulations. J. Mol. Biol. **252** (1995) 492–503
12. Monahan, A.: Nonlinear principal component analysis by neural networks: Theory and application to the lorenz system. J. Climate **13** (2000) 821–835
13. Christiansen, B.: The shortcomings of NLPCA in identifying circulation regimes. J. Climate **18** (2005) 4814–4823
14. Aggarwal, C., Wolf, J., Yu, P., Procopiuc, C., Park, J.: Fast algorithms for projected clustering. Proceedings of the 1999 ACM SIGMOD international conference on Management of data (1999)
15. Chakrabarti, K., Mehrotra, S.: Local dimensionality reduction: A new approach to indexing high dimensional spaces. In: Proceedings of the 26th VLDB Conference, Cairo,Egypt (2000) 98–115

16. Zhang, P., Huang, Y., Shekhar, S., Kumar, V.: Correlation analysis of spatial time series datasets: A filter-and-refine approach. the Proc. of the 7th Seventh Pacific-Asia Conference on Knowledge Discovery and Data Mining(PAKDD 2003) (2003)

17. Baum, L., Petrie, T., Soules, G., Weiss, N.: A maximization technique occuring in the statistical analysis of probabilistic functions of Markov chains. Ann. Math. Stat. **41** (1970) 164–171

18. Baum, L.: An inequality and associated maximization technique in statistical estimation for probabilistic functions of Markov processes. Inequalities **3** (1972) 1–8

19. Bilmes, J.: A Gentle Tutorial of the EM Algorithm and its Applications to Parameter Estimation for Gaussian Mixture and Hidden Markov Models. Thechnical Report. International Computer Science Institute, Berkeley (1998)

20. Ghahramani, Z.: An introduction to hidden Markov models and Bayesian networks. Int. J. Pattern Recognition and Artificial Intelligence **15** (2001) 9–42

21. Frydman, J., Lakner, P.: Maximum likelihood estimation of hidden Markov processes. Ann. Appl. Prob. **13** (2003) 1296–1312

22. Horenko, I., Dittmer, E., Fischer, A., Schütte, C.: Automated model reduction for complex systems exhibiting metastability. SIAM Multiscale Modeling and Simulation (2005) accepted for publication.

23. Golub, G., van Loan, C.: Matrix computations. Second edition edn. The John Hopkins University Press (1989)

24. Dempster, A.P., Laird, N.M., Rubin, D.B.: Maximum likelihood from incomplete data via the EM algorithm. J. Roy. Stat. Soc. B **39** (1977) 1–38

25. Viterbi, A.: Error bounds for convolutional codes and an asymptotically optimum decoding algorithm. IEEE Trans. Informat. Theory **13** (1967) 260–269

26. Schmidt-Ehrenberg, J., Baum, D., Hege, H.C.: Visualizing dynamic molecular conformations. In: Proceedings of IEEE Visualization 2002. (2002) 235–242

27. Schütte, C., Fischer, A., Huisinga, W., Deuflhard, P.: A direct approach to conformational dynamics based on hybrid Monte Carlo. J. Comput. Phys. **151** (1999) 146–168

Relational Subgroup Discovery for Descriptive Analysis of Microarray Data

Igor Trajkovski[1], Filip Železný[2], Jakub Tolar[3], and Nada Lavrač[1]

[1] Department of Knowledge Technologies, Jozef Stefan Institute
Jamova 39, 1000 Ljubljana, Slovenia
{igor.trajkovski, nada.lavrac}@ijs.si
[2] Department of Cybernetics, Czech Technical University in Prague,
Technicka 2, 166 27 Praha 6, Czech Republic
zelezny@fel.cvut.cz
[3] Department of Pediatrics, University of Minnesota Medical School,
420 Delaware Street, 55455 Minneapolis, USA
tolar003@umn.edu

Abstract. This paper presents a method that uses gene ontologies, together with the paradigm of relational subgroup discovery, to help find description of groups of genes differentially expressed in specific cancers. The descriptions are represented by means of relational features, extracted from gene ontology information, and are straightforwardly interpretable by the medical experts. We applied the proposed method to two known data sets: acute lymphoblastic leukemia (ALL) vs. acute myeloid leukemia and classification of fourteen types of cancer. Significant number of discovered groups of genes had a description, confirmed by the medical expert, which highlighted the underlying biological process that is responsible for distinguishing one class from the other classes. We view our methodology not just as a prototypical example of applying sophisticated machine learning algorithms to microarray data, but also as a motivation for developing more sophisticated functional annotations and ontologies, that can be processed by such learning algorithms.

1   Introduction

Microarrays are at the center of a revolution in biotechnology, allowing researchers to simultaneously monitor the expression of tens of thousands of genes. Independent of the platform and the analysis methods used, the result of a microarray experiment is, in most cases, a list of genes found to be differentially expressed. A common challenge faced by the researchers is to translate such gene lists into a better understanding of the underlying biological phenomena. Manual or semi-automated analysis of large-scale biological data sets typically requires biological experts with vast knowledge of many genes, to decipher the known biology accounting for genes with correlated experimental patterns. The goal is to identify the relevant "functions", or the global cellular activities, at work in the experiment. For example, experts routinely scan gene expression clusters to

M.R. Berthold, R. Glen, and I. Fischer (Eds.): CompLife 2006, LNBI 4216, pp. 86–96, 2006.

see if any of the clusters are explained by a known biological function. Efficient interpretation of these data is challenging because the number and diversity of genes exceed the ability of any single researcher to track the complex relationships hidden in the data sets. However, much of the information relevant to the data is contained in the publicly available gene ontologies. Including the ontologies as a direct knowledge source for any algorithmic strategy to approach such data may greatly facilitate the analysis.

Here we present an algorithm that for given multi-dimensional numerical data set, representing the expression of the genes under different conditions (that define the classes of examples) and ontology used for producing background knowledge about these genes, is able to identify groups of genes, described by conjunctions of first order features, whose expression is highly correlated with one of the classes. For example, one of the applications of this algorithm is to describe groups of genes that were selected as discriminative for some classification problem. Medical experts are usually not satisfied with having a separate description of every discriminative gene, but want to know the processes that are controlled by these genes. With our algorithm we are able to find these processes and the cellular components where they are "executed", indicating the genes from the preselected list of discriminative genes which are included in these processes.

These goals can be achieved by using the methodology of Relational Subgroup Discovery (RSD) [7]. With RSD we were able to induce set of discrimination rules between the different types (or subtypes) of cancers in terms of functional knowledge extracted from the gene ontology and information about gene interactions. In this way, we have succeeded to explain the differences between the types of cancer in terms of the functions of the genes that are differentially expressed in these types.

1.1 Analysis of Gene Expression Data

Large scale gene expression data sets include thousands of genes measured at dozens of conditions. The number and diversity of genes make manual analysis difficult and automatic analysis methods necessary. Initial efforts to analyze these data sets began with the application of unsupervised machine learning, or clustering, to group genes according to similarity in gene expression [2]. Clustering provides a tool to reduce the size of the dataset to a simpler one that can more easily be manually examined. The analysis of gene expression data for various tissue samples has enabled researchers to determine gene expression profiles characteristic of the disease subtypes. The groups of genes involved in these genetic profiles are rather large and a deeper understanding of the functional distinction between the disease subtypes might help not only to select highly accurate "genetic signatures" of the various subtypes, but hopefully also to select potential targets for drug design. Most current approaches to microarray data analysis use (supervised or unsupervised) clustering algorithms to deal with the numerical expression data. While a clustering method reduces the dimensionality of the data to a size that a scientist can tackle, it does not identify the critical background biological information that helps the researcher understand the significance of each cluster. However, that biological knowledge in terms of functional annotation of the genes is already available in public databases. Direct inclusion of this knowledge source can greatly improve the analysis, support (in term of user confidence) and explain obtained numerical results.

1.2   Gene Ontologies

One of the most important tools for the representation and processing of information about gene products and functions is the Gene Ontology (GO). It provides a controlled vocabulary for the description of cellular components, molecular functions, and biological processes. As of January 2006 (www.geneontology.org), GO contains 1681 component, 7386 function and 10392 process terms. Terms are organized in parent-child hierarchies, indicating either that one term is more specific than another (is_a) or that the entity denoted by one term is part of the entity denoted by another (part_of). Typically, such annotations are first of all established electronically and later validated by a process of manual verification.

Recently, an automatic ontological analysis approach using GO has been proposed to help solving the task of interpreting the results of gene expression data analysis [5]. From 2003 to 2005, 13 other tools have been proposed for this type of analysis and more tools continue to appear daily. Although these tools use the same general approach, identifying statistically significant GO terms that cover a selected list of genes, they differ greatly in many respects that influence in an essential way the results of the analysis. A general idea and comparison of those tools is presented in [6]. Another approach to descriptive analysis of gene expression data is [11]. They present a method that uses text analysis to help find meaningful gene expression patterns that correlate with the underlying biology as described in the scientific literature.

2   Descriptive Analysis of Gene Expression Data

The fundamental idea of this paper is as follows. First, we construct a set of discriminative genes, $G_C(c)$, for every class $c \in C$. These sets can be constructed in several ways. For example: $G_C(c)$ can be the set of k ($k > 0$) most correlated genes with class c, computed by, for example, Pearson's correlation. $G_C(c)$ can also be the set of best k single gene predictors, using the recall values from a microarray experiment (absent/present/marginal) as the expression value of the gene. These predictors can look like this: **If** $gene_i$ = present **Then** class = c. In our experiments we used a measure of correlation, $P(g, c)$, that emphasizes the "signal-to-noise" ratio in using gene g as predictor for class c. The definition and analysis of $P(g, c)$ is presented in later section.

The second step aims at improving the interpretability of G_C. Informally, we do this by identifying groups of genes in $G_C(c)$ (for each $c \in C$) which can be summarized in a compact way. Put differently, for each $c_i \in C$ we search for compact descriptions of group of genes which correlate strongly with c_i and weakly with all $c_j \in C; j \neq i$.

Searching for these groups of genes, together with their description, is defined as a separate supervised machine learning task. This secondary task is, in a way, orthogonal to the primary discovery process in that the original attributes (genes) now become training examples, each of which has a class label $c \in C$. To apply a discovery algorithm, information about relevant features of the new examples is required. No such features are usually present in the gene expression data sets themselves. However, this information can be extracted from a public database of gene annotations. For each gene we extracted its molecular functions, biological processes and cellular components where its protein products are located. Next, using GO, in the gene's background knowledge we also included its generalized GO terms and information about its interaction with other genes.

In traditional machine learning, examples are expected to be described by a tuple of values corresponding to some predefined, fixed set of attributes. Note that a gene annotation does not straightforwardly correspond to a fixed attribute set, as it has an inherently relational character. For example, a gene may be related to a variable number of cell processes, can play a role in variable number of regulatory pathways etc. This imposes 1-to-many relations which are hard to be elegantly captured within an attribute set of a fixed size. Furthermore, a useful piece of information about a gene g may for instance be expressed by the following feature:

gene g interacts with another gene whose functions
include protein binding.

In summary, we are approaching the task of subgroup discovery from a relational data domain. For this purpose we employ the methodology of relational subgroup discovery proposed in [7,13] and implemented in the RSD[1] algorithm. Using RSD, we were able to discover knowledge such as

The expression of genes coding for proteins located in the integral-to-membrane cell component, whose functions include receptor activity, has a high correlation with the BCR class of acute lymphoblastic leukemia (ALL) and a low correlation with the other classes of ALL.

The RSD algorithm proceeds in two steps. First, it constructs a set of relational features in the form of conjunctions of first order logic atoms. The entire set of features is then viewed as an attribute set, where an attribute has the value true for a gene (example) if the gene has the feature corresponding to the attribute. As a result, by means of relational feature construction we achieve the conversion of relational data into attribute-value descriptions. In the second step, groups of genes are searched, such that each group is represented as a conjunction of selected features. The subgroup discovery algorithm employed in this second step is an adaptation of the popular propositional rule learning algorithm CN2 [1].

---

[1] http://labe.felk.cvut.cz/železny/rsd/rsd.pdf

2.1 Relational Feature Construction

The feature construction component of RSD aims at generating a set of relational features in the form of relational logic atom conjunctions. For example, the feature exemplified informally in the previous section has the following relational logic form:

$$\text{interaction}(g, G), \text{function}(G, \text{protein\_binding})$$

Here, upper cases denote existentially quantified variables and g is the key term that binds a feature to a specific example (here a gene). The user specifies a grammar declaration which constraints the resulting set of constructed features. RSD accepts feature language declarations similar to those used in the inductive logic programming system Progol [9].

A remark is needed concerning the way constants (such as *protein_binding*) are employed in features. RSD extracts them automatically from the training data. For each constant-free feature, a number of different features are generated, each corresponding to a possible replacement of the combination of the indicated variables with constants. RSD then only proceeds with those combinations of constants which make the feature true for at least a pre-specified number of examples. Finally, to evaluate the truth value of each feature for each example for generating the attribute-value representation of the relational data, the first-order logic resolution procedure is used, provided by a Prolog language engine.

2.2 Subgroup Discovery

Subgroup discovery aims at finding population subgroups that are statistically "most interesting", e.g., are as large as possible and have the most unusual statistical characteristics with respect to the property of interest [12].

Rule learning, as implemented in RSD, involves two main procedures: the search procedure that performs search to find a single subgroup discovery rule, and the control procedure (the weighted covering algorithm) that repeatedly executes the search in order to induce a set of rules. Description of the two procedures, described in [7], are omitted for space constraints.

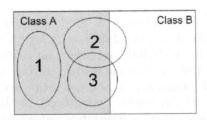

Fig. 1. Description of discovered subgroups can cover: a) individuals from only one class (1), b) individuals from other classes (2) and c) individuals covered also by other subgroups (2,3)

3 Experiments

This section presents a statistical validation of the proposed methodology. We aim at evaluating the properties of the secondary, descriptive learning task. Namely, we wish to determine if the high descriptive capacity pertaining to the incorporation of the expressive relational logic language incurs a risk of *descriptive overfitting*, i.e., a risk of discovering fluke subgroups. We thus aim at measuring the discrepancy of the quality of discovered subgroups on the training data set on one hand and an independent test set on the other hand. We will do this through the standard 10-fold stratified cross-validation regime. The specific qualities measured for each set of subgroups produced for a given class are average *precision* (PRE) and *recall* (REC) values among all subgroups in the subgroup set.

3.1 Materials and Methods

We apply the proposed methodology on two problems of predictive classification from gene expression data.

The first was introduced in [3] and aims at distinguishing between samples of ALL and AML from gene expression profiles obtained by the Affymetrix HU6800 microarray chip, containing probes for 6817 genes. The data contains 73 class-labeled samples of expression vectors. The second was defined in [10]. Here one tries to distinguish among 14 classes of cancers from gene expression profiles obtained by the Affymetrix Hu6800 and Hu35KsubA microarray chip, containing probes for 16,063 genes. The data set contains 198 class-labeled samples.

To access the annotation data for every gene considered, it was necessary to obtain unique gene identifiers from the microarray probe identifiers available in the original data. We achieved this by querying Affymetrix site[2] for translating probe ID's into unique gene ID's. Knowing the gene identifiers, information about gene annotations and gene interactions can be extracted from Entrez gene information database[3]. We developed a program script [4] in the Python language, which extracts gene annotations and gene interactions from this database, and produces their structured, relational logic representations which can be used as input to RSD.

In both data sets, for each class c we first extracted a set of discriminative genes $G_C(c)$. In our experiments we used a measure of correlation $P(g, c)$, used by [3], that emphasizes the "signal-to-noise" ratio in using gene g as predictor for class c. $P(g, c)$ is computed by the following procedure:

Let $[\mu_1(g), \sigma_1(g)]$ and $[\mu_2(g), \sigma_2(g)]$ denote the means and standard deviations of log of the expression levels of gene g for the samples in class c and samples in all other classes, respectively.

Let $P(g, c) = \frac{\mu_1(g) - \mu_2(g)}{\sigma_1(g) + \sigma_2(g)}$, which reflects the difference between the classes relative to the standard deviation within the classes. Large values of $|P(g, c)|$ indicate

---

[2] www.affymetrix.com/analysis/netaffx/

[3] ftp://ftp.ncbi.nlm.nih.gov/gene/

[4] This script is available on request to the first author.

a strong correlation between the gene expression and the class distinction, while the sign of $P(g, c)$ being positive or negative correspond to g being more highly expressed in class c or in other classes. Unlike a standard Pearson̂correlation coefficient, $P(g, c)$ is not confined to the range $[-1, +1]$. The set of informative genes for class c, $G_C(c)$ of size n, consists of the n genes having the highest $|P(g, c)|$ value. If we have only two classes, then $G_C(c_1)$ consist of genes having the highes $P(g, c_1)$ values, and $G_C(c_2)$ consists of genes having the highest $P(g, c_2)$ values.

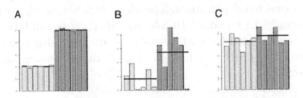

Fig. 2. Expression of three genes (A, B and C) for five patients of class 1 and five patients of class 2. The class distinction is represented by an idealized gene A, in which the expression level is uniformly low in class 1 and uniformly high in class 2. Gene B is well correlated with the class distinction. Gene C is also well correlated, but we are interested in genes that have significant difference in its expression between classes, like A and B.

For both problems we selected 50 discriminative genes[5]. The average value of correlation coefficient, $|P(g, c)|$, of selected discriminatory genes for each class/problem are listed in Table 1. The usage of the gene correlation coefficient is twofold. In the first part of the analysis, for a given class it is used for selection of discriminative genes, and in the second part as initial weight of the example-genes for the meta-mining procedure where we try to describe these discriminative genes. In the second mining task RSD will prefer to group genes with large weights, so these genes will have enough weight to be grouped in several groups with different descriptions.

After the selection of sets of discriminatory genes, $G_C(c)$ for each $c \in C$, these sets were merged and every gene coming from $G_C(c)$ was class labeled as c. Now RSD was run on these data, with the aim to find as large as possible and as pure (in terms of class labels) as possible subgroups of this population of example-genes, described by relational features constructed from GO and gene interaction data.

3.2   Results

The discovered regularities have very interesting biological interpretations.

In ALL, RSD has identified a group of 23 genes, described as: component(G, 'nucleus') AND interaction(G,B), process(B,'regulation of transcription, DNA-dependent'). The products of these genes, proteins, are located in the nucleus of

---

[5] This is a usual number of selected genes in the context of microarray data classification with SVM or voting algorithms.

Table 1. Average(AVG), maximal(MAX) and minimal(MIN) value of $\{|P(g,c)|g \in G_C(c)\}$ for each task and class c

TASK	CLASS	AVG	MAX	MIN
ALL-AML	ALL	0.75	1.25	0.61
	AML	0.76	1.44	0.59
MULTI	BREAST	1.06	1.30	0.98
CLASS	PROSTATE	0.91	1.23	0.80
	LUNG	0.70	0.99	0.60
	COLORECTAL	0.98	1.87	0.73
	LYMPHOMA	1.14	2.52	0.87
	BLADDER	0.88	1.15	0.81
	MELANOMA	1.00	2.60	0.73
	UTERUS	0.82	1.32	0.71
	LEUKEMIA	1.35	1.75	1.18
	RENAL	0.81	1.20	0.70
	PANCREAS	0.75	1.08	0.67
	OVARY	0.70	1.04	0.60
	MESOTHELIOMA	0.90	1.91	0.74
	CNS	1.38	2.17	1.21

the cell, and they interact with genes that are included in the process of regulation of transcription. In AML, RSD has identified several groups of overexpressed genes, located in the membrane, that interact with genes that have 'metal ion transport' as one of their functions.

In breast cancer, RSD has identified a group of genes (described as process (G,'regulation of transcription'), function(G,'zinc ion binding')) containing five genes (Entrez Gene id's: 4297, 51592, 91612, 92379, 115426) whose under expression is a good predictor for that class. These genes are simultaneously involved in regulation of transcription and in zinc ion binding. Zinc is a cofactor in protein-DNA binding, via a "zinc finger" domain (id 92379). Second, zinc is an essential growth factor and a zinc transporter associated with metastatic potential of estrogen positive breast cancer, termed LIV-1, has been described [4]. A separate group of genes involved in ubiquitin cycle (process(G,'ubiquitin cycle')) was identified in breast cancer, (Entrez id's: 3093, 10910, 23014, 23032, 25831, 51592, 115426). The role of ubiquitin in a cell is to recycle proteins. This is of a paramount importance to the overall cellular homeostasis, since inappropriately active proteins can cause cancer [8]. This is the example where one gene, id: 115426, was included in two groups with different descriptions.

In CNS (central nervous system) cancer, we discovered two important groups concerning neurodevelopment (description: process(G,'nervous system development'), Entrez Gene id's: 333, 1400, 2173, 2596, 2824, 3785, 4440, 6664, 7545, 10439, 50861), and immune surveillance (Entrez Gene id's: 199, 1675, 3001, 3108, 3507, 3543, 3561, 3588, 3683, 4046, 5698, 5699, 5721, 6352, 9111, 28299, 50848, 59307). The genes in the first/second group are over/under-expressed (respectively) in CNS. As for the former, reactivation of genes relevant to early development (i. e., ineffective recapitulation of embryonal or fetal neural growth at a wrong time) is a hallmark of the most rapidly growing tumors (id's 3785 and 10439 specific to neuroblastoma). The latter illustrates the common clinical observation that immune deficiency (subnormal expression of genes active in immune response shown in this work) creates a permissive environment for cancer persistence. Thus, both major themes of malignant growth are represented in this example: active unregulated growth and passive inability to clear the abnormal cells.

In addition, we subjected the RSD algorithm to 10-fold stratified cross-validation on both classification tasks. Table 2 shows the PRE and REC values (with standard deviation figures) results for the two respective classification tasks. Overall, the results show only a small drop from the training to the

Table 2. Precision-recall figures and average sizes of found subgroups of differentially expressed genes (DEG), for the ALL/AML and multi-class problem, obtained through 10-fold cross-validation

TASK	DATA	PRE(st.dev.)	REC(st.dev.)	AVG. SIZE
ALL-AML DEG	Train	0.96(0.01)	0.18(0.02)	12.07
	Test	0.76(0.06)	0.12(0.04)	
MULTI-CLASS DEG	Train	0.51(0.03)	0.15(0.01)	8.35
	Test	0.42(0.10)	0.10(0.02)	

testing set in terms of both PRE and REC, suggesting that the number of discriminant genes selected (Table 1) was sufficient to prevent overfitting. In terms of *total* coverage, RSD covered more then $\frac{2}{3}$ of the preselected discriminative genes (in both problems), while $\frac{1}{3}$ of the preselected gene were not included in any group. One interpretation is that they are not functionally connected with the other genes, but were selected by chance. This information can be used in the first phase of the classification problem, feature selection, by choosing genes that were covered by some subgroup. That will be the next step in our future work, using the proposed methodology as feature (gene) selection mechanism.

4   Discussion

In this paper we presented a method that uses gene ontologies, together with the paradigm of relational subgroup discovery, to help find patterns of expression for genes with a common biological function that correlate with the underlying biology responsible for class differentiation. Our methodology proposes to first select a set important discriminative genes for all classes and then finding compact, relational descriptions of subgroups among these genes.

Since genes frequently have multiple functions that they may be involved in, they may under some of the conditions exhibit the behavior of genes with one function and in other conditions exhibit the behavior of genes with a different function. Here subgroup discovery is effective at selecting a specific function. The same gene can be included in multiple subgroup descriptions (gene id: 115426 in breast cancer), each emphasizing the different biological process critical to the explanation of the underlying biology responsible for observed experimental results. Unlike other tools for analyzing gene expression data that use gene ontologies, which report statistically significant single GO terms and do not use gene interaction data, we are able to find a set of GO terms (the first reported group of genes, for breast cancer, is described with two GO terms), that cover the same set of genes, and we use available gene interaction data to describe features of genes that can not be represented with other approaches (the third reported group, for ALL).

However, this approach of translating a list of differentially expressed genes into subgroups of functional categories using annotation databases suffers from a few important limitations. The existing annotations databases are incomplete, only a subset of known genes is functionally annotated and most annotation databases are built by curators who manually review the existing literature. Although unlikely, it is possible that certain known facts might get temporarily overlooked. For instance, [6] found references in literature published in the early 90s, for 65 functional annotations that are yet not included in the current functional annotation databases. However, many such annotations are often made at very high-level GO terms, which limit their usefulness.

Despite the current imperfectness of the available ontological background knowledge, the presented methodology was able to discover and compactly describe several gene groups, associated to specific cancer types, with highly plausible biological interpretation. We thus strongly believe the presented approach will significantly contribute to the application of relational machine learning to gene expression analysis, given the expected increase in both the quality and quantity of gene/protein annotations in the near future.

Acknowledgment

The research of Igor Trajkovski and Nada Lavrač is supported by the Slovenian Ministry of Higher Education, Science and Technology. Filip Železný is supported by the Czech Academy of Sciences through the project KJB201210501 Logic Based Machine Learning for Analysis of Genomic Data. Jakub Tolar is supported by the Children's Cancer Research Fund, University of Minnesota Cancer Center and Department of Pediatrics.

References

1. Clark, P. & Niblett T. (1989). The CN2 induction algorithm. Machine Learning, pages 261-283.
2. Eisen, M. B., Spellman, P. T., Brown, P. O., & Botstein, D. (1998). Cluster analysis and display of genome-wide expression patterns. Proc. Natl. Acad. Sci. USA, 95:25, 14863-14868.

3. Golub, T. R., Slonim, D. K., Tamayo, P., Huard, C., Gaasenbeek, M. et al. (1999). Molecular classification of cancer: Class discovery and class prediction by gene expression monitoring. Science, 286:5439, 531-537.

4. Kasper, G. et al. (2005). Expression levels of the putative zinc transporter LIV-1 are associated with a better outcome of breast cancer patients. Int J Cancer. 20;117(6):961-73.

5. Khatri, P. et al. (2002). Profiling gene expression using Onto-Express. Genomics, 79, 266-270.

6. Khatri, P. & Draghici S. (2005). Ontological analysis of gene expression data: current tools, limitations, and open problems. Bioinformatics, 21(18):3587-3595

7. Lavrač, N., Železný F. & Flach P. (2002). RSD: Relational subgroup discovery through first-order feature construction. In Proceedings of the 12th International Conference on Inductive Logic Programming, pages 149-165.

8. Mani, A. & Gelmann E.P. (2005). The ubiquitin-proteasome pathway and its role in cancer. Journal of Clinical Oncology. 23:4776-4789

9. Muggleton, S. (1995). Inverse entailment and Progol. New Generation Computing, Special issue on Inductive Logic Programming, 13(3-4):245-286.

10. Ramaswamy, S., Tamayo P., Rifkin R. et al. (2001). Multiclass cancer diagnosis using tumor gene expression signatures, PNAS 98 (26) 15149-15154.

11. Raychaudhuri, S., Schtze,H.S. & Altman,R.B. (2003). Inclusion of textual documentation in the analysis of multidimensional data sets: application to gene expression data. Machine Learn., 52, 119-145

12. Wrobel, S. (1997). An algorithm for multi-relational discovery of subgroups. In Jan Komorowski and Jan Zytkow, editors, Proceedings of the First European Symposion on Principles of Data Mining and Knowledge Discovery, 78-87.

13. Železný, F. & N. Lavrač. (2006). Propositionalization-Based Relational Subgroup Discovery with RSD. Machine Learning 62(1-2):33-63. Springer 2006.

Applicability of Loop Recombination in Ciliates Using the Breakpoint Graph[*]

Robert Brijder[1], Hendrik Jan Hoogeboom[1], and Michael Muskulus[2]

[1] Leiden Institute of Advanced Computer Science, Universiteit Leiden,
Niels Bohrweg 1, 2333 CA Leiden, The Netherlands
`rbrijder@liacs.nl`
[2] Mathematical Institute, Universiteit Leiden,
Niels Bohrweg 1, 2333 CA Leiden, The Netherlands

Abstract. The concept of breakpoint graph, known from the theory of sorting by reversal, has been successfully applied in the theory of gene assembly in ciliates. We further investigate its usage for gene assembly, and show that the graph allows for an efficient characterization of the possible orders of loop recombination operations (one of the three types of molecular operations that accomplish gene assembly) for a given gene during gene assembly. The characterization is based on spanning trees within a graph built upon the connected components in the breakpoint graph. We work in the abstract and more general setting of so-called legal strings.

1 Introduction

Gene assembly is an involved DNA transformation process in ciliates (a large group of single cell organisms) which transforms a nucleus (the micronucleus) into a functionally different nucleus (the macronucleus). The process is accomplished using three types of DNA splicing operations, which operate on special DNA sequences called pointers. Each pointer can be seen as a breakpoint with a 'tag' which specifies how the splicing should be done, ensuring that the end result is fixed. The process however is not deterministic: for every gene in its micronuclear form, there can be several sequences of operations (called strategies) to transform this gene to its macronuclear form. For a given micronuclear gene, strategies may differ in the number of operations. It has been shown however that the number of loop recombination operations is independent of the chosen strategy [1,2], and that this number can be efficiently calculated [3,4].

In this paper we characterize for a given set of pointers D, whether or not there is a strategy that applies loop recombination (called string negative rule in the formal model that we use) on exactly these pointers. This result depends heavily on the reduction graph, which was introduced at CompLife '05 [4], and is motivated by the breakpoint graph in the theory of sorting by reversal [5,6,7]

---

[*] This research was supported by the Netherlands Organization for Scientific Research (NWO) project 635.100.006 'VIEWS'.

M.R. Berthold, R. Glen, and I. Fischer (Eds.): CompLife 2006, LNBI 4216, pp. 97–106, 2006.

since it adopts the concept of reality-and-desire for DNA sequences with break-points. More specifically, we define a graph, called the pointer-component graph, 'on top of' the reduction graph, thereby depicting the distribution of pointers over the connected components of the reduction graph [3,4]. We show that one can apply loop recombination on the pointers in D exactly when D forms a spanning tree in the pointer-component graph. Also, we characterize in which order the pointers of D can possibly be applied in strategies. Due to space constraints, proofs of the results are omitted, but can be found in an extended version of this paper [8].

2   String Pointer Reduction System

Three (almost) equivalent formal models for gene assembly were considered in [9,10,2]. In this section we briefly recall the one that we will use in this paper: the string pointer reduction system. For a detailed motivation and other results concerning this model we refer to [2].

We fix $\kappa \geq 2$, and define the alphabet $\Delta = \{2, 3, \ldots, \kappa\}$. For $D \subseteq \Delta$, we define $\bar{D} = \{\bar{a} \mid a \in D\}$ and $\Pi = \Delta \cup \bar{\Delta}$. The elements of Π will be called *pointers* (the pointers represent specific sequences of DNA where recombination will take place). We use the 'bar operator' to move from Δ to $\bar{\Delta}$ and back from $\bar{\Delta}$ to Δ. Hence, for $p \in \Pi$, $\bar{\bar{p}} = p$. For a string $u = x_1 x_2 \cdots x_n$ with $x_i \in \Pi$, the *inverse* of u is the string $\bar{u} = \bar{x}_n \bar{x}_{n-1} \cdots \bar{x}_1$. For $p \in \Pi$, we define **p** to be p if $p \in \Delta$, and $\bar{p}$ if $p \in \bar{\Delta}$, i.e., **p** is the 'unbarred' variant of p. The *domain* of a string $v \in \Pi^*$ is $dom(v) = \{\mathbf{p} \mid p$ occurs in $v\}$. A *legal string* is a string $u \in \Pi^*$ such that for each $p \in \Pi$ that occurs in u, u contains exactly two occurrences from $\{p, \bar{p}\}$. For a pointer p and a legal string u, if both p and $\bar{p}$ occur in u, then we say that both p and $\bar{p}$ are *positive* in u; if on the other hand only p or only $\bar{p}$ occurs in u, then both p and $\bar{p}$ are *negative* in u.

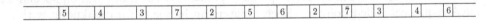

Fig. 1. Sequence of breakpoints represented by legal string u

A legal string is a representation of a sequence of pointers (breakpoints). Legal string $u = 543725627346$ corresponds to the sequence of pointers in Figure 1. Each gene in the micronucleus in ciliates can be represented by such a legal string. For example, legal string 3445675678932289 corresponds to the micronuclear form of the gene that corresponds to the actin protein in the stichotrich *Sterkiella nova* (see [11,2,12]). Gene assembly transforms each gene in micronuclear form to their macronuclear form by three splicing operations which operate on the pointers. These three operations are formally defined on legal strings through the string pointer reduction system, each defined on a specific pattern of the pointers.

The string pointer reduction system consists of three types of reduction rules operating on legal strings. For all $p, q \in \Pi$ with $\mathbf{p} \neq \mathbf{q}$:

- the *string negative rule* for p is defined by $\mathbf{snr}_p(u_1 p p u_2) = u_1 u_2$,
- the *string positive rule* for p is defined by $\mathbf{spr}_p(u_1 p u_2 \bar{p} u_3) = u_1 \bar{u}_2 u_3$,
- the *string double rule* for p, q is defined by $\mathbf{sdr}_{p,q}(u_1 p u_2 q u_3 p u_4 q u_5) = u_1 u_4 u_3 u_2 u_5$,

where $u_1, u_2, \ldots, u_5$ are arbitrary (possibly empty) strings over Π. We also define $Snr = \{\mathbf{snr}_p \mid p \in \Pi\}$, $Spr = \{\mathbf{spr}_p \mid p \in \Pi\}$ and $Sdr = \{\mathbf{sdr}_{p,q} \mid p, q \in \Pi, \mathbf{p} \neq \mathbf{q}\}$ to be the sets containing all the reduction rules of a specific type.

Let $S \subseteq \{Snr, Spr, Sdr\}$. Then a composition φ of reduction rules from S is called an *(S-)reduction*. Let u be a legal string. We say that φ is a *reduction of u*, if φ is a reduction and φ is applicable to (defined on) u. We call φ *successful for u* if $\varphi(u) = \lambda$, the empty string. We say that u is *successful in S* if there is a successful S-reduction of u. We say that a linearly ordered set $L = (p_1, \ldots, p_n)$ over $dom(u)$ is the *Snr-order* of φ, if $\varphi = \varphi_{n+1} \mathbf{snr}_{\tilde{p}_n} \varphi_n \mathbf{snr}_{\tilde{p}_{n-1}} \cdots \varphi_2 \mathbf{snr}_{\tilde{p}_1} \varphi_1$ for some (possible empty) $\{Spr, Sdr\}$-reductions $\varphi_1, \varphi_2, \ldots, \varphi_{n+1}$ and $\tilde{p}_i \in \{p_i, \bar{p}_i\}$ for $1 \leq i \leq n$ and $n \geq 0$.

Example. Let $u = \bar{4}377\bar{4}3$. Then $\varphi_1 = \mathbf{sdr}_{\bar{4},3} \mathbf{spr}_7$ is a successful $\{Spr, Sdr\}$-reduction of u. However, both $\varphi_2 = \mathbf{snr}_3 \mathbf{spr}_7$ and $\varphi_3 = \mathbf{snr}_8$ are *not* reductions of u.

Since for every (non-empty) legal string there is an applicable reduction rule, it follows that for every legal string u there is a successful reduction of u.

The *domain* of a reduction rule ρ, denoted by $dom(\rho)$, is defined by $dom(\mathbf{snr}_p) = dom(\mathbf{spr}_p) = \{\mathbf{p}\}$ and $dom(\mathbf{sdr}_{p,q}) = \{\mathbf{p}, \mathbf{q}\}$ for $p, q \in \Pi$. For a composition $\varphi = \varphi_n \cdots \varphi_2 \varphi_1$ of reduction rules $\varphi_1, \varphi_2, \ldots, \varphi_n$, the *domain*, denoted by $dom(\varphi)$, is defined by $dom(\varphi) = dom(\varphi_1) \cup dom(\varphi_2) \cup \cdots \cup dom(\varphi_n)$. Note that if φ is a reduction of u, then $dom(\varphi) = dom(u) \backslash dom(\varphi(u))$.

Example. The domain of $\varphi = \mathbf{snr}_2 \mathbf{spr}_{\bar{4}} \mathbf{sdr}_{7,5} \mathbf{snr}_{\bar{9}}$ is $dom(\varphi) = \{2, 4, 5, 7, 9\}$.

3   Reduction Graph

In this section we recall the definition of reduction graph and some results concerning this graph. First we give the definition of *pointer removal operations* [3]. For a subset $D \subseteq \Delta$, the *D-removal operation*, denoted by rem_D, is the mapping on Π^* that removes the occurrences in $D \cup \bar{D}$ from a string. Formally, it is the homomorphism $\Pi^* \rightarrow \Pi^*$ such that $\varphi(p) = p$ if $p \notin D \cup \bar{D}$, and $\varphi(p) = \lambda$ if $p \in D \cup \bar{D}$.

Example. Let $u = 543725627346$ be a legal string. Then for $D = \{4, 6, 7, 9\}$, we have $rem_D(u) = 532523$. In the remaining examples we will keep using this legal string u.

The next lemma is an easy consequence of Lemma 8 from [3]. It is used to prove Theorem 12. It cannot be extended to Snr rules: if $\mathbf{snr}_p$ is applicable to $rem_D(u)$, then it is not necessarily applicable to u.

Lemma 1. *Let u be a legal string and $D \subseteq dom(u)$. There is a $\{Spr, Sdr\}$-reduction φ of u with $dom(\varphi(u)) = D$ iff there is a successful $\{Spr, Sdr\}$-reduction of $rem_D(u)$.*

We now recall the reduction graph by an example. A reduction graph is an undirected graph with two (not necessarily disjoint) sets of edges called *reality edges* and *desire edges*. Moreover, all vertices, except for two distinct vertices s and t, are labelled by an element from Δ. Recall that the physical representation of legal string $u = 543725627346$ is given in Figure 1. Figure 2 shows the reduction graph $\mathcal{R}_u$ of this u. Notice that in this figure, the linear ordering of the pointers in u (cf. Figure 1) remains intact. The graph is defined in such a way that (1) each pointer of u appears twice (in unbarred form) as labels of vertices in the graph to represent both sides of the pointer in Figure 1, (2) the reality edges (depicted as 'double edges' to distinguish them from the desire edges) represent the segments between the pointers, (3) the desire edges represent which segments should be glued to each other when operations are applied on the corresponding pointers. Positive pointers are connected by crossing desire edges (cf. pointer 7 in Figure 2), while negative pointers are connected by parallel desire edges.

Notice that the reduction graph uses the intuition of the breakpoint graph known from the theory of sorting by reversal, see e.g. [6,5], since it connects the segments of Figure 1 through desire edges in such a way that it represents how the segments will be concatenated to each other when splicing operations are applied. Since the exact identities of the vertices are irrelevant for our purposes, in graphical depictions of reduction graphs, such as Figure 2, we will instead use their labels. We refer to [3,4] for a more detailed motivation and a formal definition. In Figure 3 (left-hand side) we have rearranged the vertices of Figure 2.

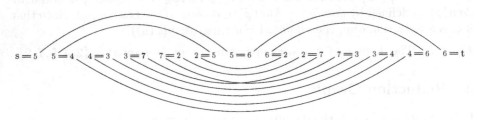

Fig. 2. The reduction graph $\mathcal{R}_u$ of u from the Example

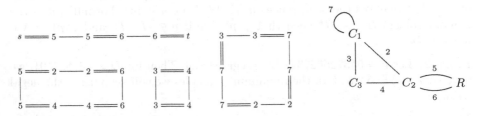

Fig. 3. The reduction graph $\mathcal{R}_u$ (left) and pointer-component graph $\mathcal{PC}_u$ (right) of u from the Example

Each reduction graph has a connected component with a linear structure containing both the source and the target vertex. This connected component is called the *linear component* of the reduction graph. The other connected components are called *cyclic components* because of their structure.

4 Pointer-Component Graphs

In this section we introduce the pointer-component graph built on the components of the reduction graph. We reformulate two results from [3] for use on this new graph.

A *(undirected) multigraph* is a structure $G = (V, E, \epsilon)$, where E is a finite set of edges and $\epsilon : E \to \{\{x, y\} \mid x, y \in V\}$ is the *endpoint mapping*. We allow $x = y$, and $\epsilon(e) = \{x, x\} = \{x\}$ then corresponds to e being a 'loop' on vertex x. We write $o(G) = |V|$, and for $E' \subseteq E$, we denote multigraph $(V, E', \epsilon|E')$ by $G|_{E'}$, where $\epsilon|E'$ is the restriction of ϵ to E'. Isomorphism, denoted by $\approx$, is defined as usual.

If it is clear from the context which legal string u is meant, we will denote by ζ the set of connected components of the reduction graph of u. We now define a graph on ζ that we will use throughout the rest of this paper. The graph represents how the labels of the vertices of a reduction graph are distributed among its connected components. We show that this graph is particularly useful in determining which sets D of pointers correspond to strategies that apply string negative rules on exactly the pointers of D.

Definition 2. Let u be a legal string. The *pointer-component graph of u (or of $\mathcal{R}_u$)*, denoted by $\mathcal{PC}_u$, is a multigraph (ζ, E, ϵ), where $E = dom(u)$ and ϵ is, for $e \in E$, defined by $\epsilon(e) = \{C \in \zeta \mid C \text{ contains vertices labelled by } e\}$.

Note that for each $e \in dom(u)$, there are exactly two desire edges connecting vertices labelled by e, thus the four vertices with this label can be found in at most two components. Therefore, ϵ is well defined.

Example. Consider $\mathcal{R}_u$ shown on the left-hand side of Figure 3. Let us define C_1 to be the cyclic component with vertices labelled by 7, C_2 to be the cyclic component with vertices labelled by 5, C_3 to be the third cyclic component, and R to be the linear component. Then $\zeta = \{C_1, C_2, C_3, R\}$. The pointer-component graph $\mathcal{PC}_u$ of u is shown on the right-hand side of Figure 3.

One of the motivations for the reduction graph is the easy determination of the number of string negative rules needed in each successful reduction. We can use the definition of pointer-component graph to reformulate Theorem 26 of [3]. Recall that $o(\mathcal{PC}_u)$ equals the number of connected components of $\mathcal{R}_u$.

Proposition 3. *Every successful reduction of a legal string u has exactly $o(\mathcal{PC}_u) - 1$ string negative rules.*

Example. Since $\mathcal{PC}_u$ has four vertices, by Proposition 3, every successful reduction φ of u has exactly three string negative rules. For example $\varphi = \mathbf{snr}_6 \ \mathbf{snr}_4$

$\mathbf{snr}_2 \, \mathbf{spr}_{\bar{7}} \, \mathbf{sdr}_{5,3}$ is a successful reduction of u. Indeed, φ has exactly three string negative rules. Alternatively, $\mathbf{snr}_6 \, \mathbf{snr}_4 \, \mathbf{snr}_3 \, \mathbf{spr}_2 \, \mathbf{spr}_5\mathbf{spr}_7$ is also a successful reduction of u, with a different number of ($\mathbf{spr}$ and $\mathbf{sdr}$) operations.

The effect of each of the three reduction rules on the level of reduction graphs can be described by a single operation, the reduction function. The transition from $\mathcal{R}_u$ to $\mathcal{R}_{\varphi(u)}$ is achieved by deleting the vertices labelled by elements of $dom(\varphi)$ and replacing paths via the deleted vertices by a single reality edge ([3]). In particular, none of the operations can join or split components of $\mathcal{R}_u$. In fact, only Snr rules change the number of components when deleting the last vertices in one of the components. The reduction function can be reformulated on pointer-component graphs. The p-reduction function deletes edge p from $\mathcal{PC}_u$ and removes the vertices that become isolated in the process.

Definition 4. For each edge p, we define the p-*reduction function* rf_p, for multi-graph $G = (V, E, \epsilon)$, by $rf_p(G) = (V', E', \epsilon|E')$, where $E' = E \backslash \{p\}$ and $V' = \{v \in V \mid v \in \epsilon(e) \text{ for some } e \in E'\}$ if $E' \neq \varnothing$, and $V' = \{\varnothing\}$ otherwise.

The next result is now a reformulation of Theorem 17 in [3].

Proposition 5. *Let u be a legal string, and let φ be a reduction of u. Then $(rf_{p_n} \cdots rf_{p_2} \, rf_{p_1})(\mathcal{PC}_u) \approx \mathcal{PC}_{\varphi(u)}$, where $dom(\varphi) = \{p_1, p_2, \ldots, p_n\}$.*

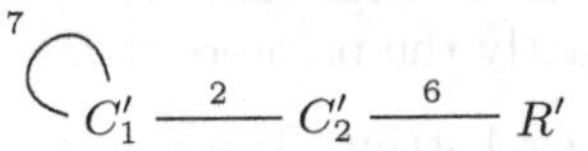

$$\overset{7}{\underset{C_1'}{\bigcirc}} \overset{2}{\rule{1cm}{0.4pt}} C_2' \overset{6}{\rule{1cm}{0.4pt}} R'$$

Fig. 4. Pointer-component graph PC_1 from the Example

Example. We have $(\mathbf{snr}_4 \, \mathbf{sdr}_{5,3})(u) = 62\bar{7}726$. The pointer-component graph PC_1 of this legal string is shown in Figure 4. It is easy to see that the graph obtained by applying $(rf_5 \, rf_4 \, rf_3)$ to $\mathcal{PC}_u$ (Figure 3) is isomorphic to PC_1.

5   Spanning Trees in Pointer-Component Graphs

We consider spanning trees in pointer-component graphs, and we show that there is an intimate connection between these trees and applicable strategies of string negative rules. The Snr rules in a reduction can be 'postponed' without affecting the applicability. Thus we can separate each reduction into a sequence without Snr rules, and a tail of Snr rules. We often use this 'normal form' for notational convenience. First, we characterize reductions φ without Snr rules by comparing the pointer-component graph *after* applying φ to u with the graph *before* applying φ to u. By Propositions 3 and 5, such a reduction φ results in the removal of edges in the pointer-component graph, whereas the number of vertices is not changed.

Lemma 6. *Let φ be a reduction of legal string u, and let $D = dom(\varphi(u))$. Then φ is a $\{Spr, Sdr\}$-reduction iff $\mathcal{PC}_{\varphi(u)} \approx \mathcal{PC}_u|_D$.*

The next result shows that every pointer-component graph is connected.

Theorem 7. *The pointer-component graph of any legal string is connected.*

If a legal string u is successful in $\{Snr\}$, then, by Proposition 3, the number of edges of $\mathcal{PC}_u$ is one less than the number of vertices. Thus, by Theorem 7, it is a tree. The argument can be reversed.

Lemma 8. *Legal string u is successful in $\{Snr\}$ iff $\mathcal{PC}_u$ is a tree.*

By Lemma 6 and Lemma 8 the pointers on which string negative rules are applied in any successful reduction form a spanning tree of $\mathcal{PC}_u$.

Theorem 9. *Let u be a legal string, and let $D \subseteq dom(u)$. If there is a successful reduction φ of u, where the domain of Snr rules in φ equals D, then $\mathcal{PC}_u|_D$ is a tree.*

In the next few sections we show that the reverse implication of the previous theorem also holds. This will require considerably more effort than the forward implication. The reason for this is that it is not obvious that when $\mathcal{PC}_u|_D$ is a tree, there is a reduction φ_1 of u such that $D = dom(\varphi_1(u))$. We will use the pointer removal operation to prove this.

First, we consider a special case of the previous theorem. For a legal string u, we denote the set of loops $\{e \in E \mid |\epsilon(e)| = 2\}$ of $\mathcal{PC}_u = (\zeta, E, \epsilon)$ by $snrdom(u)$. Since a loop can never be part of a tree, if $\mathbf{snr}_p$ or $\mathbf{snr}_{\bar{p}}$ is in a (successful) reduction of u, then $p \in snrdom(u)$. We will show at the end of Section 7 that the reverse implication also holds. Hence, the name $snrdom(u)$ is explained: the pointers $p \in snrdom(u)$ are exactly the pointers for which $\mathbf{snr}_p$ or $\mathbf{snr}_{\bar{p}}$ occurs in some (successful) reduction of u.

Example. We saw that $\varphi = \mathbf{snr}_6\,\mathbf{snr}_4\,\mathbf{snr}_2\,\mathbf{spr}_7\,\mathbf{sdr}_{5,3}$ is a successful reduction of u. By Theorem 9 and Figure 3, $\mathcal{PC}_u|_{\{2,4,6\}}$ is a tree. Also, we have $2, 4, 6 \in snrdom(u)$. Both statements are clear from Figure 3 where $\mathcal{PC}_u$ is depicted.

6   Merging and Splitting Components

We consider the effect of pointer removal operations on pointer-component graphs. It turns out that these operations correspond to the merging and splitting of connected components of the underlying reduction graph. First, we formally introduce the merging operation. Intuitively, the p-merge rule 'merges' the two endpoints of edge p into one vertex, and therefore the resulting graph has exactly one vertex less than the original graph.

Definition 10. For each edge p, the p-*merge rule*, denoted by $merge_p$, is a rule applicable to (defined on) multigraph $G = (V, E, \epsilon)$ with $p \in E$ and $|\epsilon(p)| = 2$. It is defined by $merge_p(G) = (V', E', \epsilon')$, where $E' = E \backslash \{p\}$, $V' = (V \backslash \epsilon(p)) \cup \{v'\}$ with $\{v'\} \cap V = \varnothing$, and $\epsilon'(e) = \{h(v_1), h(v_2)\}$ iff $\epsilon(e) = \{v_1, v_2\}$ where $h(v) = v'$ if $v \in \epsilon(p)$, otherwise it is the identity.

Again, we allow both $v_1 = v_2$ and $h(v_1) = h(v_2)$ in the previous definition.

One of the most surprising aspects of this paper is that the pointer removal operation is crucial in the proofs of the main results. The next theorem compares $\mathcal{PC}_u$ with $\mathcal{PC}_{rem_{\{p\}}(u)}$ for a legal string u and $p \in dom(u)$. We distinguish three cases: either the number of vertices of $\mathcal{PC}_{rem_{\{p\}}(u)}$ is one less, is equal, or is one more than the number of vertices of $\mathcal{PC}_u$. The proof of this theorem (omitted here) shows that the first case corresponds to merging two connected components of $\mathcal{R}_u$ into one connected component, and the last case corresponds to splitting (the inverse of merging) one connected component of $\mathcal{R}_u$ into two connected components.

Theorem 11. *Let u be a legal string, and let $p \in dom(u)$.*

- *If $p \in snrdom(u)$, then $\mathcal{PC}_{rem_{\{p\}}(u)} \approx merge_p(\mathcal{PC}_u)$*
  (and therefore $o(\mathcal{PC}_{rem_{\{p\}}(u)}) = o(\mathcal{PC}_u) - 1$).
- *If $p \notin snrdom(u)$, then $o(\mathcal{PC}_u) \leq o(\mathcal{PC}_{rem_{\{p\}}(u)}) \leq o(\mathcal{PC}_u) + 1$.*

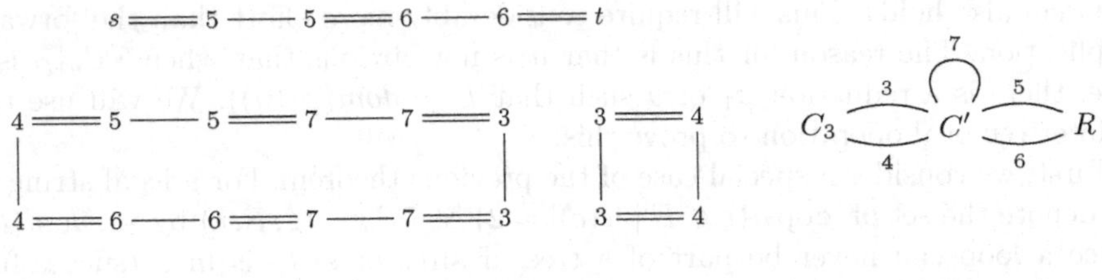

Fig. 5. Both $\mathcal{R}_{rem_{\{2\}}(u)}$ (left) and $\mathcal{PC}_{rem_{\{2\}}(u)}$ (right) from the Example

Example. By Figure 3, $2 \in snrdom(u)$. Therefore, by Theorem 11, we have $\mathcal{PC}_{rem_{\{2\}}(u)} \approx merge_2(\mathcal{PC}_u)$. Indeed, this is transparent from Figures 3 and 5, where $\mathcal{PC}_u$ and $\mathcal{PC}_{rem_{\{2\}}(u)}$ are depicted, respectively. For convenience $\mathcal{R}_{rem_{\{2\}}(u)}$ is also given in Figure 5. Now, by Theorem 11, it follows that $\mathcal{R}_{rem_{\{2,7\}}(u)}$ has two or three cyclic components. One can verify that $\mathcal{R}_{rem_{\{2,7\}}(u)}$ has two cyclic components.

7   Applicability of the String Negative Rule

In this section we characterize for a given set of pointers D, whether or not there is a strategy that applies string negative rules on exactly these pointers. It is one of the main results of this paper and follows from the results of the previous section, Lemma 1 and Proposition 3. It improves Theorem 9 by characterizing exactly which string negative rules can be applied together in a successful reduction of a given legal string.

Theorem 12. *Let u be a legal string, and let $D \subseteq dom(u)$. There is a successful reduction φ of u, where the domain of Snr rules in φ equals D iff $\mathcal{PC}_u|_D$ is a tree.*

Since there are many well known and efficient methods for determining spanning trees in a graph, it is easy to determine, for a given set of pointers D, whether or not there is a successful reduction applying string negative rules on exactly the pointers of D (for a given legal string u).

Example. By Theorem 12 and Figure 3, there is a successful reduction $\varphi = \varphi_2\,\varphi_1$ of u, for some $\{Spr, Sdr\}$-reduction φ_1 and $\{Snr\}$-reduction φ_2 with $dom(\varphi_2) = \{2, 3, 5\}$. Indeed, we can take for example $\varphi = \mathbf{snr}_5\ \mathbf{snr}_2\ \mathbf{snr}_3\ \mathbf{spr}_7\ \mathbf{sdr}_{4,6}$.

It is easy to see that we can apply Theorem 12 also to a subset of the pointers that form the domain of Snr operations in a successful reduction, or to reductions in general. We then obtain, for legal string u and $D \subseteq dom(u)$, there is a (successful) reduction φ of u such that for all $p \in D$, φ contains either $\mathbf{snr}_p$ or $\mathbf{snr}_{\bar{p}}$ iff $\mathcal{PC}_u|_D$ is acyclic. In particular, for a single pointer $D = \{p\}$, we conclude that $\mathbf{snr}_p$ or $\mathbf{snr}_{\bar{p}}$ occurs in a reduction iff edge p is not a loop in $\mathcal{PC}_u$. In other words, when p is an element of $snrdom(u)$.

8   The Order of Loop Recombination

According to Theorem 12 a set D of pointers can occur as the domain of Snr rules in a successful reduction of a legal string u exactly when the graph $\mathcal{PC}_u|_D$ is a tree. This result can be strengthened to incorporate the order in which the Snr rules are applied. We show that in a successful reduction φ we can only apply Snr rules in orderings determined by the tree $\mathcal{PC}_u|_D$, where D is the domain of Snr rules in φ. These orderings are similar topological orderings in a directed acyclic graph, however, here we order the edges instead of the vertices. If T is a tree with root R, then an *edge-topological ordering of* T is an linear ordering $(e_1, e_2, \ldots, e_n)$ of the edges such that if the path from the root R to e_i passes e_j, then $i \leq j$. Thus, the edge-topological orderings of T are exactly those sequences of edges obtained by iteratively removing leaves from T.

Example. Consider tree $\mathcal{PC}_u|_{D_1}$ with $D_1 = \{2, 3, 5\}$. Taking R as the root of $\mathcal{PC}_u|_{D_1}$, it follows that $(3, 2, 5)$ is the only possible edge-topological ordering of $\mathcal{PC}_u|_{D_1}$.

Theorem 13. *Let u be a legal string, and let L be an linearly ordered set over $dom(u)$. There is a successful reduction φ of u with Snr-order L iff $\mathcal{PC}_u|_L$ is a tree and L is an edge-topological ordering of $\mathcal{PC}_u|_L$ with the linear component R of $\mathcal{R}_u$ as root.*

Example. Consider tree $\mathcal{PC}_u|_{D_2}$ with $D_2 = \{2, 4, 6\}$. Since $L = (4, 2, 6)$ is an edge-topological orderings of tree $\mathcal{PC}_u|_{D_2}$ with root R, by Theorem 13 there is a successful reduction φ of u with Snr-order L. Indeed, we can take for example $\varphi = \mathbf{snr}_6\ \mathbf{snr}_2\ \mathbf{spr}_7\ \mathbf{snr}_4\ \mathbf{sdr}_{5,3}$.

We say that two reduction rules ρ_1 and ρ_2 can be applied in *parallel* to u if both $\rho_2\,\rho_1$ and $\rho_1\,\rho_2$ are applicable to u (see [13]). As a corollary to Theorem 13, we have, for $p, q \in dom(u)$, $\mathbf{snr}_{\bar{p}}$ and $\mathbf{snr}_{\bar{q}}$ can eventually be applied in parallel when reducing u iff p and q are incomparable in a spanning tree of $\mathcal{PC}_u$ containing both edges p and q.

9   Conclusion

This paper shows that one can efficiently determine the possible sequences of loop recombination operations that can be applied in the transformation of a given gene from its micronuclear to its macronuclear form. Formally, one can determine the orderings of string negative rules that can be present in successful reductions of u. This is a characterization in terms of a graph defined on the (components of the) reduction graph. Future research could focus on similar characterizations for the string positive rules and the string double rules. However, this would require other concepts, since the pointer-component graph does not retain information regarding positiveness or overlap of pointers, notions crucial for the applicability of the other two operations.

References

1. Ehrenfeucht, A., Petre, I., Prescott, D., Rozenberg, G.: Circularity and other invariants of gene assembly in ciliates. In Ito, M., et al., eds.: Words, Semigroups, and Transductions, World Scientific, Singapore (2001) 81–97
2. Ehrenfeucht, A., Harju, T., Petre, I., Prescott, D., Rozenberg, G.: Computation in Living Cells – Gene Assembly in Ciliates. Springer Verlag (2004)
3. Brijder, R., Hoogeboom, H., Rozenberg, G.: Reducibility of gene patterns in ciliates using the breakpoint graph. Theor. Comput. Sci. **356** (2006) 26–45
4. Brijder, R., Hoogeboom, H., Rozenberg, G.: The breakpoint graph in ciliates. In Berthold, M., et al., eds.: CompLife. Volume 3695 of LNCS., Springer (2005) 128–139
5. Pevzner, P.: Computational Molecular Biology: An Algorithmic Approach. MIT Press (2000)
6. Setubal, J., Meidanis, J.: Introduction to Computional Molecular Biology. PWS Publishing Company (1997)
7. Bergeron, A., Mixtacki, J., Stoye, J.: On sorting by translocations. In Miyano, S., et al., eds.: RECOMB. Volume 3500 of LNCS., Springer (2005) 615–629
8. Brijder, R., Hoogeboom, H., Muskulus, M.: Strategies of loop recombination in ciliates. LIACS Technical Report 2006-01, [arXiv:cs.LO/0601135] (2006)
9. Ehrenfeucht, A., Petre, I., Prescott, D., Rozenberg, G.: String and graph reduction systems for gene assembly in ciliates. Math. Struct. in Comput. Sci. **12** (2002) 113–134
10. Ehrenfeucht, A., Harju, T., Petre, I., Prescott, D., Rozenberg, G.: Formal systems for gene assembly in ciliates. Theor. Comput. Sci. **292** (2003) 199–219
11. Prescott, D., DuBois, M.: Internal eliminated segments (IESs) of oxytrichidae. J. Euk. Microbiol. **43** (1996) 432–441
12. Ciliates IES MDS database, http://oxytricha.princeton.edu/dimorphism/
13. Harju, T., Li, C., Petre, I., Rozenberg, G.: Parallelism in gene assembly. In Ferretti, C., et al., eds.: DNA 10. Volume 3384 of LNCS., Springer (2004) 138–148

High-Throughput Identification of Chemistry in Life Science Texts

Peter Corbett and Peter Murray-Rust

Unilever center for Moleclular Sciences Informatics, Lensfield Road, Cambridge,
CB2 1EW

Abstract. OSCAR3 is an open extensible system for the automated annotation
of chemistry in scientific articles, which can process thousands of articles per
hour. This XML annotation supports applications such as interactive browsing
and chemically-aware searching, and has been designed for integration with
larger text-analysis systems. We report its application to the high-throughput
analysis of the small-molecule chemistry content of texts in life sciences, such
as PubMed abstracts.

1 Introduction

Chemical knowledge is an important component of bio-literature, as show by the
recent prominence of ontolgies and resources on bio-informatics sites. ChEBI[1] (at
the European Bioinformatics Institute) lists 7592 compounds of relevance to
bioscience. The NCBI has recently provided PubChem as the repository of chemical
data created in the NIH's Molecular Libraries Roadmap. Its importance has been
shown by the addition of the ZINC database of commercially available compounds,
bringing PubChem's size to about 5 million compounds and covering almost all those
in common use.

Currently few publishers actively add chemical semantics to their products.
Recently Nature Chemical Biology has produced connection tables (the formal
computer representation of the structure) and made them available to PubChem. For
almost all articles, however, the identity of the chemistry in the text has had to be
extracted by human experts.

Natural language processing (NLP) of the biological, biochemical and biomedical
literature is a well-developed area with a lot of activity. There is considerable activity
from both a number of commercial entities (eg. Temis, Linguamatics, PubGene) and
academic projects (eg. PennBioIE[2], FlySlip[3], GENIA[4]). There are a number of
freely-usable web tools based on NLP technology (eg. MEDIE[5], Info-Pubmed[6],
iHOP[7], EBIMed[8], Textpresso[9]). The biological NLP community is assisted by a
variety of publically-available resources. Hand-annotated corpora have been made
available by the PennBioIE and GENIA groups. Large amounts of the literature are
available in the form of PubMed/MEDLINE, which is commonly used as a corpus,
for example by the web tools mentioned above, and other resources such as the Gene

M.R. Berthold, R. Glen, and I. Fischer (Eds.): CompLife 2006, LNBI 4216, pp. 107–118, 2006.
© Springer-Verlag Berlin Heidelberg 2006

Ontology and other members of the Open Biomedical Ontologies family (which includes ChEBI). Finally, the field is assisted by competitive evaluations like BioCreAtIvE[10] and the TREC genomics track[11].

Development of natural language methodologies for chemistry lags behind that of the biochemical world. We have been able to find several works in the area[12-20], but not the same level of activity as is observed in the bioscience.

In this paper we describe an open source system, OSCAR3 which can identify much of the chemical terminology in articles, and extract molecular connection tables with useful recall and precision. OSCAR3 is being developed as a part of the SciBorg project which aims to use deep parsing of text (described in [21]) to analyse chemistry papers. Here we present the architecture and strategy of OSCAR3 as an open extensible modular framework for individual and high-throughput chemical text processing. The source code to OSCAR3 is on sourceforge.net[22].

1.1 Chemical Language in Bioscience

Among the useful information in manuscripts is
1. mention of chemical compounds.
2. details of synthesis (in vivo and in vitro) of compounds.
3. proof of structure (spectra and analytical data).
4. methods and reagents in bioscience bio-protocols.
5. properties of compounds.
6. reactions and their properties, both in enzymes and enzyme-free systems.

We consider both abstracts and full-text articles (a typical example on "Proton-sensing G-protein-coupled receptors" [23], (reproduced with permission of Nature))

Buffers and pH Experiments were carried out in a physiological salt solution (PSS) containing 130 mM NaCl, 0.9 mM NaH2PO4, 5.4 mM KCl, 0.8 mM MgSO4, 1.0 mM CaCl2, 25 mM glucose. This solution was buffered either with HEPES alone (20 mM) or HEPES/EPPS/MES (8 mM each; HEM-PSS), to cover a wider pH range. HEPES-buffered PSS was used in all experiments unless HEM-PSS is specifically mentioned. HEPES is 4-(2-hydroxyethyl) piperazine-1-ethanesulphonic acid...

This type of chemistry is very well understood and has a simple generic vocabulary which has not changed over decades. Unlike much bioscience, where ontological tools are an essential part of reconciling the domain-dependent approaches, much chemistry has an implicitly agreed abstract description. The problems are primarily reconciling syntax and semantics. This is because chemists use abbreviated methods of communicating data, relying on trained readers to add information from the context. We have reviewed current problems of machine-understanding of chemistry in a typical chemistry journal[24] and requirements for parsing chemistry in life sciences [25].

2   OSCAR3

OSCAR3 is being developed as part of the SciBorg system for the deep parsing and analysis of scientific texts [21] and Fig 1. shows its role. Scientific articles in bespoke

XML are converted to a canonical XML ("SciXML", a schema developed by the Cambridge natural language group). A selection of modules (including OSCAR3) then annotate various concepts in the text. These annotations are collected together in a standoff annotation document – a separate document that contains pointers to elements in the source text, allowing the integration of the results from a range of different parsers, and for earlier parsers to pass information to later parsers. The role of OSCAR3 in the analysis of full text has informed many of our design and prioritisation decisions. For example, it will be easier for the SciBorg system to deal with false positives generated by OSCAR3 than with false negatives. Also, it is important for OSCAR3 to be able to recognise as many entities as possible, even if it cannot assign any semantics to many of them.

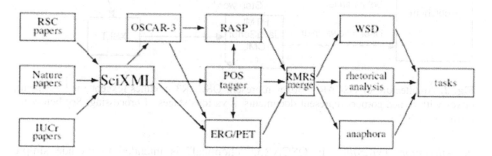

Fig. 1. Architecture of SciBorg. POS = Part Of Speech, RASP and ERG/PET are parsers for English text, RMRS is a format for representing the semantics of parsed language, WSD = Word Sense Disambiguation

However, OSCAR3 can also be used in contexts other than the SciBorg framework. For example, we have been developing a set of web tools for analysing, viewing and searching texts using OSCAR3 in an essentially standalone manner. OSCAR3 could also potentially be incorporated into a different framework – for example, integrating OSCAR3 with biological NLP tools such as those mentioned in the introduction. To expedite this, we have been developing OSCAR3 with thought for modification, extensibility and integratability, making components modular and supporting them with XML data files. For example, it should be possible to configure OSCAR3 to recognise many GO terms by simple lookup, to recongise names of computation chemistry codes and to configure OPSIN to parse many semi-systematic names without editing the OSCAR3 source code.

The overall architecture of OSCAR3 is shown in Figure 2. The preferred input format is SciXML, but rudimentary facilities exist for producing SciXML from plain text or HTML, allowing documents to be fed to OSCAR3 from the world-wide web via a Javascript bookmarklet. Abstracts may also be fetched from PubMed. An initial module – the recogniser - finds chemical names, data and other entities in the text. The names can be systematic (eg. *propan-2-ol*), trivial (*morphine, water*), semisystematic (*diacetylmorphine*), acronyms and other abbreviations, (*DMSO, 5-HT*) or formulae ($C_6H_{12}O_6$, *EtOAc*). We also consider names of groups or other fragments (*methyl*) and names in plural, verb or adjective form (*azidomethanones,*

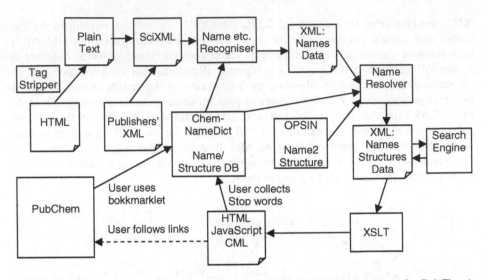

Fig. 2. Architecture of OSCAR3. Boxes represent OSCAR3 modules (except for PubChem), boxes with turned corners represent documents at various stages of processing. See below for further details.

demethylation, pyrazolic). In OSCAR3, "chemical" is intended to include simple polymers as well as small molecules, but not complex biopolymers such as proteins and nucleic acids. Elements are listed as a separate class (as it is often nontrivial to determine whether "carbon" (for example) refers to carbon atoms within a molecule or to the pure substance). "Data" refers to conventional representations of experimental results – e.g. "[α]22D +10.0 (c 1.00, MeOH)", which represents the optical rotation of a substance. "Other entities" are other nonwords that occur in chemical free text, for example "C(1)-N(6)" which refers to a particular bond in a molecule.

A second module – the resolver - examines the chemical names and attempts to assign structures to them. Two forms of output are produced – a standoff annotation document (for the rest of the SciBorg framework), and an enhanced SciXML document that incorporates the annotations inline while preserving all of the formatting, bibliographic, metadata and other markup that was in the source document.

Prior to name recognition, the text must be tokenised, by splitting on whitespace and recognising full stops and other similar characters. A particular difficulty is posed by hyphens, as these sometimes occur within chemical names (eg *tert-butyl peroxide*) and sometimes divide names from other words (eg "*hexane-ethyl acetate*" is a common phrase used to describe a mixture of the two solvents). Zamora and Blower [17] only considered hyphens with two alphabetic characters on either side to be word boundaries; we also have lists of strings that denote non-word-boundaries if they come before the hyphen (eg. *tert-* is often a part of chemical names), and strings that denote word boundaries if they occur after the hyphen (eg –containing, -induced).

To cope with the variety of forms of names and data to be recognised, several methods are used in parallel for name recognition. OSCAR3 keeps an internal lexicon of chemical names and structures – we have initially populated this using ChEBI[1]

This collection is by no means comprehensive, but it does cover many key solvents, reagents and biomolecules. This lexicon can be extended at run-time.

Names are also recognised using a naïve-Bayesian method based on overlapping 4-Grams[26]. Wilbur et al. [13] have reported the use of a simple 4-Gram based approach to recognise chemical names to obtain high precision and recall on their test data. However, Vasserman [12] found that their simple approach gave poor results in the more difficult context of PubMed abstracts, and experimented with a variety of smoothing algorithms for the 4-Gram models to improve the performance of their classifier. Neither approach has been used to find the bounds of chemical names in free text. Here, we use modified Knesser-Ney smoothing[27] to produce a refined 4-Gram model. Further improvements are obtained by applying a penalty to the scores of words in a standard English dictionary, and by rejecting names that did not possess a suffix from a list of 49 common chemical suffixes. A lexicon of stop words exists to catch the most common errors. There are rules for which words to group together to make multiword chemical names – for example a group name can add onto the front of a chemical name, as in *ethyl acetate*. Finally, a set of cascaded regular expressions is used to recognise chemical data[28], chemical formulae and other forms of notation.

Structures are assigned to chemical names *via* two methods – lookup (using the lexicon above), and parsing of systematic nomenclature (see below). The structures are stored as SMILES, InChIs (International Chemical Identifier – an algorithmically-generated, canonical identifier for chemical compounds developed by IUPAC) and CML (Chemical Markup Language).

After parsing, the enhanced SciXML can then be rendered into HTML using an XSLT stylesheet, and displayed in a browser. Javascript routines in the HTML are used to feed the molecular structures back to the server when the mouse is moved over them to produce and display a structural diagram. The pages are indexed by a search engine based on Apache Lucene, which indexes the compounds by their InChI identifiers, allowing them to be searched for specific compounds. The search engine also contains the full structures of all of the compounds indexed; these may be queried using substructure and similarity searches. This produces a list of InChIs which are then used to retrieve documents relevant to the query. As well as retrieving the documents, it is possible to list all of the compounds occurring in the documents, sorted by their frequencies of occurrence.

The browser-based architecture also provides a useful way of retraining OSCAR3. If a name does not have an associated structure, the structure (and synonyms) may be fetched from PubChem. We have not incorporated PubChem into OSCAR3 wholesale, partly out of a concern for efficiency, and partly as not all of the names in PubChem are accurate (see CIDs 26, 311) or appropriate (see CID 446220) Likewise, if a word has been misclassified as a chemical name, it can be fed to a list of the list of stop words, so that future occurrences of the word will not be classed as chemical. The stop words are also used to retrain the Bayesian classifier, further improving its accuracy. These techniques are especially useful in conjunction with the sorted compound lists described above, as this allows for common compounds and errors to be identified and dealt with appropriately, reducing the amount of effort required to adapt OSCAR3 to new domains.

2.1 Chemical Names

The conversion of systematic chemical names to structures is a surprisingly difficult task: the official description[29] of the nomenclature is not presented in a formal way, and so it falls to software authors to convert the descriptions provided into working code. This is not a trivial problem and the various commercial offerings on the market compete on the range of names that they can parse.

It is clear that OSCAR requires a chemical name parser, and there are a number of reasons for us to write our own. The first is a sheer practicality – an open-source parser made in-house is easier to integrate with other systems, and it is legally easier to deploy and distribute the results. More importantly, having such a parser will allow ourselves and others to experiment with parsing many of the ways nomenclature is used in the chemical literature. We have therefore developed OPSIN – an Open Parser for Systematic IUPAC Nomenclature.

A number of parsing schemes have been described in the literature. The earliest approach we are aware of was reported by Vander Stouw et al.[30-31]. Later, Kirby et al.[30] fitted names to a formal context-free grammar using a modified SLR parser. In building a commercial parser for CambridgeSoft, Brecher expressed a frustration with the formal grammars, and appears to have chosen a less formal approach.[32]

The published descriptions of these parsers are not sufficiently detailed to allow a reconstruction of their work: for example Vander Stouw [30] reports the use of a "special routine" to disambiguate the possible meanings of the token "hex", with little further explanation. These omissions are at least partly due to a lack of space in which to present full details; to overcome this, we have deposited the source code for OPSIN on sourceforge.net under an open-source license, along with the rest of OSCAR3.

Previously, we have worked on an ad-hoc approach to nomenclature parsing. However, this ran into severe difficulties owing to ambiguity. Therefore, we are currently pursuing a hybrid approach. Rather than using context-free grammars, we chose to partially interpret the names using finite-state grammars, which are less expressive but more tractable. A series of informal rules is then used to construct a full interpretation, thus escaping the limitations of the finite-state parsing.

The input to the parser is a complete systematic name: the current parser cannot detect and remove extraneous verbiage (e.g. *glacial* in *glacial* acetic acid) from its input, nor does it request additional information if it is only given a partial name to work with. There are a number of stages of processing – key ones are shown below.

The first step (not shown in Figure 3) breaks the chemical name into its constituent words, by splitting on spaces, and assigns those words to particular roles. A set of regular expressions are used to detect whether the compound is represented as a one-word compound (eg. ethanol), an ester (ethyl acetate), an acid (ethanoic acid), a salt (sodium acetate) or one of a number of other possibilities. Having done this, the various words are labelled as being root-like (eg "ethanol"), substituent-like (eg "ethyl") or a simple token ("acid").

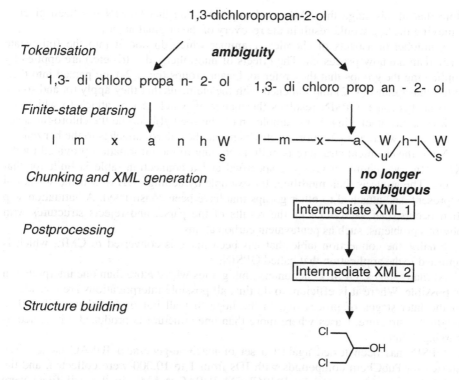

Fig. 3. Steps in the parsing of chemical names by OPSIN. Key to "finite-state parsing": l = locant group, m = multiplier, x = simple substituent, a = alkane name stem, n = "n-" (indicating a straight-chain alkane), h = hyphen, s = suffix, W = Hantzsch-Widman ring system.

The second stage (*tokenisation*) breaks the individual words into a list of multi-character tokens. Tokens come from two sources – a set of lists of tokens, along with data on what they mean, and a set of regular expressions.

In the third stage, *finite-state parsing*, roles are assigned to the various tokens – some, such as "chloro", are unambiguous. Others, such as "hex" and "ane", could have multiple roles – each of these are represented by a 1-letter symbolic code. The first could act as a multiplier (as in hexachloro-, "m" in Figure 3) or to indicate six carbon atoms in a row (as in hexane), whereas the latter could indicate an alkane (hexane again, "a" in Figure 3) or a six-membered aliphatic ring (as in dioxane, "W" in Figure 3). The various possibilities for each token in the word are tabulated, and a finite automaton generated from a regular expression (representing the grammar of chemical names) is used to select a valid path (if present) through the word. At this stage, an essentially 'flat' parse is generated – further processing is needed to construct a full parse tree.

The next stage is to group together the tokens into chunks, each representing the "root group" of the compound, or a substituent. These chunks also include metadata such as brackets and locants and multipliers.

Once the chunk structure of the name has been determined, an XML representation of the name is constructed, with one XML element per token in the chemical name.

Note that at this stage the XML marks-up the name that OPSIN has been given – removing the tags would result in the recovery of the original name.

A number of aspects of chemical grammar which do not fit into the finite-state formalism are now processed. The effects of multipliers (di-, tri- etc.) are applied by duplicating the groups that they refer to, locant groups (eg. 1,2-) are parsed into their individual locants (1, 2) and matched with the elements that they apply to, and so on. Also at this stage, OPSIN resolves the matter of which groups attach to which, by looking at the bracketing. For example, in (2-chloroethyl)benzene the chloro- attaches to the ethyl, whereas in 2-chloro-1-ethylbenzene the chloro- attaches to the benzene.

Next, the chemical structure is built, according to the information provided by the XML. In each chunk, a group is specified and a connection table is built for that group. The group is then modified, for example by adding –OH to it if the suffix –ol is present, and attached to other groups that have been constructed. A validation step then occurs, that sanity-checks the results of the parse, and rejects structures with obvious problems, such as pentavalent carbon atoms.

Finally, the connection table that has been made is converted to CML, which is returned to the application that called OPSIN.

At many stages, OPSIN encounters ambiguities where more than one interpretation is possible. Where it is efficient to do this, all possible interpretations are considered in the later stages of processing, in the hope that all but one of these will fail to produce a structure. Cases where more than one structure is produced are treated as parsing failures.

OPSIN has been tested against a set of machine-generated IUPAC names. The entries for PubChem compounds with IDs from 1 to 10,000 were collected, and the IUPAC names (designated by PUBCHEM_IUPAC_NAME in the .sdf files) were harvested, along with the associated InChIs. This gave 8183 names (average length: 58 characters) with associated InChI pairs. The names were passed to OPSIN, and where OPSIN produced a structure for the name, the structure was converted to an InChI and compared against the published InChI. From the 8183 names, OPSIN produced 4475 correct InChIs (54.7%), 162 incorrect InChIs (2.0%) and did not produce a result for the remaining 3546 of the names (43.3%). This test run took 442.5 seconds – equal to 18 parse attempts per second.

There are a number of areas where OPSIN is currently lacking in functionality. One of these is in the handling of "generic" nomenclature, where the chemical structure is underspecified by the name – for example in "aminopyrazoles" where the amino group attaches to the pyrazole is not specified. Often, these appear as plurals, and are used to specify a class of chemical compounds. Currently, OPSIN will reject plurals that are handed to it, but if they are re-cast as singulars by removing the terminal 's', OPSIN will attempt to parse them. Often, where these names differ from fully-specified names in that locants are missing, the current OPSIN will assume that "default" locants are meant – for example aminopyrazole is treated as 1-aminopyrazole. It should be noted that to the best of our knowledge none of the other parsers are able to produce underspecified structures from chemical names either. Parsing these remains as one of our research objectives.

Features not currently implemented include the handling of tri- and higher poly-cyclic nomenclature (eg. tricyclo[5.4.0.0^{2,9}]undecane), fused aromatic nomenclature (eg. thieno[3,2-b]furan – however fused systems with trivial names such as naphtha-alene are included, as is the common system cyclopenta[a]phenanthrene, which is included as a special case, as it has a special locant numbering for compatibility with steroid nomenclature), and many forms of nomenclature specific to complex inorganic chemicals and natural products. Another problem is that the bracketing that is used to show which group attaches to which is frequently left out – for example, in 2-chloroethylbenzene it is not trivial to determine whether the chloro group attaches to the ethyl group or to the benzene. We are therefore looking into ways in which this information can be inferred. Possible approaches involve looking for common fragments of chemical names (eg. chloromethyl-, dimethylamino-), or by seeing which interpretation requires the fewest assumptions later on (for example if the chloro group above attaches directly to the benzene, the ethyl group has three inequivalent places where it could attach. If the chloro group attaches via the ethyl group, the symmetry of the benzene ring removes this problem).

3 Evaluation

An initial informal evaluation was carried out the 1100 PubMed abstracts in the section of the BioIE[2] corpus dealing with cytochrome P450 biochemistry, suggesting an average precision of around 75%. These took about 1300 seconds to parse. After further development OSCAR3 was re-tested on unseen abstracts.

Batches of ca. 200 abstracts were fetched from PubMed using five different search terms. "Smith" was used to get an essentially random collection of abstracts. "Bacterial metabolism" was used to represent a fairly large field, "veterinary toxicology" for a small one. "Porcine skin" and "grapefruit" were also selected for their ability to give lots of interesting abstracts.

From them, abstracts containing five or more chemicals (or other entities to be annotated by OSCAR3) were selected. For each batch, the following procedure was followed: the abstracts were auto-annotated by OSCAR3, and a gold standard was created by correcting the annotations by hand. Per-abstract precision and recall ("P1", "R1") were calculated for the concepts annotated. Two annotations matched if they were in the same place, the same length and had the same type (compound, group, element etc.). Currently the name resolution is not advanced enough for formal testing. A second, larger, batch of abstracts (ca 500 where available, not overlapping with the first) was fetched using the same search term, and auto-annotated with OSCAR3. The most commonly occurring compounds were then tabulated, and exactly five minutes was spent adding mis-recognised words (eg. *swine, prostate*) as stop-words, starting with the most common mistakes and working down the list in sequence. The first batch was re-annotated with OSCAR3, and precision and recall ("P2", "R2") were recalculated. The parser was then retrained, and precision and recall ("P3", "R3") were recalculated again. Finally, exactly five minutes was spent selecting various abbreviations that had been mis-recognised as formulae (eg. IC50, P450, H5N1, CI), and final precision and recall ("P4", "R4") were calculated. This procedure demonstrated that the second batch of abstracts could act as training data for the first, without the need for laborious hand-annotation of the entire batch.

Table 1. Results from evaluation of OSCAR3. P1 etc. as defined above.

Search term	Smith	Veterinary Toxicology	Bacterial Metabolism	Porcine Skin	Grapefruit
Abstracts fetched	168	91	185	196	180
Abstracts selected	48	31	63	64	129
Abstracts in batch 2	432	147	483	483	234
P1	60.8%	67.5%	67.7%	72.1%	65.0%
R1	69.6%	74.0%	80.8%	69.7%	72.0%
P2	62.9%	70.5%	69.9%	74.0%	66.7%
R2	69.3%	74.0%	80.8%	69.8%	72.0%
P3	63.0%	71.4%	70.5%	73.7%	66.9%
R3	69.3%	74.0%	80.8%	69.2%	72.0%
P4	64.1%	74.3%	71.6%	75.3%	70.5%
R4	69.6%	74.0%	80.8%	69.1%	72.0%

From this work, it is clear that there are a number of areas in which OSCAR3 could be improved. Some changes should be trivial to implement – for example the list of allowable chemical suffixes was missing a few key entries (-am, -fil, -id, -vir, -oids) that were common in drug names. More training data would also help.

Other changes are not so straightforward. Acronyms and other abbreviations, for example, are a real problem. The regex-based chemical formula recogniser mis-classifies many of these as chemicals (some fortuitously do represent chemicals eg. BP (benzopyrene), but most represent non-chemicals eg. HIV), and many acronyms/abbreviations for chemical names are missed by the parser. Fortunately, many of these acronyms are defined in the text of the abstract – in the SciBorg framework, we see finding these and using them to handle acronyms appropriately as being a task for word sense disambiguation (or anaphora/coreference resolution) modules. Likewise, words like In, No and At are commonly mis-labelled as chemical elements – these errors could be removed with the aid of a part-of-speech tagger.

Currently OSCAR3 uses a medium-size lookup for names, but we believe that almost all common trivial and drug names are to be found in PubChem whose name, synonym and connection table data are freely downloadable and could be configured with OSCAR3.

References

1. de Matos, P., Ennis, M., Guedj, M., Degtyarenko, K., Apweiler, R. ChEBI – Chemical Entities of Biological Interest, Nucleic Acids Res., Database Summary Paper 646.
2. http://bioie.ldc.upenn.edu
3. http://www.cl.cam.ac.uk/users/av308/Project_Index/index.html
4. http://www-tsujii.is.s.u-tokyo.ac.jp/GENIA
5. http://www-tsujii.is.s.u-tokyo.ac.jp/medie
6. http://www-tsujii.is.s.u-tokyo.ac.jp/info-pubmed
7. http://www.ihop-net.org/UniPub/iHOP/
8. http://www.textpresso.org/
9. http://www.ebi.ac.uk/Rebholz-srv/ebimed/index.jsp
10. http://pdg.cnb.uam.es/BioLINK/BioCreative.eval.html

11. http://ir.ohsu.edu/genomics/
12. Vasserman, A.: Identifying Chemical Names in Biomedical Text: An Investigation of the Substring Co-occurrence Based Approaches. Proceedings of the Student Research Workshop at HLT-NAACL, 2004.
13. Wilbur, J. W., Hazard, G. F., Divita, G., Mork, J. G., Aronson, A. R., Browne, A. C.: Analysis of Biomedical Text for Chemical Names: A Comparison of Three Methods. Proc AMIA Symp 1999, 176-80.
14. Chowdhury, G. G., Lynch, M. F., Semantic Interpretation of the Texts of Chemical Patent Abstracts. 1. Lexical Analysis and Categorization. Journal of Chemical Informatics and Computer Science 32 (1992) 463-467.
15. Chowdhury, G. G., Lynch, M. F., Semantic Interpretation of the Texts of Chemical Patent Abstracts. 2. Processing and Results. Journal of Chemical Informatics and Computer Science 32 (1992) 468-473.
16. Al, C. S., Blower, P. E. Jr., Ledwith, R. H., Extraction of Chemical Reaction Information from Primary Journal Text. Journal of Chemical Informatics and Computer Science 30 (1990), 163-169.
17. Zamora, E. M., Blower, P. E. Jr.: Extraction of Chemical Reaction Information from Primary Journal Text Using Computational Linguistics Techniques. 1. Lexical and Syntactic Phases. Journal of Chemical Informatics and Computer Science 24 (1984), 176-181.
18. Zamora, E. M., Blower, P. E. Jr.: Extraction of Chemical Reaction Information from Primary Journal Text Using Computational Linguistics Techniques. 2. Semantic Phase. Journal of Chemical Informatics and Computer Science 24 (1984), 181-188.
19. Postma, G. J., van der Linden, B., Smits, J. R., Kateman, G.: TICA: A System for the Extraction of Data from Analytical Chemical Text. Chemometrics and Intellegent Laboratory Systems, 9 (1990) 65-74.
20. Cooper, J. W., Boyer, S., Nevidomsky, A., Coden, A. R.: Automatic discovery and annotation of organic chemical names in patents, 229th ACS National Meeting 2005.
21. Copestake, A., Corbett, P. T., Murray-Rust, P., Rupp, C. J., Siddharthan, A., Teufel, S., Waldron, B.: An Architecture for Language Technology for Processing Scientific Texts, submitted for UK e-Science All Hands Meeting 2006.
22. http://sourceforge.net/projects/oscar3-chem.
23. Ludwig, M.-G., Vanek, M., Guerini, D., Gasser, J. A., Jones, C. E., Junker, U., Hofstetter, H., Wolf, R. M., Seuwen, K.: Proton-sensing G-protein-coupled receptors, Nature 425 (2003), 93-98.
24. Murray-Rust, P., Mitchell, J. B. O., Rzepa, H. S.: Communication and re-use of chemical information in bioscience, BMC Bioinformatics 6 (2005), 180.
25. Murray-Rust, P., Mitchell, J. B. O., Rzepa, H. S.: Chemistry in Bioinformatics, BMC Bioinformatics 6 (2005), 141.
26. Townsend, J., Copestake, A., Murray-Rust, P., Teufel, S., Waudby, C., Language Technology for Processing Chemistry Publications, in Proceedings of the fourth UK e-Science All Hands Meeting, 2005.
27. Chen, S. F., Goodman, J.: An empirical study of smoothing techniques for language modeling. Computer Speech and Language 13 (1999), 359-394.
28. Townsend, J. A., Adams, S. E., Waudby, C. A., de Souza, V. K., Goodman, J. M., Murray-Rust, P.: Chemical documents: machine understanding and automated information extraction, Organic & Biomolecular Chemistry 2 (2004), 3294.

29. A Guide to IUPAC Nomenclature of Organic Chemistry, Recommendations 1993 (including Revisions, Published and hitherto Unpublished, to the 1979 Edition of Nomenclature of Organic Chemistry), IUPAC (1993)
30. Vander Stouw, G. G., Naznitsky, I., Rush, J. E.: Procedures for Converting Systematic Names of Organic Compounds into Atom-Bond Connection Tables. Journal of Chemical Documentation 7 (1967) 165-169.
31. Vander Stouw, G. G., Elliott, P. M., Isenbert, A. C.: Automated Conversion of Chemical Substance Names into Atom-Bond Connection Tables. Journal of Chemical Documentation 14 (1974), 185-193.
32. Cooke-Fox, D. I., Kirby, G. H., Rayner, J. D.: Computer Translation of IUPAC Systematic Organic Chemical Nomenclature. 1. Introduction and Background to a Grammar-Based Approach, J. Chem. Inf. Comp. Sci. 29 (1989) 101.
33. Brecher, J.: Name=Struct: A Practical Approach to the Sorry State of Real-Life Chemical Nomenclature, J. Chem. Inf. Comp. Sci. 39 (1999) 943.

Beating the Noise: New Statistical Methods for Detecting Signals in MALDI-TOF Spectra Below Noise Level

Tim O.F. Conrad[1], Alexander Leichtle[2], Andre Hagehülsmann[3],
Elmar Diederichs[1], Sven Baumann[2], Joachim Thiery[2], and Christof Schütte[1]

[1] Free University Berlin, Department of Mathematics, Germany
[2] Institute of Laboratory Medicine, Clinical Chemistry and Molecular Diagnostics,
University Hospital Leipzig, Germany
[3] Microsoft Research, Cambridge, UK

Abstract

Background: The computer-assisted detection of small molecules by mass spectrometry in biological samples provides a snapshot of thousands of peptides, protein fragments and proteins in biological samples. This new analytical technology has the potential to identify disease associated proteomic patterns in blood serum. However, the presently available bioinformatic tools are not sensitive enough to identify clinically important low abundant proteins as hormons or tumor markers with only low blood concentrations.

Aim: Find, analyze and compare serum proteom patterns in groups of human subjects having different properties such as disease status with a new workflow to enhance sensitivity and specificity.

Problems: Mass data acquired from high-throughput platforms frequently are blurred and noisy. This complicates the reliable identification of peaks in general and very small peaks even below noise level in particular.[1] However, this statement is only valid for single or few spectra. If the algorithm has access to a large number of spectra (e.g. $N > 1000$), new possibilities arise, one of such being a statistical approach.

Approach: Apply signal preprocessing steps followed by statistical analyses of the blurred data and the region below the typical noise threshold to identify signals usually hidden below this "barrier".

Results: A new analysis workflow has been developed that is able to accurately identify, analyze and determine peaks and their parameters even below noise level which other tools can not detect. A Comparison to commercial software[2] has clearly proven this gain in sensitivity. These additional peaks can be used in subsequent steps to build better peak patterns for proteomic pattern analysis. We belive that this new approach will foster identification of new biomarkers having not been detectable by most algorithms currently available.

---

[1] It is due to the fact that it is not possible to distinguish noise from signals if these two components fully overlay.

[2] ClinProTools 2.0, Bruker Daltronics: manufacturer of mass spectrometry instruments and accessories.

M.R. Berthold, R. Glen, and I. Fischer (Eds.): CompLife 2006, LNBI 4216, pp. 119–128, 2006.
© Springer-Verlag Berlin Heidelberg 2006

1 Introduction

With the advantages of proteomic technologies it became possible to assess small and low abundant molecules in biological fluids. These peptide or protein patterns (often referred to as "fingerprints") can indicate status or progress of a specific disease. Several studies have shown the potential of such patterns for early detection of different types of cancer [1, 2].

After reducing sample complexity by biochemical separation techniques [3], the protein fingerprinting of biological samples consists of three main steps: first, generating of mass spectra, that is presenting the complete spectrum of peptides and proteins in the sample. Second, features (usually peaks) in the resulting spectra are detected that can be used to discriminate between groups of individuals with various phenotypes (eg. gender, age or disease). Third, the features have to be tested in independent studies to confirm and to detect the underlying molecules.

Mass spectrometry is a high-throughput profiling technique able to fulfill the first part of the task[3]. The second step is usually done by machine learning algorithms and statistical approaches which are used to analyze the data obtained from mass spectrometry and to detect phenotype specific patterns.

Todays mass spectrometry based protein fingerprinting techniques rely on the analysis of spectra from complex biological protein mixtures (e.g. serum) obtained from high-throughput platforms in clinical settings. An unsolved bioinformatic problem is the highly sensitive detection of peaks (potential features) within this "crossfire of influences". Embedded in systematic and random noise introduced during acquisition of data it is very difficult and questionable to detect peaks and correctly determine their parameters, such as location, height or width. Most methods simply ignore any signals below an estimated noise threshold and potentially lose many signals hidden in this region.

In this study, we propose a new statistical driven approach that allows to analyze noise and identify signals below the commonly used signal-to-noise threshold[4]. This is done by sophisticated preprocessing steps and statistical analysis of all potential signals in a large number of spectra by identifying even smallest features. Compared to commercial software[5] (see Tab.1) our approach - in the cases tested - is about 20000 times more sensitive without loss of specificity. Additionally peaks identified can be used in subsequent steps to build better patterns for proteomic fingerprinting analysis. We belive that this will foster identification of new biomarkers having not been detectable by most algorithms currently available.

---

[3] In this study we used the "Matrix-assisted Laser Desorption/Ionization - Time-Of-Flight - Mass Spectrometry" (MALDI-TOF MS). For a recent review and a good introdutcion to this topic see e.g.[4] and references therein.

[4] This method only regards peaks if their height is above a certain value determined by a noise-estimation step. A common setting for the minimum peak height is three times the estimated noise level.

[5] ClinProTools 2.0, Bruker Daltronics: manufacturer of mass spectrometry instruments and accessories.

2 Material and Methods

2.1 Spectra Preprocessing

The raw spectrum acquired by a TOF mass-spectrometer (see Fig.1 left) is a mixture of the real signal and noise. The noise itself consists of a low-frequency baseline and high-frequency chemical and random noise. Preprocessing of TOF spectra includes suppression of noise and enhancing of the real signal - it is therefore a crucial step prior to the actual signal extraction. The next sections describe steps performed to prepare the raw signal enabling subsequent reliable peak detection and analyses.

Baseline Correction. The baseline is an exponential like offset dependent on the m/z value (mass-to-charge; x-value). It is mainly caused by clusters of matrix components and small molecular fragments originating from degradation processes, desorption and collisions in the acceleration phase. A baseline correction is performed to remove this rather low-frequency noise from the spectrum.

Following [5–7] we use a morphological TopHat filter. Mathematical morphology is the analysis of spatial structures and is used here to eliminate certain spatial structures within the signal, in our case the baseline. Its simplicity and rapidity make it extremely handy for application to large amounts of data. Fig. 1 illustrates this method. Note that this technique does not produce negative intensity values as many other popular methods depending on polynomial fitting [8], piecewise linear regression [9] or convex hulls [10] do.

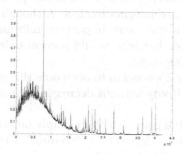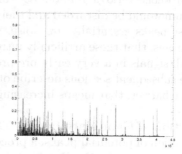

Fig. 1. Application of the TopHat Filter: the opening is substracted from the raw signal (left) yielding the filtered version (right)

Smoothing. Smoothing or denoising the raw signal X tries to separate the Gaussian contribution from the undisturbed signal S and generally yields better results in subsequent steps of the analysis workflow, since some general assumptions about smoothness can be taken. We define the Gaussian distributed component of the signal as noise. Consider the problem of denoising a raw signal $X \in \mathbb{R}$ having additive noise n with zero mean:

$$x_i = s_i + n_i \quad i = 1..N, \quad n \sim \mathcal{N}(0, \sigma^2) \tag{1}$$

Since the noise described above occurs on much faster time scales as the signal, we use a multiresolution analysis on X [11] based on a time-invariant discrete orthogonal wavelet transformation [12].

Opposed to other denoising algorithms, such as moving average or low-pass filter (e.g. Savitzky-Golay), this approach utilizes the multi-scale nature of the signal and therefore has better energy conservation properties, that is, the amplitude of the signal decreases less through denoising.

Normalization. Inter-spectrum normalization is the process of removing systematic variations between spectra. Many different techniques exist such as "Inverse Normalization" [13] or "Logarithmic Normalization" [14]. Our implementation follows the idea of the most frequently used method which is global normalization with respect to the average *total ion current* (TIC[6])[7] [16, 17] with an important extension: from the set of spectra to be normalized all TIC values are computed, outliers removed and the remaining highest value (instead of the average) is used for the actual computation.

2.2   Peak Identification in Single Spectra

Most peak detection algorithms have in common that they use threshold driven detection techniques. That is, a peak will only be regarded if it is higher than a predetermined signal-to-noise threshold depending on the calculated noise level (see e.g. [18]).

As shown exemplarily in Fig. 2, by assuming a noise level of 50[8] and using a signal-to-noise ratio of 3[9] about 85% of the 1332 potential peaks in this particular spectrum would be discarded and their assigned information lost. Although most of these peaks essentially *are* noise, some might carry important information. This means, that these artificially introduced "barriers" would prevent detection of small signals in a very early pre-processing stage.

The subsequent sections describe our new approaches to overcome this signal-to-noise barrier, that means increasing sensitivity without decreasing specificity.

Detection of Candidate Peaks. The initial peak detection simply determines the location of potential peaks, a process often referred to as "seeding". Utilizing the properties of the TopHat filter (see Sec. 2.1) it is sufficient to detect interception points of the curve with the X-axis. These points define start- and end points of potential peaks and are stored in a database for further analyses.

The advantage of this approach is that even smallest peaks are considered for consecutive steps. However, a deliberate validation algorithm must be applied to this set of candidate peaks to distinguish real peaks from noise and detect and deconvolute overlapping peaks.

---

[6] The TIC is the sum of the area of all peaks in a spectra.

[7] The normalization by the ratio $R = \frac{\text{"TIC of spectrum"}}{\text{"Average TIC of all spectra"}}$ is reported to be superior to other methods tested [15, 6].

[8] Different noise-estimators compute values ranging from 50 to 150.

[9] A commonly used value to get reliable results.

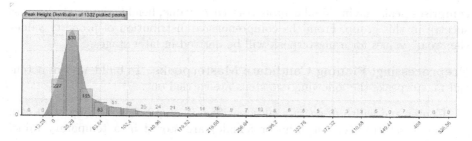

Fig. 2. Histogram of peak heights in a randomly chosen spectrum of the study described in [3]. Only peaks smaller than 500 are displayed.

Analyzing Candidate Peaks. In mass spectra of complex protein mixtures (such as serum) most of the peaks detected are broadened and/or highly convoluted. This results for example from molecular fragments having very similar masses and thus partly overlay, from poor machine resolution, or different isotopic forms of the same molecule. Therefore, a successful peak detection algorithm needs to deconvolute those peaks. A widely accepted method assumes a "blurred peak" to be a mixture of Gaussians and tries to resolve it back into its original components. A commonly used technique is "Maximum Entropy" that has been originally developed for clarifying blurred images (see [19]). Based on this idea we have developed an approach to separate and evaluate the assumed mixture of Gaussians. The key steps for each candidate peak found are as follows:

1. Determine number of Gaussian components by density estimation using the "Greedy Expectation Maximization" algorithm [20]. This algorithm has been shown to have a very good performance even on large mixtures often found in peaks at higher masses (> 3000Da). (For a comparative study see [21].)
2. To account for isotopic forms a de-isotoping step is carried out by fitting a mass dependent pre-calculated model (see [22] for details) if more than one Gaussian component is found. If successfull, the peaks involved are tagged as belonging to the same molecule(-fragment). If the quality of the fit is too poor the peak is split according to the number of Gaussians found and for each of the new parts step 1 is performed again.
3. Determine and store the parameters (height, width, center, area, shape quality[10]) of this peak.

2.3 Peak Assignment Across Spectra

In order to identify a particular peak across spectra a list of so called "masterpeaks" is maintained per spectra group (e.g. male or healthy). A masterpeak

---

[10] This is achieved by geometrical hashing [23]. The hashing algorithm returns a discrete value $c \in \{0, 1 \ldots 5\}$, indicating the class the hashing algorithm has assigned to this shape. $c = 0$ means "noise" and $c = 1..5$ peak, where $c = 5$ is assigned if the peak looks "perfect". The categories are trained a priori.

comprises peaks having similar properties (m/z value, height, shape, etc.) across spectra in this group. From the comprehensive distribution of property values the "real" values for a masterpeak will be derived in later stages.

Preprocessing: Finding Candidate Masterpeaks. To build a set of potential masterpeaks the following two steps are carried out:

1. Center and width of every peak identifyed in step 2.2 in a group of spectra under scrutiny (e.g. healthy or female) are stored in a temporary table, ordered by their center.
2. Candidate clusters of these peaks are built with respect to the centers and the width of these peaks. Peaks belong to the same cluster if they overlap in at least one point. Since we can simply "march" through this ordered set of peaks with a linear number of comparisons, the complexity is $\mathcal{O}(n)$. Alternatively, complete-linkage hierarchical clustering could be performed [24] to build the clusters which is computationally more expensive.

Masterpeak Property Determination. We now have a set of candidate clusters often containing more than one "real" group of similar peaks. This step is going to resolve these groups by a Bayesian Clustering approach. From the clusters found in this step all properties such as center, height or width are derived. From the law of large numbers we know that the average values will converge to the real values. The probabilistic object that underlies this approach is a distribution on partitions of integers[11] known as the (weighted) *Chinese restaurant process* (CRP) [27–29].

The CRP can be best described by a process where N customers sit down in a Chinese restaurant with an infinite number of tables $C_1, C_2, \ldots$ and each table has an infinite number of seats. Suppose customers arrive sequentially. Per definition the first guest sits down at the first table. The $n + 1$th subsequent customer x_{n+1} sits at a table drawn from the following distribution:

$$p(\text{occupied table } i \mid \text{previous customers}) = \frac{c_i}{\alpha+n} \cdot R \cdot s(x_{n+1}|x_j, j \in C_i)$$
$$p(\text{new table } \mid \text{previous customers}) = \frac{\alpha}{\alpha+n}$$

where c_i is the number of customers already sitting at table C_i, R is a rescaling factor and $\alpha > 0$ is a parameter defining the CRP. Obviously, the choice of the similarity function $s(\cdot)$ is crucial and is explained in the following paragraphs.

Let $C_i(A)$ be the average value of a property A of a set of peaks "sitting" at table i. (For example, $C_2(center)$ would be the average center of peaks at the second table.) Let $x_j(A)$ be the value of property A of peak j. $s(\cdot)$ has the following properties:

1. The distance of the center of a peak to the average center of an existing group of peaks can not be further away than 2 Da.
2. $s(\cdot)$ is the likelihood of x_j belonging to "table" C_i depending on how similar the properties of x_j are to the peaks already at "table" C_i.

---

[11] Interestingly, the partition after N steps has the same structure as draws from a Dirichlet process [25, 26].

This results in:

$$s(x) = \begin{cases} 0 & \text{if } |C_i(\text{center}) - x_j(\text{center})| > 2 \\ \sum_{A \in \mathcal{PP}} k_{i,A}(x(A)) & \text{otherwise} \end{cases}$$

$\mathcal{PP}$ being the set of peak properties and $k_{i,A}(\cdot)$ is constructed as follows:

1. Kernel Density Estimation (KDE)[12] [30] is performed on $x_j(A), j \in C_i$.
2. This resulting density is transformed through interpolation to the continuous function $k_{i,A}(\cdot)$.

The resulting sub-clusters (tables) are processed further in order to merge together similar groups and stored in the database.

3 Results and Discussion

Although having been designed for large amounts of spectra, we conducted first experiments with a small set of samples. To obtain a first "proof-of-principle" and to test the overall performance of our workflow we spiked a subset of human serum samples[13] with a peptide mix[14]. We split 16 different samples into five groups each. Before sample pretreatment and measurement each of the groups was spiked with one of the following concentrations: 121.21nMol/L, 0.76nMol/L, 0.30nMol/L, 3.03pMol/L, 0.075pMol/L, resulting in 320 spectra (64 for each concentration group due to 4-fold spotting).

We then processed each resulting raw spectrum as described above. For each of the five resulting concentration groups we evaluated the masterpeaks found. Subsequently we validated whether masterpeaks originating from the spiked peptides were identified by the algorithms and checked the deviation of the determined centers to the postulated ones. Note that at this stage no analysis of other detected peaks has been performed. Table 1 summarizes the findings:

1. The algorithms successfully detect peaks even for very small concentrations at pMol/L level. This is exemplarily shown for the hormones Angiotensin, Bombesin and ACTH clip 18-39 which can be detected in a very low and biologically relevant concentration range ($\sim$3 pMol/L). Peaks for Angiotensin and Bombesin are not detected by commercial software[15]. Therefore, in these examples, our algorithm is at least 20000 times more sensitive than a commercial algorithm using a signal-to-noise threshold.

---

[12] Following the Parzen Window approach with Gaussian Kernel.

[13] The protocol used for preprocessing and (magnetic bead) fractionation has been described in [3].

[14] Protein calibration standard mix (Part No.: 206355 & 206196 purchased from Bruker Daltronics, Leipzig, Germany).

[15] ClinProTools 2.0, Bruker Daltronics. Parameters used: Signal-to-Noise Level: 3, Peak Detection Algorithm: Centroid.

Table 1. Results of the spiking experiments (see text for explanation). Note that no calibration has been performed for the Proteomics.NET platform.

Substance	Plat-form[3]	Theor. Center	Center found (Concentration of spiked peptide mix as stated below)				
			121.21nMol/L	0.76nMol/L	0.30nMol/L	3.03pMol/L	0.075pMol/L
Angiotensin II	P.NET	1047.20	1047.1	-[1]	-[1]	-[1]	-[1]
Angiotensin II	CPT	1047.20	1046.9	-[2]	-[2]	-[2]	-[2]
Angiotensin I	P.NET	1297.51	1297.6	1298.0	1299.3	1299.2	-[1]
Angiotensin I	CPT	1297.51	1297.2	-[2]	-[2]	-[2]	-[2]
Bombesin	P.NET	1620.88	1620.9	1618.1	1617.2	1617.2	-[1]
Bombesin	CPT	1620.88	1620.6	-[2]	-[2]	-[2]	-[2]
ACTH clip 18-39	P.NET	2466.73	2466.8	2465.8	2465.8	2466.2	-[1]
ACTH clip 18-39	CPT	2466.73	2466.2	-[2]	2466.1	2465.9	-[2]
Somatostatin 28	P.NET	3149.61	3149.5	-[1]	-[1]	-[1]	-[1]
Somatostatin 28	CPT	3149.61	3149.0	-[2]	-[2]	-[2]	-[2]
Insulin	P.NET	5734.56	5734.3	-[1]	-[1]	-[1]	-[1]
Insulin	CPT	5734.56	5734.2	-[2]	-[2]	-[2]	-[2]

[1]: No significant masterpeak in this range at this concentration found, [2]: No significant peaks in this range at this concentration found, [3]: CPT: Bruker ClinprotTools, P.NET: Our Proposed Platform: Proteomics.NET.

2. The proposed methods are able to detect peak centers accurately since shifts and noise in the spectra largely are cancelled out after averaging by the cluster partition process described above[16]. The centers found are determined more precise than the commercial software does.

4   Conclusion

We have presented new methods for preprocessing MALDI-TOF MS spectra and detecting and evaluating peaks in these spectra. These steps lead to an enhanced sensitivity in the overall peak detection process. In a proof-of-concept setting, our results show that our algorithms with the statistical driven approach are able to detect spiked peptides in serum spectra in a concentration as low as 3.03pMol/L. We believe that this new approach will promote the sensitivity of proteome pattern diagnostics in laboratory medicine.

Acknowledgements

The study presented here is supported by a grant from Microsoft Research Ltd. to C. Schütte, Berlin and J. Thiery, Leipzig. We are grateful to Dr. Markus

---

[16] Note that the spectra have not been calibrated and therefore systematic shifts can occure. This is due to the fact that none of the calibration methods from the literature having been applied to this highly complex dataset has shown stable and reliable results. We belive that new ideas and approached have to be developed to successfully tackle this problem.

Kostrzewa (Bruker Daltonics, Leipzig, Germany) for technical support and helpful discussions. This work was also supported by a grant from the Sächsische Aufbaubank (SAB) to J. Thiery, Leipzig.

References

1. K. R. Kozak, F. Su, J. P. Whitelegge, K. Faull, S. Reddy, and R. Farias-Eisner. Characterization of serum biomarkers for detection of early stage ovarian cancer. *Proteomics*, 5(17):4589–4596, Nov 2005.
2. S. Becker, L. H. Cazares, P. Watson, H. Lynch, O. J. Semmes, R. R. Drake, and C. Laronga. Surfaced-enhanced laser desorption/ionization time-of-flight (SELDI-TOF) differentiation of serum protein profiles of BRCA-1 and sporadic breast cancer. *Ann Surg Oncol*, 11(10):907–914, Oct 2004.
3. S. Baumann, U. Ceglarek, G. M. Fiedler, J. Lembcke, A. Leichtle, and J. Thiery. Standardized approach to proteome profiling of human serum based on magnetic bead separation and matrix-assisted laser desorption/ionization time-of-flight mass spectrometry. *Clin Chem*, 51(6):973–980, Jun 2005.
4. Glen L Hortin. The MALDI TOF Mass Spectrometric View of the Plasma Proteome and Peptidome. *Clin Chem*, Apr 2006.
5. E. J. Breen, F. G. Hopwood, K. L. Williams, and M. R. Wilkins. Automatic poisson peak harvesting for high throughput protein identification. *Electrophoresis*, 21(11):2243 – 2251, 2000.
6. A. C. Sauve and T. P. Speed. Normalization, baseline correction and alignment of high-throughput mass spectrometry data. In *Procedings Gensips 2004*, 2004.
7. C. Gröpl, A. Hildebrandt, O. Kohlbacher, E. Lange, S. Lövenich, and M. Sturm. OpenMS - Software for Mass Spectrometry. Poster presented at the MBI Workshop on Computational Proteomics and Mass Spectrometry 2005, Ohio State University, 2005.
8. V. Mazet, D. Brie, and J. Idier. Baseline spectrum estimation using half-quadratic minimization. In *Proceedings of the European Signal Processing Conference*, Vienna, Autriche, September 2004.
9. M. Wagner, D. Naik, and A. Pothen. Protocols for disease classification from mass spectrometry data. *Proteomics*, 3(9):1692–1698, Sep 2003.
10. Q. Liu, B. Krishnapuram, P. Pratapa, X. Liao, A. Hartemink, and L. Carin. Identification of differentially expressed proteins using maldi-tof mass spectra. In *Conference Record of the Thirty-Seventh Asilomar Conference on Signals, Systems and Computers*, volume 2, pages 1323– 1327, Nov 2003.
11. A.K. Louis, P. Maass, and A. Rieder. *Wavelets: Theorie und Anwendungen*. B.G. Teubner, Stuttgart, 2nd edition, 1998.
12. G. P. Nason and B. W. Silverman. The stationary wavelet transform and some statistical applications. In *Lecture Notes in Statistics*, volume 103, pages 281–300, 1995.
13. E. F. Petricoin, A. M. Ardekani, B. A. Hitt, P. J. Levine, V. A. Fusaro, S. M. Steinberg, G. B. Mills, C. Simone, D. A. Fishman, E. C. Kohn, and L. A. Liotta. Use of proteomic patterns in serum to identify ovarian cancer. *Lancet*, 359(9306):572–577, Feb 2002.
14. L. Li, H. Tang, Z. Wu, J. Gong, M. Gruidl, J. Zou, M. Tockman, and R. A. Clark. Data mining techniques for cancer detection using serum proteomic profiling. *Artif Intell Med*, 32(2):71–83, Oct 2004.

15. J. L. Norris, D. S. Cornett, J. A. Mobley, S. A. Schwartz, H. Roder, and R. M. Caprioli. Preparing maldi mass spectra for statistical analysis: A practical approach. In *Proceedings of the 53rd ASMS Conference on Mass Spectrometry and Allied Topics*, San Antonio, TX, June 2005.
16. E. T. Fung and C. Enderwick. ProteinChip clinical proteomics: computational challenges and solutions. *Biotechniques*, Suppl:34–8, 40–1, Mar 2002.
17. K. A. Baggerly, J. S. Morris, J. Wang, D. Gold, L.-C. Xiao, and K. R. Coombes. A comprehensive approach to the analysis of matrix-assisted laser desorption/ionization-time of flight proteomics spectra from serum samples. *Proteomics*, 3(9):1667–1672, Sep 2003.
18. R. N. McDonough and A. D. Whale. *Detection of Signals in Noise*. Academic Press, San Diego, 2nd edition, 1995.
19. S. Guiasu and A. Shenitzer. The principle of maximum entropy. *The Mathematical Intelligencer*, 7(1):42–48, 1985.
20. J. J. Verbeek, N. Vlassis, and B. Kröse. Efficient greedy learning of gaussian mixture models. *Neural Comput*, 15(2):469–485, Feb 2003.
21. P. Paalanen, J.-K. Kamarainen, J. Ilonen, and H. Kälviäinen. Representation and discrimination based on gaussian mixture model probability densities - practices and algorithms. Technical Report 95, Lappeenranta University of Technology, Department of Information Technology, 2005.
22. Jingfen Zhang, Wen Gao, Jinjin Cai, Simin He, Rong Zeng, and Runsheng Chen. Predicting molecular formulas of fragment ions with isotope patterns in tandem mass spectra. *IEEE/ACM Transactions on Computational Biology and Bioinformatics*, 02(3):217–230, 2005.
23. H. J. Wolfson and I. Rigoutsos. Geometric hashing: An overview. *IEEE Computational Science & Engineering*, 4(4):10–21, /1997.
24. R. Tibshirani, T. Hastie, B. Narasimhan, S. Soltys, G. Shi, A. Koong, and Q.-T. Le. Sample classification from protein mass spectrometry, by 'peak probability contrasts'. *Bioinformatics*, 20(17):3034–3044, Nov 2004.
25. T. S. Ferguson. A bayesian analysis of some nonparametric problems. *The Annals of Statistics*, 1:209–230, 1973.
26. D. Blackwell and J. MacQueen. Ferguson distributions via polya urn schemes. *The Annals of Statistics*, 1:353–355, 1973.
27. D.J. Aldous. *Exchangeability and related topics*, volume 1117 of *Lecture Notes in Math - Ecole d'ete de probabilites de Saint-Flour*. Springer, Berlin, 1983.
28. H. Ishwaran and L. F. James. Generalized weighted chinese restaurant process for species sampling mixture models. *Statistica Sinica*, 3:1211–1235, 2003.
29. A. Y. Lo. Weighted chinese restaurant processes. *Cosmos*, 1(1):107–111, 2005. World Scientific Publishing Company.
30. S. J. Scheather. Density estimation. *Statistical Science*, 19(4):588–597, 2004.

Dynamic Complexity of Chaotic Transitions in High-Dimensional Classical Dynamics: Leu-Enkephalin Folding

Dmitry Nerukh, George Karvounis, and Robert C. Glen

Unilever Centre for Molecular Informatics, Department of Chemistry,
Cambridge University, Cambridge CB2 1EW, UK
Tel.: +44 (0)1223 763078; Fax: +44 (0)1223 763076.
dn232@cam.ac.uk

Abstract. Leu-Enkephalin in explicit water is simulated using classical molecular dynamics. A β-turn transition is investigated by calculating the topological complexity (in the "computational mechanics" framework [J. P. Crutchfield and K. Young, Phys. Rev. Lett., **63**, 105 (1989)]) of the dynamics of both the peptide and the neighbouring water molecules. The complexity of the atomic trajectories of the (relatively short) simulations used in this study reflect the degree of phase space mixing in the system. It is demonstrated that the dynamic complexity of the hydrogen atoms of the peptide and almost all of the hydrogens of the neighbouring waters exhibit a minimum precisely at the moment of the β-turn transition. This indicates the appearance of simplified periodic patterns in the atomic motion, which could correspond to high-dimensional tori in the phase space. It is hypothesized that this behaviour is the manifestation of the effect described in the approach to molecular transitions by Komatsuzaki and Berry [T. Komatsuzaki and R.S. Berry, *Adv. Chem. Phys.*, **123**, 79 (2002)], where a "quasi-regular" dynamics at the transition is suggested. Therefore, for the first time, the less chaotic character of the folding transition in a realistic molecular system is demonstrated.

Keywords: folding, enkephalin, dynamic complexity, computational mechanics.

1 Introduction

Self-organisation in molecular systems is ubiquitous (micelles, membranes, crystal structures etc.). This process is vital for the proper functioning of biomolecules, the most prominent example being protein folding. Despite the dynamic character of the process, the majority of the literature is mostly devoted to the structural aspects of protein folding; a significantly lower number of publications are devoted to folding dynamics and very few (if any) works exist that consider the problem from the self-organisation point of view. There are difficulties in simulating protein folding which derive mostly from the size of the molecular systems and ultimately, the complexity of both their structure and dynamics. However, research in the field has evolved and it

M.R. Berthold, R. Glen, and I. Fischer (Eds.): CompLife 2006, LNBI 4216, pp. 129–140, 2006.

is now possible to investigate computational models of sufficient quality and size over a realistic time scale covering, in some cases, a complete folding event.

Protein folding is currently described using the notion of a "thermodynamic folding funnel"[1] which generally concentrates on the structural aspects of folding while explaining some aspects of the dynamics (particularly the selection of favoured transition structures). However, there is evidence that the funnel approach may fail to explain the folding process, its mechanisms and speed[2]. Therefore, we believe that direct analysis of the folding dynamics using the apparatus of non-linear dynamical systems and dynamic chaos is one of the most promising ways of moving towards understanding the protein folding phenomenon.

Within this framework, the study of dynamic complexity gives new insights into the details of the dynamics of molecular systems[3]. In particular, it is shown that the methodology we have developed is capable of distinguishing between different regimes of molecular motion which have different chaoticity[4]. This suggests that the computation of complexity may help to understand the features that characterise self-organisation of molecules in time frames much shorter than those predicted by e.g. the thermodynamic folding funnel approach.

The methodology we have developed is based on the computational mechanics theory developed by Crutchfield et al[5,6,7,8]. By "discovering" dynamic patterns rather than by postulating them (like, for example, in Fourier analysis), this theory is designed to elucidate the property of emergence in dynamic systems. That is, to uncover qualitatively new behaviour when it appears. This is exactly the essence of self-organisation and, to our view, a perception currently missing in the analysis of dynamic molecular systems.

Another aspect long recognized as fundamental to protein folding studies is of course the role and importance of water. There are a growing number of publications proving both the decisive role of water[9-14] and also the very different dynamics of water molecules at different locations with respect to the biomolecule[15]. This, on the one hand complicates the situation (since a significant number of water molecules now needs to be included into simulations of the essential "core" of the system) and on the other hand, indicates new directions of research to explain phenomenologically the observed folding events.

In a recent publication of a computer simulation of the folding of Leu-Enkephalin, we reported some key observations of the dynamics of the water shell around the peptide during folding[16]. Specifically, during the short time period of the folding event, at the moment of folding when i and $i+3$ C_α atoms come together (this nomenclature is described in reference 15), the water network around the peptide undergoes significant collective changes. By making and breaking water-water and peptide-water hydrogen-bonds the whole network becomes more mobile at the moment of folding, in contrast to having a more "frozen" dynamics before and after the folding event.

Even though the formation of β-turns in proteins is well defined (through the distance between i and $i+3$ C_α atoms) there is no explanation of the driving force for the molecule to undergo the transformation. There is no apparent reason why a protein fluctuating in the conformation close to a β-turn for a relatively long time suddenly and quickly enters into the transition regime and acquires the specific folded configuration. In the present study we link this behaviour to a general property of dynamic systems – the phase space mixing or, loosely speaking, their chaoticity. By calculating the dynamic complexity of the system's trajectories we demonstrate a substantial decrease of chaoticity specifically at the moment of transition. We hypothesise that this phenomenon can be attributed to a general theory of transitions in molecular systems that predicts the non-chaotic character of the system's dynamics at these transitions[17].

2 The System

All-atom classical molecular dynamics simulations of a Leu-Enkephalin molecule in explicit water have been carried out. Analysis of the dynamics of the atoms of the peptide, as well as the first solvation shell and some of the second solvation shell waters has been performed. The same analysis for a small group of water molecules in the bulk was performed for testing purposes.

The Leu-Enkephalin molecule (NH^{3+}-TYR-GLY-GLY-PHE-LEU-COO⁻), Fig. 1, was built using Sybyl6.9[18] in a linear conformation in a zwitterionic form (with N-terminal NH^{3+} and C-terminal COO⁻). It was solvated in a rectangular box of 5826 SPC^{19} water molecules and energy minimized using the steepest descent method. The GROMOS96[20] force field was used in all simulations. Four simulations of 3ns (of 0.002 ps time steps) were performed with weak coupling to a water bath of constant temperature at 300 K (peptide and solvent individually), with a coupling time of $0.1ps^{21}$. Pressure coupling was also applied to a pressure bath with reference pressure of 1 bar and a coupling time of 0.1ps. All bond-lengths of the peptide were constrained using the SHAKE[22] algorithm. The SETTLE[23] algorithm was used to constrain the bond lengths and angles of the SPC waters. The simulations were performed using the GROMACS[19] package. A 1 nm cut-off distance for both van der Waals and Coulomb potentials was used. The system was allowed to equilibrate for the first 100ps before collecting data for analysis.

A β-turn was defined for 4 consecutive residues if the distance between the C_α atom of residue i and the C_α atom of residue $i+3$ was less than 7Å and if the two central residues are not helical[24]. There are two possible β-turns in Leu-Enkephalin: 1-4 and 2-5. As has previously been reported in the literature[25] and observed in this study, the β-turn in this molecule is not stable, however, it survives for relatively long periods of time and is created very quickly. This kind of event in this small peptide is probably consistent with similar events found during protein folding.

We have chosen four moments of β-turn formation for preliminary inspection. We would like to stress that this formal β-turn definition is not exhaustive. In terms of both structure and dynamics, in all four cases we observed somewhat different results. These differences are reflected in the configuration of the side chains, dynamics of the C_α distances, etc. Of course, each simulation starts from a different starting point and would be expected to be different. Two out of the four events exhibited the effect of formation of the β-turn and we have selected one transition for detailed analysis. This transition was chosen as it shows the most pronounced effect in dynamic complexity.

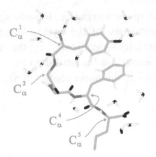

Fig. 1. A Leu-Enkephalin molecule with hydrogen-bonded waters at the moment of transition. The colours represent (from grey to black): H, C, N, O atoms.

3 The Method: Statistical Complexity of Deterministic Chaotic Systems

3.1 General Idea

The complexity measure used in this work is described in detail in ref. 3. Here we give an outline of the theory and specific characteristics used to quantify dynamic complexity.

In the computational mechanics framework[5,6,7] symbolic dynamics is considered i.e. the signal consists of discrete symbols assigned to discrete time steps. Let a set of symbols corresponding to each time step t_i form a sequence S. To calculate the statistical complexity, S is decomposed into a set of left s_i^l (past) of length l and right s_i^r (future) of length r halves joined together at time points t_i. Consider all equivalent left subsequences s_k^l. Collect a set of all right subsequences following this unique left subsequence. Each right subsequence has its probability conditioned on the particular left one: $\Pr\left(s^r \mid s_k^l\right)$. The equivalence relation between any two left subsequences is defined as follows: two unique left subsequences s_i^l and s_j^l are equivalent if their right distributions are the same up to some tolerance value δ:

$$\Pr\left(s^r \mid s_i^l\right) = \Pr\left(s^r \mid s_j^l\right) + \delta . \tag{1}$$

A set of all equivalent left subsequences forms an "equivalence class". The equivalence classes have their own probabilities ($\Pr\left(A_i\right)$) calculated from the probabilities of the constituent left subsequences.

The importance of the notion of equivalence classes is that they represent the states of the system that define the dynamics at future moments – the "causal states" (here equal to "equivalence classes" with corresponding probabilities). The time evolution of the system can be viewed as traversing from one causal state to the other with a transition probability equal to $\Pr\left(s^r \mid s_i^l\right)$. The set of the causal states together with the transition probabilities constitute a so called "ε-machine". ε-machines represent the minimal computation necessary to reproduce the dynamics of the system.

A number of useful mathematical properties of the ε-machine are proved[8]. Among others is the fact that the ε-machines are monoids (semi-groups with an identity element). From this it follows that causal states are (i) sufficient statistics; (ii) unique; (iii) minimal. These and other properties serve as a fundamental basis for defining a complexity measure that stands out when compared to the majority of current approaches to quantify complexity of dynamic systems.

The statistical complexity is defined as the informational size of the ε-machine. The measure of this is the Shannon entropy of the causal states:

$$C_\mu \equiv -\sum_{A_i} \Pr(A_i) \log_2 \Pr(A_i),$$
(2)

where A_i are causal states. In contrast to Kolmogorov complexity, this measure provides a zero complexity for *both* extremes – a constant signal and a purely random process. The maximum value of complexity lies somewhere in between these two limits.

For the purposes of the present study "topological complexity" is important. It is defined as the size of the ε-machine:

$$C \equiv \log_2 N,$$
(3)

where N is the number of causal states A_i. For the reasons explained further, the highest of the transition probabilities (1) of causal states also carry important information about the dynamic patterns of the signal. To estimate it for the entire ε-machine, the highest transition probabilities averaged over all states were calculated:

$$D \equiv \frac{1}{N} \sum_i \max_j (\Pr_i^j),$$
(4)

where $\Pr_i^j$ is a probability of a transition j of a state A_i.

3.2 Application to Molecular Dynamics

We are currently developing a general theory of dynamic complexity of molecular Hamiltonian systems. In this framework we define the complexity of the system as a statistical (or topological) complexity of a signal comprising of the realisations of the system's phase-space trajectories symbolised using a particular partitioning method. This general definition can be exploited to investigate the properties of molecular systems using the results of chaotic non-linear dynamics from which some nontrivial implications follow.

However, such an analysis requires the information on the full-dimensional trajectories, possibly of infinite lengths. This is, of cause, impossible to obtain in a simulation like the one used in the present work. Therefore, a simpler approach is employed, namely the one-atom projections of the full-dimensional trajectories are used and the length of the signal was chosen such that it is long enough to accumulate a useful statistic and short enough not to average out interesting effects of the system's dynamics. Pieces of trajectory of 4ps long were used to calculate the values of complexity. These values were then plotted at the times corresponding to the middle of the interval used. The procedure was repeated to obtain the time evolution of the complexity value of the system (Fig. 4,5).

Initially we used the three-dimensional Cartesian coordinates of the atoms as the signals to calculate the complexities. This approach, however, has the following disadvantages. The domain of the coordinates is very large compared to the area a trajectory covers during the time periods studied. In other words, the atoms only drift slowly in small regions (different for all atoms) of the allowed coordinate area. This is a major obstacle for the complexity method described above since the method is statistical in nature. Therefore, it requires a representative sampling of the whole domain of the signal.

Instead we used the velocities of the individual atoms as the signals. These quickly fluctuate around the origin and each undergoes a significant number of fluctuations during the time period of the investigation. Another advantage is that they are of comparable value for different atoms and that makes their comparison possible on the same grounds.

3.3 Symbolisation

The velocities were symbolised according to the following scheme. In spherical coordinates, the φ and the θ angles were partitioned into n and $n/2$ regions respectively and radius r was partitioned into m regions. The resulting spherical sectors were assigned a number (in an arbitrary order). A symbol was assigned to each value of the velocity depending upon which sector it lies.

We have tested the partitions for the values of n from 8 to 16, and m equal to 1 and 2. Overall, particular combinations of indices n and m were not essential; the total number of symbols ($n \times \frac{n}{2} \times m$) was found to be the key parameter. The latter did influence the value of complexity significantly and the appearance of the complexity minimum during the β-turn formation was only observed for a narrow interval of the number of symbols (discussed in the "Results and Discussion" section). Indirectly, the number of symbols defines the time scale of the analysis: for a fixed time interval; small partitioning extracts more information on a fine-scale, while coarse partitioning is only sensible for very long run times.

3.4 Dynamic Scenarios

There are particular reasons that we observe relations between the values of the parameters used for calculating complexities and the dynamics of the system.

If a trajectory consists of slowly changing values this leads to a situation where the symbolic sequence is comprised of a long series of repeating symbols. In this case, the majority of the transitions occur with the same symbol (for a history length much smaller than the repetition length) and the ε-machine consists of the states corresponding to each symbol with the prevailing probability of transition from each state to the same state. In this situation, the complexity reflects essentially the statistics on the symbols and contains very little information on the dynamical patterns. Analysis of the highest transition probability (4) allows us to recognize this situation: the value of D in this case is high, reflecting the fact that most transitions are to the same causal state. The topological complexity in this case is close to the logarithm of the number of symbols used N_S: $C = \log_2 N_S$. We call this situation "slow" dynamics.

We should emphasize that the (relatively) short length of the signals introduces a specific feature to computational mechanics, different from its conventional meaning. Namely, when the signal is highly chaotic, that is when almost all subsequences entering probability (1) are different, the ε-machine reconstruction algorithm does not recognise them as belonging to the same causal state, as should happen for infinitely long signals. Therefore, the large size of the ε-machine (i.e. the topological complexity (3)) in this case signifies the high chaoticity (high degree of the phase space mixing) of the signal. The statistical complexity (2), however, can be of low value, depending on a particular distribution of subsequencies. This may contradict the normal interpretation of low complexity of a highly random (less structured) signal and should be taken into account in the analysis.

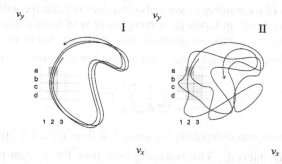

Fig. 2. Schematic representation of the trajectories of different chaoticity and symbolic sequences they produce. In case I the trajectory generates three identical symbolic sequences: {a2,b2,c2,d2}. In case II the sequences started with the same symbol a2, but because of high chaoticity generate three different sequences: {a2,b2,b3,c3}, {a2,b2,c2,d2}, {a2,b1,c1,d2}.

For chaotic signals generated by the atomic trajectories (evidence of the chaotic character of molecular trajectories is given in e.g. ref. 26, 27 through the calculation of positive Lyapunov exponents), particular importance can be assigned to the lengths of the left and right subsequence entered in the formula for probabilities (1), in other words, the memory of the system. The degree of chaoticity can be revealed by analysing the change in topological complexity C with increasing memory length. For signals having persistent dynamic structures the value of C will not change significantly, whereas for a chaotic signal, because of high sensitivity to the initial conditions, C will quickly rise with the length of the memory since quickly diverging trajectories will generate more and more new unique subsequences, i.e. causal states (Fig. 2).

4   Results and Discussion

The complexities obtained for the atoms investigated can be classified into two categories. All hydrogen atoms and some water oxygen atoms belong to the first category, the rest of the atoms – to the second category. The latter is characterised by the "slow" dynamics described in the previous chapter. Besides the apparent prevalence of a long series of the same symbol, this is also supported by the relatively high value of D (4). For both classes its value does not significantly change with time and lies in the range from 0.8 to 0.9 for "slow" dynamics and from 0.6 to 0.65 for "fast" dynamics.

For the "fast" atoms (mostly the hydrogens of the system), the main result of this work is observed: we observe low topological complexity values exactly at the moment of β-turn formation, being high at all other times (Fig. 4,5). It should be stressed that this effect is only visible in a narrow range of the complexity algorithm parameters: the number of symbols from 8 to 36 and the memory length – 3-4 ps. For too many symbols or too long a memory length the dynamics appear "random" (chaotic) for all time periods. For too few symbols, when the symbolisation is too coarse, or for too short a memory length, the dynamics shows a value typically lower than the logarithm of the number of symbols (that does not change with time).

Even though the "gap" in topological complexity is in most cases self-evident on the graphs for sets of atoms, to quantify the minimum value of complexity the averaged complexities $\overline{C}$ at each time t_i for u number of atoms was calculated:

$$\overline{C_i} = \frac{1}{u}\sum_{l=1}^{u}\frac{1}{b}\sum_{k=i-b/2}^{i+b/2}C_{lk}, \tag{5}$$

where C_{lk} is a topological complexity for atom l at time t_k and b (typically equal to 6) is the averaging interval. The averaging over time for b data points is needed because the "gap" in complexity for different atoms was sometimes misaligned for 1-2 data points, especially for the peptide's atoms (Fig. 4). The averaging thus emphasises an effect that would otherwise be difficult to recognise.

Typical ε-machines obtained for the signals investigated are shown in Fig 3. The apparently less sophisticated structure can be seen for the transition moment of 1657

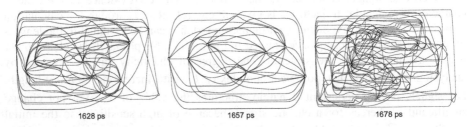

1628 ps   1657 ps   1678 ps

Fig. 3. ε-machines of a representative atomic trajectory at the moments before, at (1657ps), and after the β-turn formation. Dots represent the causal states, arcs – the transitions between the states. A state can lead to several other states with corresponding probabilities.

ps. It should be stressed that the ε-machine structure reflects the dynamic patterns present in the signal, not the statistical features of the symbols comprising the signal. This is important because it proves that the dynamics does not become "slow" at the moment of transition, which is also confirmed by the value of D as discussed above.

The original and averaged topological complexity of the peptide's atoms and its evolution in time are shown in Fig. 4. Not all atoms have the complexity minimum at exactly the transition time, having a small difference of a few ps. This makes the minimum in the complexity less pronounced in the summary graph of the original complexities (lower plot), but is clearly emphasised by the averaging procedure (upper plot). This is in contrast to the water atoms where the minimum falls exactly at the moment of transition (Fig. 5). The reasons for this may be that the peptide's atoms interact with each other much more strongly and the transition path has more complicated character, whereas the weaker interactions of water, still being involved in the dynamics of transition, serve as a "bath". Another possibility may be that in the full-dimensional space the low complexity value is observed for the wider period covering all the atoms, while for the projections studied here, it falls at slightly different times.

The complexities of the water atoms are shown in Fig. 5. Here the minimum in the complexity is most clearly visible for both the original values as well as for the averaged values of C.

To check if this behaviour is not simply a result of appropriately selected water molecules, we have performed the following test. At different moments (1616, 1640, 1657, 1672, and 1692 ps) a group of about 50 water molecules belonging to the first solvation shell of the peptide was chosen. The complexities of the atoms of these molecules were then calculated and plotted in a similar fashion to Fig. 5. If the effect (the decrease in complexity) was only the result of proximity of the water molecules to the peptide, it would be observed in all 5 test calculations at the chosen times. In reality it was only observed at 1657 ps, the moment of β-turn formation, therefore strongly implying that the dynamics of the system is the reason for this behaviour.

For the plots presented in Fig. 5 the complexities of 111 water atoms were calculated. Only one atom showed a high value at 1657 ps (excluded from the plot for clarity). This one atom is specific in that it is part of a peptide – water – peptide hydrogen-bonding bridge at the moment of transition. This, most probably, makes its dynamics special and different from the majority of water molecules neighbouring the peptide.

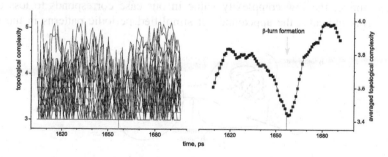

Fig. 4. Topological complexities of the peptide's atoms. A symbolisation alphabet of size 8 and a history length of 3 ps were used. The β-turn transition is at 1657 ps.

Another subtle observation worth mentioning here is a much lower spread of values of complexities at the moment of transition, not only from above, but also from below (Fig. 5). In other words, the lines become on average slightly closer to each other. This is in line with our further explanation of the phenomenon.

We would like to point out the connections between the phenomenon found here and our previous investigations. In ref. 16 we reported the change in the dynamics of the water network around Leu-Enkephalin during the β-turn formation. We observed that the water network exhibits a "freezing" behaviour before and after the transition with high mobility and the disruption of hydrogen-bonding at the moment of transition. We now see evidence of a completely different characteristic that directly investigates the dynamics of the atoms and that also demonstrates a fundamentally different character of the dynamics of the water before, at, and after the transition.

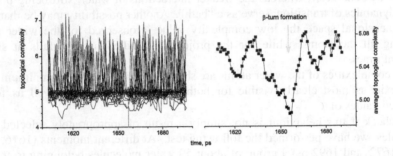

Fig. 5. Topological complexities of the water atoms. A symbolisation alphabet of size 32 and a history length of 4 ps were used. The β-turn transition is at 1657 ps.

Also, these results correlate well with our calculations of the complexity of a simpler (model) molecular system: a zwitterion that showed similar changes in statistical complexity during the transition between different conformational states[4].

The authors of ref. 17 suggest that the dynamics of the states between the transitions is significantly chaotic while at the moment of the transition the dynamics becomes semi-chaotic or quasi-regular i.e. the system can maintain approximate constants of motion and possess fully deterministic dynamics. We hypothesise that the effect reported in this work is the manifestation of this phenomenon. Indeed, as discussed above, the low complexity value in our case corresponds to less chaotic motion. This indicates the appearance of simplified periodic patterns in the atomic

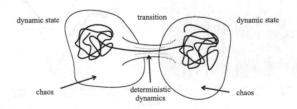

Fig. 6. Schematic illustration of the transitions between different phase-space regions of the system

motion, which could correspond to high-dimensional tori in the phase space. As a schematic illustration of the dynamics, the transition can be visualised as passing through a narrow "tunnel" connecting two states. In this situation the phase-space flow should "straighten" (or rather confined to the tori) in order to be transferred from one basin to the other, Fig. 6.

5 Conclusion

Summarising, the complexity of the dynamics of the Leu-Enkephalin – water system has been investigated paying particular attention to the moment of β-turn formation. The dynamic complexity of the atomic velocities has been calculated and its evolution during the fold transition has been analysed. It is found that exactly at the moment of transition, the complexity shows a significant drop that in this framework indicates a low degree of phase space mixing (less chaotic character) of the dynamics of the system.

Work is planned to extend this complexity analysis to other molecules and elementary folding events such as the formation of α helices. We also plan to link this behaviour to what we think may be a general characteristic of biomolecular systems during their self-organisation, that is changes in the complexity of the dynamics of the system.

Acknowledgement. The work is supported by the Isaac Newton Trust and Unilever.

References

1. N.D. Socci, J.N. Onuchic, P.G. Wolynes, Protein folding mechanisms and the multidimensional folding funnel, *Proteins*, **32** (2), 136 (1998).
2. S.V. Krivov and M. Karplus, Hidden complexity of free energy surfaces for peptide (protein) folding, *Proc. Nat. Acad. Sci.*, **101** (41), 14766 (2004).
3. D. Nerukh, G. Karvounis, and R.C. Glen, Complexity of classical dynamics of molecular systems. Part I: methodology, *J. Chem. Phys.*, **117**(21), 9611 (2002); Complexity of classical dynamics of molecular systems. Part II: finite statistical complexity of a water-Na⁺ system, *ibid*, 9618.
4. D. Nerukh, G. Karvounis, and R.C. Glen, Quantifying the complexity of chaos in multi-basin multidimensional dynamics of molecular systems, *Complexity*, **10** (2), (2004).
5. J. P. Crutchfield and K. Young, Inferring statistical complexity Phys. Rev. Lett., **63**, 105 (1989).
6. J. P. Crutchfield and K. Young, Computation at the Onset of Chaos, in *Entropy, Complexity, and Physics of Information, SFI Studies in the Sciences of Complexity, VIII*, edited by W. Zurek (Addison-Wesley, Reading, Massachusetts (1990)).
7. J. P. Crutchfield, The Calculi of Emergence: Computation, Dynamics, and Induction, Physica D **75**, 11 (1994).
8. Cosma Rohilla Shalizi, *Causal Architecture, Complexity and Self-Organization in Time Series and Cellular Automata*, PhD thesis, University of Wisconsin at Madison, 2001.
9. D. Beck, D. Alonso, V. Daggett, A microscopic view of peptide and protein salvation, Biophys. Chem. **100**, 221 (2003).

10. J. M. Sorenson, G. Hura, A. K. Soper, A. Petsemlidis and T. Head-Gordon; Determining the Role of Hydration Forces in Protein Folding, J. Phys. Chem. B **103** (26), 5413 (1999).
11. A.E.Garcia and G. Hummer, Water penetration and escape in proteins, *PROTEINS: Struct, Funct. and Genet.* **38**, 261-272 (2000).
12. S. Dennis, C.J. Camacho and S. Vajda, Continuum electrostatic analysis of preferred solvation sites around proteins in solution, PROTEINS: Struct, Funct. and Genet. **38**, 176 (2000).
13. R. L. Baldwin, Relation between peptide backbone solvation and the energetics of peptide hydrogen bonds, Biophys. Chem. **101-102**, 203 (2002).
14. S. M. Bhattacharyya, Z. Wang and A. H. Zewail, Dynamics of Water near a Protein Surface, J. Phys. Chem. B **107** (107), 13218 (2003).
15. P.W. Fenimore, H. Frauenfelder, B.H. McMahon, and R.D. Young, Bulk-solvent and hydration-shell fluctuations, similar to - and -fluctuations in glasses, control protein motions and functions, *Proc. Nat. Acad. Sci.*, **101** (40), 14408 (2004).
16. G. Karvounis, D. Nerukh, and R.C. Glen, Water network dynamics at the critical moment of a peptide's beta-turn formation: an MD study, *J. Chem. Phys.*, **121** (10), 4925 (2004).
17. T. Komatsuzaki and R.S. Berry, Chemical Reaction Dynamics: Many-Body Chaos and Regularity, Adv. Chem. Phys. **123**, 79 (2002).
18. Sybyl [molecular modeling package], version 6.8. St Louis (MO): Tripos Associates; (2000).
19. H. J. C. Berendsen, J.P.M. Postma, W.F. van Gunsteren, J. Hermans, Interaction Model for Water in Relation to Protein Hydration, in Intermolecular Forces, edited by B. Pullman, (D. Reidel Publishing Company, Dordrecht, 1981), pp. 331-342.
20. W.R.P. Scott, P.H. Hunenberger, I.G. Tironi, A.E. Mark, S.R. Billeter, J. Fennen, A.E. Torda, T. Huber, P. Kruger, and W.F van Gunsteren, The GROMOS Biomolecular Simulation Program Package, *J. Phys. Chem.*, **A 103**, 3596 (1999).
21. W.F. van Gunsteren and H.J.C. Berendsen, Computer simulation of molecular dynamics, *Angew. Chem. Int. Ed. Engl.*, **29**, 992 (1990).
22. J. P. Ryckaert, G. Ciccotti, and H. J. C. Berendsen, Numerical integration of the cartesian equations of motion of a system with constraints: molecular dynamics of n-alkanes, *J. Comput. Phys.*, **23**, 327 (1977).
23. S. Miyamoto and P. A. Kollman, An analytical version of the SHAKE and RATTLE algorithm for rigid water models, *J. Comp. Chem.*, **13**, 952 (1992).
24. E.G. Hutchinson, J.M. Thornton, PROMOTIF--A program to identify and analyze structural motifs in proteins, *Prot. Sci.*, 5, 212 (1996).
25. D. van der Spoel, H. J. C. Berendsen, Molecular dynamics simulations of Leu-enkephalin in water and DMSO, *Biophys J.*, **72**, 2032 (1997).
26. M. Braxenthaler, R. Unger, D. Auerbach, J.A. Given, and J. Moult, Chaos in protein dynamics, *Proteins: structure, function, and genetics*, **29**, 417 (1997).
27. Huai-bei Zhou, Chaos in Biomolecular Dynamics, *J. Phys. Chem.*, **100**, 8101 (1996).

Solvent Effects and Conformational Stability of a Tripeptide

Maxim V. Fedorov[1], Stephan Schumm[2], and Jonathan M. Goodman[1]

[1] Unilever Centre for Molecular Science Informatics, Department of Chemistry,
University of Cambridge, Lensfield Road, CB2 1EW Cambridge, UK
[2] Unilever Food and Health Research Institute, Unilever R&D Vlaardingen, Olivier
van Noortlaan 120 3133 AT Vlaardingen, The Netherlands

Abstract. In this work we are trying to gain an insight on the molecular
mechanisms of the salt effects on conformational stability of proteins with
use of fully atomistic Molecular Dynamics simulations techniques. Such
'in silico' approach allows us to obtain quite realistic data on the time
and scale resolutions that are unavailable for both 'in vitro' and 'in vivo'
experimental techniques. We investigated a trialanine peptide which is
the one of the simplest examples of biomolecules, bearing the essential
features of proteins.

1 Introduction

Studies of biomolecular hydration have more than a hundred year history. This
has lead to many achievements and a number of books [1,2,3,4,5,6,7]. Netherthe-
less, there are still many standing questions in this area. One of the questions is:
"what is the influence of different ions and their concentrations to the stability
of a macromolecular solute?"

In general, the addition of salts into biomolecular solution has three main
effects that may influence the macromolecular conformation:

1. Debye-Huckel screening effect;
2. Specific interaction with charges by ion-pair formation (or ion binding);
3. Disruption of water structure which consequently results in an increase or
   decrease in the hydrophobic interaction of proteins and biomolecules

If the **first** (Debye-Huckel screening) has the major contribution, the effect of
various ions will be determined **only** by the ionic strength of solution.

If the **second** factor is dominant, the effect of different ions should follow the
electroselectivity series of the salts toward anion (cation)-exchange resins [8,9,10].

The importance of the **third** factor can be determined by comparing the
effects of different ions with the *Hofmeister* series [11,12].

The Hofmeister series is different from electroselectivity series. For alkali
halides the electroselectivity series has almost the inverse order to the Hofmeister
series.

M.R. Berthold, R. Glen, and I. Fischer (Eds.): CompLife 2006, LNBI 4216, pp. 141–149, 2006.
© Springer-Verlag Berlin Heidelberg 2006

To investigate these ideas we performed long range fully atomistic molecular dynamics (MD) simulations of the alanine tripeptide (AA3) in ionic solutions with fully atomistic protein force-field OPLSAA-2001 and an explicit water model TIP5P-EW [13]. We have chosen this tripeptide as a model system because it is the one of the simplest examples of biomolecules which also contains the essential features of proteins. To investigate the specific effects of anion series we simulated water solutions of sodium salts of fluorine, chlorine, bromine and iodine with different molar concentrations: $0.20\,M$, $0.50\,M$, $1.00\,M$ and $2.00\,M$.

2   Materials and Methods

2.1   Procedure for MD Simulations

In this work we used the Gromacs MD software package [14,15]. For our simulations we have employed OPLSAA-2001 [16] fully atomistic force field. In all simulations we placed the tripeptide in a periodic dodecahedron box of TIP5P-EW water [13]. The TIP5P-EW water is slightly modified version of TIP5P water [17]. For simulations of bulk water solution we used 1201 water molecules in the box. For simulations of ionic solutions of AA3 we randomly substituted some of the water molecules by the corresponding ions to provide a finite concentration of ions. We used the following concentration range: 0.20; 0.50; 1.00 and 2.00 mole per litre. We generated *in silico* model of the tripeptide using the Molden software [18]. The initial conformation of the tripeptide corresponded to a segment of an ideal alpha-helix with central (Φ,Ψ) dihedral angles equal to $(-57°,-47°)$. Electrostatic interactions were treated with use of Particle Mesh Ewald (PME) summation technique. The ion parameters were taken from the OPLSAA standard set of ion parameters [16].

The first step in the computational procedure was the minimisation of the potential energy of each system by using a version of the steepest descent algorithm [14,15]. Then for each system we performed a single nanosecond equilibration run with NVT conditions and constrained positions of the tripeptide atoms to equilibrate solvent molecules in the box. Then we performed 27 ns run for the following equilibration of the system with NPT ensemble where the pressure and the temperature were maintained at 1 atmosphere and 300 K by coupling the system to a heat bath via the Berendsen thermostat [19]. Such long preliminary simulations provided a good equilibration of the system. Then we performed an additional 27 ns production run to collect the data. We collected the data each 0.2 *ps*. For integration of the Newton's equations of motion we used the velocity Verlet algorithm with a timestep of 2 *fs*.

For preprocessing and analysis of the MD data we mainly used the GROMACS analysis tool [14,15]. We used this collection of different programs and subroutines for calculation of solvent accessible area, dihedral angles, etc. We calculated the peptide solvent accessible area by the *g_sas* program from the GROMACS analysis tool. We used the solvation probe of radius 1.4 Å. For further preprocessing of the obtained arrays of parameters we used the OCTAVE

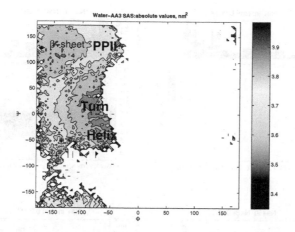

Fig. 1. Water Surface Accessible Area for AA3 with regard to the Ramachandran angles Φ and Ψ, nm^3

software which is a GNU analogue of the commercial Matlab software. All pictures in this work were created using the GNUPLOT 4.0 software.

3   Results and Discussion

For all the systems we calculated the density of conformational states projected on the plane of Ramachandran (Φ,Ψ) angles of the central AA3 residual. In the case of bulk water solution, the most favoured conformation is the Poly-Proline II (PPII) conformation $\approx (-80°, 140°)$. There are also another secondary maxima on this map which correspond to the beta-sheet and alpha-helix conformations. Such distribution is in a good correspondence with experimental results [20,21,22] and previous simulation data [23].

Table 1. Ratio of compact conformations of the tripeptide as a percentage for different salt concentrations. For comparison, we also give here this ratio for bulk water solution (0.0 M concentration).

Salt	0.0 M	0.20 M	0.50 M	1.00 M	2.00 M
NaF	12.2	21.9	52.6	33.8	15.3
NaCl	12.2	17.1	14.5	9.0	9.0
NaBr	12.2	19.5	16.1	21.2	14.8
NaI	12.2	22.5	15.2	20.2	10.9

To reveal a possible relationship between conformation of the tripeptide and its solvent accessible area (SAS) we plotted on the Fig. 1 the absolute values of SAS of AA3 for bulk water solution projected on the plane of Ramachandran

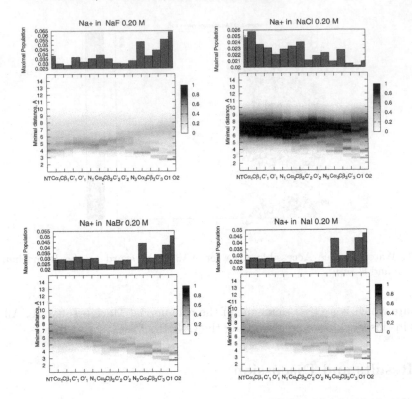

Fig. 2. AA3 - Cation minimal distance distributions for 0.20 M salt solution showing distribution of minimal distance between ion and specified peptide atoms. The top panels show the frequency values of the most frequent distance.

(Φ,Ψ) angles of the central AA3 residual. As is clear from the figure, the turn and helical conformations have smaller SAS than the beta-sheets and PPII conformations. The result is a general consequence from the polypeptide geometry - the turn and helical conformations are more compact, and, therefore, have smaller SAS with compare to 'extended' conformations.

This distribution of SAS with regard to the dihedral angles gives us a possibility to make a crude classification of the tripeptide conformations. Thus, to generalise the data we calculated the ratios (in percents) of 'folded' and 'extended' conformations for all investigated systems which are given in the Tab. 1. We determine these rates as following: any conformation of AA3 with (Φ,Ψ) belonging to the region ($[-110° : -20°],[-70° : +50°]$) we label as a 'compact' conformation. All other conformations we label as 'extended' ones. Of course, such separation of the density space is somewhat arbitrary but it helps one to reveal the most important trends in the conformational changes of the tripeptide.

As one may see from this table, the number of compact conformations varies with tye and concentration of ions. Fluorine seems to be the best stabiliser of the

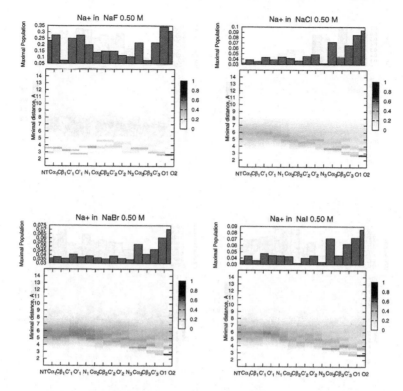

Fig. 3. AA3 - Cation minimal distance distributions for 0.50 M salt solution showing distribution of minimal distance between ion and specified peptide atoms. The top panels show the frequency values of the most frequent distance.

compact conformation until the solubility limit of NaF (app. 1.0 M). Other salts behave differently, but a general trend for them corresponds to electroselectivity hypothesis rather than Hofmeister. In average, the highly chaotropic anions (Br- and I-) seem to be better stabilisers than the chlorine anion which possess almost neutral (kosmo)chaotropic properties. These results are in line with experimental observations of Goto and coworkers [8,9,10]: anion effects on amphiphilic peptides follow the electroselectivity series.

To gain an insight on the specific tripeptide-ion interactions on the atomistic level we performed the following calculations: for each MD time-frame we calculated minimal atom-atom distances for all heavy atoms of the tripeptide and nearest ions. Then we calculated the frequency distributions (histograms) for all these minimal distances. The results for peptide-cations distributions are shown on the Figs. 2–3. The results for peptide-anions distributions are shown on the Figs. 4–5. On these pictures we show these data only for 0.20 M and 0.50 M solutions as the most interesting cases. We plotted them as two-dimensional density maps with regard to the distance and atom name where the density is indicated

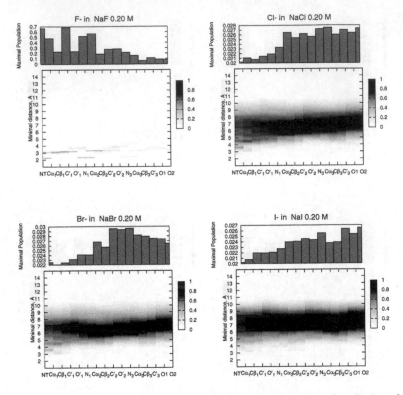

Fig. 4. AA3 - Anion minimal distance distributions for 0.20 M salt solution showing distribution of minimal distance between ion and specified peptide atoms. The top panels show the frequency values of the most frequent distance.

by the colour intensity. These distributions have quite different amplitudes of the highest peaks. Therefore, for the sake of conveniency we normalise them on the height of highest peak for any given system. To have an idea about the absolute values of the highest peaks for each atom-ion distribution we plotted them as separate plots on the top sides of the maps.

As one may see from these pictures the ions in water have very peculiar behaviour. For example, the cations in 0.50 M sodium fluoride solutions and in 0.20 M sodium chloride solution tend to attract the peptide charged groups of the same charge! A similar phenomena takes place for iodine, bromine and chlorine anions in 0.20 M solution. It has happened because of the high polarisability of water - the charged groups formate an ordered solvation shell which leads to high charge density fluctuations near these groups. For the fluorine anions the mechanism is different - these small anions can be in direct contact with positively charged groups and formate quite stable 'zwitterionic' complexes which may attract the sodium ions.

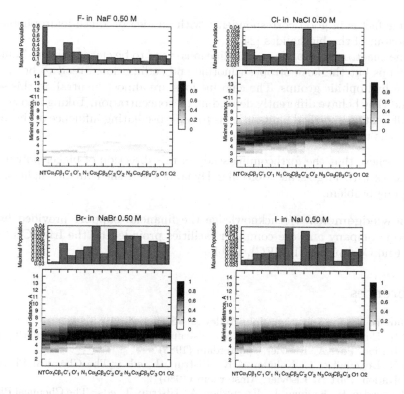

Fig. 5. AA3 - Anion minimal distance distributions for 0.50 M salt solution showing distribution of minimal distance between ion and specified peptide atoms. The top panels show the frequency values of the most frequent distance.

4   Conclusion

In this article, we presented the results from Molecular Dynamics simulations of the alanine tripeptide in different electrolyte solution. This leads to the conclusions:

1. The charge-charge and dipole-charge interactions in water are not well described by any continuum models. Indeed, for some solutions we have a strange picture : - the ions seem to attract the peptide charged groups of the same charge. This effect is known as a 'repolarisation' phenomena [1,2,3,4] and it happens in highly polarisable medias such as aqueous solutions. This fact could serve as an explanation of the fact that the conventional electrostatic theories often fail to describe ion-protein interactions.

2. Generally, in case of comparatively small ionic radius (fluorine and chlorine anions), the ions tend to 'solubilise' highly charged groups (NH3+, COO- groups and peptide backbone in this case). In the case of the bigger anions, both anions and cations tend to be closer to the hydrophobic peptide groups.

This fact is in good correspondence with an idea of 'quasi-hydrophobic' behaviour of the big halides [24].

3. The chaotropes (bromine and iodine ions) tend to be close to the hydrophobic groups but the kosmotropes (fluorine and sodium ions) tend to be close to the hydrophilic groups. The chlorine ions are almost 'neutral' in this sense, and they behave differently depending on concentration. Taking into account different geometrical limits it leads to some oscillating influence of the anions on the tripeptide conformation.

We believe that the forthcoming study in this direction of bigger polypeptide complexes with large-scale Molecular Dynamics will shed further light on this intriguing problem.

Acknowledgement. We acknowledge the financial support provided by the Unilever company and the computer facilities provided by the Irish Centre for High-End Computing (ICHEC), Dublin.

References

1. Marcus, Y.: Ion solvation. John Wiley and Sons Ltd. (1985)
2. R.R. Dogonadze, E. Kalman, A.K., Ulstrup, J., eds.: The Chemical Physics of Solvation. Part A. Elsevier, Amsterdam (1985)
3. R.R. Dogonadze, E. Kalman, A.K., Ulstrup, J., eds.: The Chemical Physics of Solvation. Part B. Elsevier, Amsterdam (1986)
4. Dogonadze, R., Kalman, E., Kornyshev, A., Ulstrup, J., eds.: The Chemical Physics of Solvation. Part C. Elsevier, Amsterdam (1988)
5. Westhof, E., ed.: Water and Biological Macromolecules. The Macmillan press Ltd (1993)
6. Price, N.C., Dwek, R.A., Wormald, M., Ratcliffe, G.R.: Principles and problems in physical chemistry for biochemists. Oxford University press (2001)
7. Hirata, F., ed.: Molecular theory of solvation. Kluwer Academic Publishers, Dordrecht,Netherlands (2003)
8. Goto, Y., Takahashi, N., Fink, A.L.: Mechanism of acid-induced folding of proteins. Biochemistry **29** (1990) 3480–3488
9. Goto, Y., Aimoto, S.: Anion and ph-dependent conformational transition of an amphiphilic polypeptide. Journal of Molecular Biology **218** (1991) 387–396
10. Goto, Y., Hagihara, Y.: Mechanism of the conformational transition of melittin. Biochemistry **31** (1992) 732–738
11. Cacace, M.G., Landau, E.M., Ramsden, J.J.: The hofmeister series: salt and solvent effects on interfacial phenomena. Quarterly Reviews of Biophysics **30** (1997) 241–277
12. Karlstrom, G.: On the effective interaction between an ion and a hydrophobic particle in polar solvents. a step towards an understanding of the hofmeister effect? Physical Chemistry Chemical Physics **5** (2003) 3238–3246
13. Rick, S.W.: A reoptimization of the five-site water potential (TIP5P) for use with ewald sums. Journal of Chemical Physics **120** (2004) 6085–6093
14. Berendsen, H.J.C., Vanderspoel, D., Vandrunen, R.: Gromacs: A message-passing parallel molecular-dynamics implementation. Computer Physics Communications **91** (1995) 43–56

15. Lindahl, E., Hess, B., van der Spoel, D.: Gromacs 3.0: a package for molecular simulation and trajectory analysis. Journal of Molecular Modeling **7** (2001) 306–317
16. Kaminski, G.A., Friesner, R.A., Tirado-Rives, J., Jorgensen, W.L.: Evaluation and reparametrization of the opls-aa force field for proteins via comparison with accurate quantum chemical calculations on peptides. Journal of Physical Chemistry B **105** (2001) 6474–6487
17. Mahoney, M.W., Jorgensen, W.L.: A five-site model for liquid water and the reproduction of the density anomaly by rigid, nonpolarizable potential functions. Journal of Chemical Physics **112** (2000) 8910–8922
18. Schaftenaar, G., Noordik, J.H.: Molden: a pre- and post-processing program for molecular and electronic structures. Journal of Computer-Aided Molecular Design **14** (2000) 123–134
19. Berendsen, H.J.C., Postma, J.P.M., Vangunsteren, W.F., Dinola, A., Haak, J.R.: Molecular-dynamics with coupling to an external bath. Journal of Chemical Physics **81** (1984) 3684–3690
20. Woutersen, S., Hamm, P.: Structure determination of trialanine in water using polarization sensitive two-dimensional vibrational spectroscopy. Journal of Chemical Physics B **104** (2000) 11316–11320
21. Woutersen, S., Hamm, P.: Isotope-edited two-dimensional vibrational spectroscopy of trialanine in aqueous solution. Journal of Chemical Physics **114** (2001) 2727–2737
22. Schweitzer-Stenner, R., Eker, F., Huang, Q., Griebenow, K.: Dihedral angles of trialanine in d2o determined by combining ftir and polarized visible raman spectroscopy. Journal of the American Chemical Society **123** (2001) 9628–9633
23. Mu, Y.G., Kosov, D.S., Stock, G.: Conformational dynamics of trialanine in water. 2. comparison of amber, charmm, gromos, and opls force fields to nmr and infrared experiments. Journal of Chemical Physics B **107** (2003) 5064–5073
24. Lynden-Bell, R.M., Rasaiah, J.C.: From hydrophobic to hydrophilic behaviour: A simulation study of solvation entropy and free energy of simple solutes. Journal of Chemical Physics **107** (1997) 1981–1991

Grid Assisted Ensemble Molecular Dynamics Simulations of HIV-1 Proteases Reveal Novel Conformations of the Inhibitor Saquinavir

S. Kashif Sadiq, Stefan J. Zasada, and Peter V. Coveney

Centre for Computational Science, Department of Chemistry,
University College London, Christopher Ingold Laboratories,
20 Gordon Street, London, WC1H 0AJ

Abstract. Drug resistant mutations have severely limited the success of HIV therapy. Here we provide insight into the molecular basis of drug resistance in HIV-1 protease with the inhibitor saquinavir. We employ protocols consisting of chained molecular dynamics simulations that allow preparation of desired mutants from an available wildtype structure. By conducting ensembles of molecular dynamics simulations we report differing frequencies of adoption of four stable conformations of the P2 subsite of saquinavir. The P2 subsite hydrogen bonds more frequently with the catalytic aspartic acid dyad in the wildtype, whilst preferring to bind with the flaps of the protease in three chosen mutants. Previously such simulations have been demanding to perform on computational grids due to the difficulty in tracking large numbers of simulations. Using the Application Hosting Environment, a lightweight grid middleware solution, we present a simple way to construct chained ensembles of simulations seamlessly across multiple grid resources.

1 Introduction

1.1 HIV-1 Protease Drug Resistance

The development of drug resistant mutations by the human immuno deficiency virus (HIV) remains one of the greatest challenges in the struggle against AIDS. The protease encoded by HIV is responsible for the cleavage and subsequent maturation of the matrix (gag) and enzymatic (pol) viral polyprotein precursors into their constituent proteins. The protease is a symmetric dimer (each monomer has 99 amino acids) that encloses a pair of catalytic aspartic acid residues in the active site (Fig. 1). The active site is bound by a pair of highly flexible flaps that allow the substrate access to the aspartic acid dyad [2]. Due to its crucial role in the maturation of the virus, the enzyme has been a key target for antiretroviral inhibitors and an example of structure assisted drug design [3,4]. Unfortunately, the high replication rate of the virus and low fidelity of the reverse transcription process have led to the emergence and proliferation of drug resistant mutations in various enzymes of HIV [5]. HIV-1 protease in particular, also exhibits tolerance to a significant quantity of non-drug resistant

M.R. Berthold, R. Glen, and I. Fischer (Eds.): CompLife 2006, LNBI 4216, pp. 150–161, 2006.

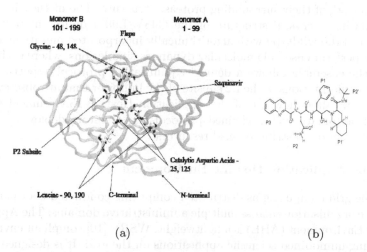

(a)   (b)

Fig. 1. (a) Dimeric three dimensional structure of HIV-1 protease complexed with the inhibitor saquinavir (extracted from 1HXB). (b) The chemical structure of saquinavir showing the P1-P3 and P1'-P2' subsites (nomenclature taken from [1]) with polar atoms labelled O^1 - O^5 and N^1 - N^6.

mutations as part of its natural variability [6]. Comparisons of resolved crystal structures of HIV-1 protease supports the stability of tertiary structure to many mutations [7], whilst computational research has shown large flexibility of the enzyme [8]. Treatment with saquinavir, the first HIV protease inhibitor, has led to the emergence of G48V and L90M single mutants and the G48V/L90M double mutant as primary mutations [5], where G48V denotes glycine at residue position 48 mutating to valine on each monomer and L90M similarly denotes mutation from leucine to methionine at position 90. Experimental studies of saquinavir binding with these mutants have shown a 13.5-, 3- and 419-fold drop respectively in binding affinity with respect to the wildtype [9]. The P2 subsite of saquinavir has been shown across two 1 ns simulations to adopt different conformations in the wildtype and G48V mutant [1], hydrogen bonding to residue 48, which lies near the tip of the flaps, in the wildtype but rotating away from the flaps in the G48V mutant. Here we extend the study of such comparative features by performing an ensemble of fifteen molecular dynamics simulations on each of four saquinavir complexes, namely the wildtype, the G48V and L90M single mutants and the G48V/L90M double mutant. However, as no crystal structures exist for the G48V and L90M systems, it is necessary to employ mutational protocols that reproduce these mutants from a wildtype protease sequence with available crystal structure. It is also important that such mutational algorithms are followed by suitable multi-step equilibration protocols to ensure that the desired mutant structure is an accurate representation of the actual structure. Here, we first present a system dependent equilibration protocol composed of a chained sequence of molecular dynamics simulations, including steps that allow conformational sampling and relaxation of the incorporated mutations within the framework of their surrounding protease structure. The method is validated by comparing a crystal structure of the G48V/L90M double mutant (1FB7) with that of the wildtype with algorithmically incorporated mutations (1HXB). We then perform ensemble molecular dynamics simulations, where each simulation in the ensemble follows a defined equilibration protocol, reporting on the different conformations of the P2 subsite of the inhibitor that are obsereved. Furthermore, we show how use of the Application Hosting Environment (AHE) [1] can both automate such a chained protocol and facilitate deployment of such simulations across distributed grid resources.

1.2   The Application Hosting Environment

We define grid computing as distributed computing conducted transparently by disparate organisations across multiple administrative domains. The Application Hosting Environment (AHE) is a lightweight, WSRF [10] compliant environment for hosting unmodified scientific applications on the grid. It is designed to allow scientists to quickly and easily run unmodified, legacy applications on a grid resource, manage the transfer of files to and from the grid resource and monitor the status of the application. The philosophy of the AHE is based around the fact that very often a group of researchers will all want to access the same application, but not all of them will possess the skill or inclination to install the application on a remote grid resource. In the AHE, an expert user installs the application and configures the AHE server, so that all participating users can share the same application. The AHE is designed to operate across multiple administrative domains seamlessly, but can also be used to provide a uniform interface to applications deployed on both local HPC machines and remote grid resources. The AHE is based on a number of pre-existing grid technologies, principally GridSAM [2] and WSRF::Lite [3]. WSRF::Lite is a Perl implementation of the OASIS Web Services Resource Framework specification, built using the Perl SOAP::Lite [4] web services toolkit. The AHE delegates job submission to GridSAM, which provides a web services interface to grid resources running a variety of backend Distributed Resource Managers (DRM), including Globus [5], Condor [6] and Sun Grid Engine [7]. Jobs submitted to GridSAM are described using Job Submission Description Language (JSDL) which GridSAM in turn uses to submit the job to a local resource. The complexity of the AHE client is kept to a minimum by storing all details of how the application is run on a central service. The design of the client is such that it can easily be installed by an end user without requiring intervention from a system administrator, and without the need for any supporting libraries to be install. The AHE client

[1] http://www.realitygrid.org/AHE
[2] http://gridsam.sourceforge.net
[3] http://www.sve.man.ac.uk/research/AtoZ/ILCT
[4] http://www.soaplite.com
[5] http://www.globus.org
[6] http://www.cs.wisc.edu/condor
[7] http://gridengine.sunsource.net

Table 1. Equilibriation protocol for molecular dynamics simulation of HIV-1 protease incorporating relaxation of mutated amino acid residues

Eq Step	Procedure	Sim Duration (ps)	Force Constant (kcal/mol)			
			wt	G48V	L90M	G48V/L90M
eq0	minimization: 700 iterations		25A	25A	25A	25A
eq1	50K - 100K	10	25A	25A	25A	25A
eq2	100K - 300K	20	25A	25A	25A	25A
eq3*	NVT**	200	25A	25A	25A	25A
eq4	NVT	50	-***	-	25B	25B
eq5	NVT	50	20A	20A	20B	20B
eq6	NVT	50	15A	15A	15B	15B
eq7	NVT	50	10A	10A	10B	10B
eq8	NVT	50	5A	5A	5B	5B
eq9	NVT	800	0	0	0	0
Total Sim Duration (ns)			1.23	1.23	1.28	1.28

A = all non-hydrogen protease atoms restrained

B = class 'A' except atoms of all amino acids within 5 Å of and including L90M mutations

* This step prevents premature flap collapse [11]

** NVT ensemble temperature maintained using Langevin thermostat with coupling coefficient of 5 /ps

*** Absence of the force constraint denotes exclusion of this step in the protocol

package comes with both GUI and command line clients. With the GUI, the user clicks through the steps of a wizard in order to prepare and launch their simulation. The command line clients provide the same functionality, but they have the advantage that they can be called from scripts, meaning that they can be used to create complex multi-application workflows, where the output of one application is used as the input of another. In order to run an application using the AHE command line clients, firstly the user must issue the ahe-listapps command to find the end point of the application factory of the application she wants to run. Next she issues the ahe-prepare command to create a new WS-Resource to manage the state of her application instance. Finally she issues the ahe-start command, which will parse her application configuration file to find any input files that need to be staged to the remote grid resource, stage the necessary files, and launch the application. The user can then use the ahe-monitor command to check on the progress of her application and, once completed, the ahe-getoutput command to stage the output files back to her local machine.

2 Methods

Here we first describe the molecular dynamics methods used to carry out the simulations followed by a description of the grid assisted implementation of such simulations, outlining the emerging bottleneck in the management of multiple input and output files. We finally describe how such a bottleneck could be removed by conducting molecular dynamics simulations using the AHE.

2.1   Molecular Dynamics Methodology

The 1HXB (2.3 Å resolution) crystal structure was used as the starting point for all molecular dynamics simulations. This structure contains dimeric wild-type HIV-1 protease complexed with saquinavir. The missing hydrogen atoms of saquinavir were inserted using the PRODRG tool [12]. Gaussian 98 [13] was used to perform geometric optimisation of the inhibitor at the Hartree Fock level with 6-31G** basis functions. The Restrained Electrostatic Potential (RESP) proce-dure, which is also part of the Gaussian package, was used to calculate the partial atomic charges. The forcefield parameters for the inhibitor were completely de-scribed by the General Amber Force Field (GAFF) [14]. GAFF has been used before in a comparison between saquinavir and a second generation inhibitor [15]. Mutations on the protease were incorporated using VMD [16] which also inserted all missing hydrogens on the protease. The standard AMBER forcefield for bio-organic systems (ff99) [17] was used to describe the protein parameters. The Leap module [18] in the AMBER7 software package [19] was then used to combine each apo-protease system with the inhibitor. Catalytic aspartic acids were modelled as charged and four Cl^- counter-ions were added to electrically neutralise each system, which was then solvated using atomistic TIP3P water [20] in a cubic box with a 10 Å buffer around the complex. The size of each prepared system was 31845, 31860, 31841 and 31856 atoms for the wildtype G48V, L90M and G48V/L90M systems respectively. The molecular dynamics package NAMD2 [21] was used for all simulations. The equilibration protocol was adapted from Perryman *et al.* [2] with several important modifications and is presented in Table 1. The long range Coulombic interaction was handled using the particle mesh Ewald summation method (PME) [22]. A non-bonded cutoff distance of 12 Å was used for all simulations. The SHAKE algorithm [23] was employed on all atoms covalently bonded to a hydrogen atom, allowing for an integration timestep of 2 fs. The equilibration process was repeated from the same initial conditions of minimised energy to construct an ensemble of fifteen equilibration runs for each protease system. Each simulation within an ensemble varied only in the initial velocities attributed to the constituent atoms, which in each case were randomised in a way that reproduced the Maxwell-Boltzmann distribution for a given temperature. The conformation of the P2 subsite in each run was then analysed for all four systems.

2.2   Grid Accelerated Implementation of Simulations

The duration of each molecular dynamics simulation was approximately 1.3 ns. This equated to approximately 330 cpuhrs/sim of wall-clock time with each job using 32 processors for optimized performance with NAMD2. In total, 60 simulations were performed giving a total wall-clock requirement of approxi-mately 20,000 cpuhrs. As the simulations were effectively independent of each other, the scope for accelerated turnover was evident given enough computa-tional power, without which, such a study would be infeasible over a realistic timescale. The simulations were deployed over several resources, namely, the UK

National Grid Service, the UK National supercomputing resources of CSAR and HPCx and USA TeraGrid clusters. Whilst use of such resources greatly parallelizes implementation of 'ensemble' molecular dynamics simulations, a new bottleneck emerges in the management of the many input and output files that result not only from the multiplicity of simulations but also from the multi-step requirement of each equilibration protocol. Further complexity is added in the management of such files across several administrative domains each with heterogenous computational architectures. In this case, job submission scripts were created on each of the resources used giving some degree of automation, coupled with monitoring and 'shepherding' of files across the various machines. In addition to this, as the multi-step simulation workflow is a sequential pipeline, each script was designed in such a way as to feed the output files of one step into the input of the next step of the simulation. Ideally however, much time could be saved given a suitable interface that provides a single point of access for multiple computational resources, which allows for the automated launching of multiple simulations each with multi-step components as well as automatic file transfer and retrieval. In the following section, we describe how such functionality is achieved using the AHE.

2.3   Chained Simulations Using the AHE

By calling the AHE command line clients described above from a shell or Perl script, the user is able to create complex application workflows, starting one application execution using the output files from a previous run. Since the AHE takes care of application execution and moving files between machines, once the workflow script has been started, no user interaction is required. By hosting the NAMD2 application in the AHE, the equilibration protocol described above can be quickly implemented using the AHE to submit separate NAMD jobs for each step, and to manage the transfer of input and output files. To do so, firstly the NAMD2 configuration files for each step in the equilibration protocol must be set up, with the output files of each step configured as the inputs of the next. To make the workflow easier to automate using the AHE, the input files should follow a naming convention (for example, eq1.conf, eq2.conf etc). The chained simulation can be run by creating a Perl script to execute the desired equilibration chain on a remote grid resource which calls the AHE command line clients. The script would execute the ahe-prepare command followed by the ahe-start command sequentially for each step of the equilibration protocol. This has the effect of preparing a WS-Resource to manage the step, staging input files necessary for the step, and executing the application. The script then polls the AHE server at regular intervals using the ahe-monitor command until the simulation step has completed. Once complete, the script stages the files back to the local machine and uses them to initiate the next step of the equilibration protocol. The script terminates after sequentially executing all desired steps in the chained protocol. Following the termination of the script, all output files from the separate NAMD2 jobs will have been staged back to the local machine and, depending on how the chaining script has been configured, can be placed in separate directories ready for analysis.

3   Results and Discussion

We report here on the validation of the mutation and equilibration protocols used in the preparation of the molecular dynamics simulations. We then report on the novel conformations of the P2 subsite exhibited by saquinavir and the differential molecular dynamics of the three above mentioned HIV-1 protease mutants with respect to the wildtype complexed with saquinavir.

3.1   Validation of Mutation and Equilibration Protocol

The mutational protocol was validated by root mean squared deviation (RMSD) analysis of the 1HXB double mutant derivative with the 1FB7 crystal structure which had the same sequence. The RMSD of the protease backbone was very small (0.48 Å) and the RMSD values for the backbone and side chain atoms of mutated amino acids were all less than 0.5 Å and 1.5 Å respectively. As these values are very small, the initial placement of algorithmically incorporated mutants is sufficiently accurate. The mutational relaxation component of the equilibration protocol in this case is not essential but nonetheless provides additional opportunity for further relaxation of mutated amino acids.

3.2   Multiple Conformations of the P2 Subsite

We ran an ensemble of 15 equilibration simulations, each of approximately 1.3 ns for every protease/saquinavir complex. These showed that multiple conformations of the P2 subsite can exist in each system. The conformations are characterised by the association of the P2 subsite with distinct residues within the active site of the protease (Fig. 2). The initial crystallographic conformation, Cx, is orientated with the hydrogens of the P2 subsite facing towards the flaps. In Ci, the P2 subsite has rotated so that the hydrogens point downwards, away from the flaps. We identify this as the conformation associated with the G48V mutation reported by previous authors [1]. Conformation C1 is characterised not only by the rotation of the P2 subsite but by further subsequent motion, forming a strong hydrogen bond with one of the oxygens belonging to the catalytic aspartic acid side chain of monomer B (residue D125). In all instances of C1, evolution occurs directly from Ci with a mean duration of 322 ps for the intermediate. C2 evolves directly from Cx, and is characterised by the P2 subsite being pulled up towards the flaps to form a strong hydrogen bond with the peptide carbonyl oxygen of residue 148. In C3, which in all but one case is preceded by an intermediate C2 conformation with a mean duration of 210 ps, the asparagine arm of P2 rotates out of the active site, while the P2 subsite anchors into a hydrophilic well composed of residues R8, G127, A128 and D129. Finally, C4 is exclusively preceded by Ci with a mean duration of 237 ps and characterised by the inward rotation of the P2 subsite to form a hydrogen bond

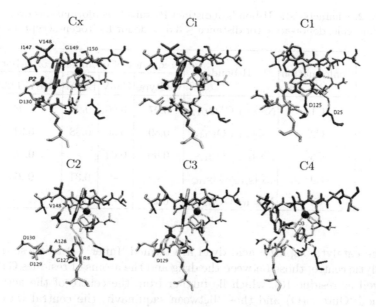

Fig. 2. Distinct conformations of the P2 subsite (red) of saquinavir (white). Sub-regions of the active site are classified as 'Inner': D25 and D125 (blue), 'Outer': G127 to D130 and R8 (yellow) and 'Top': I47 to I50 and I147 to I150 (green). Cx represents the minimised X-ray structure, Ci the intermediate structure and C1-C4 the four stable structures of the P2 subsite.

with the neighbouring carbonyl oxygen (O^3) of the drug. Mutually exclusive hydrogen bonds between the N^4 atom of the P2 subsite and various polar atoms of either the protease or the inhibitor (see Table 2) are also a feauture of each conformation. The frequency of occurrence of each ligand-protease hydrogen bond, including those of the tetrahedrally coordinated water molecule between the ligand and the flaps, as well as that of hydrogen bond characterising C4, was then calculated over the last 1 ns of these extended simulations using the criteria of a maximum donor-acceptor distance of 3.5 Å and a minimum donor-hydrogen-acceptor angle of 150°. The results for the P2 subsite H-bonds characterising each conformation are shown in Table 2. In all cases, conformations C1 - C4 are shown to persist, exhibiting a significant frequency of occurrence of characteristic H-bonds and small mean donor-acceptor distances (< 3 Å). It is then clear from these results that C1-C4 describe distinct and stable conformations of the P2 subsite over a 2 ns timescale. Ci is a strong candidate for being a transient state over the 2 ns timescale as it is a direct precursor of C1 and C4. To test this, we extended simulations for all occurrences of the conformation. The ratio of persistent to transient occurrences of Ci, as measured after the extended 2 ns simulations, was 4:3, for all instances that were stable after the initial 1.3 ns of equilibration, confirming the status of Ci as an intermediate. We classify all hydrogen bonds between ligand and protease into three sub-regions based on their location within the active site. Those between saquinavir and the atoms

Table 2. Characteristic H-bonds of distinct P2 subsite conformations over an extended 2ns timescale: donor-acceptor distance ≤ 3.5 Å, donor-hydrogen-acceptor angle $\geq 150°$

Conformation	H-bond	Frequency of occurrence of H-bonds			
		wildtype	G48V	L90M	G48V/L90M
C1	$N4_{SAQ}\text{-}OD2_{D125}$	0.97	0.96	0.94	0.94
C2	$N4_{SAQ}\text{-}O_{V148}$	0.89	0.96	0.88	0.91
C3	$N4_{SAQ}\text{-}O_{G127}$	0.90	0.64	-	0.80
C4	$N4_{SAQ}\text{-}O3_{SAQ}$	-	-	0.31	0.27
Ci	$N4_{SAQ}\text{-}OD2_{D130}$	0.22	-	0.05	-

of the catalytic aspartic acid dyad are termed 'Inner' (I) as these describe the catalytic centre, those between the drug and the atoms of residues G127 to D130 as well as residue R8, which lie further from the centre of the active site, are termed 'Outer' (O) and those between saquinavir, the central water molecule and the flaps of the protease from I47 to I50 and I147 to I150 are termed 'Top' (T), as shown in Fig. 2. The mean number of H-bonds across each of these sub-regions in all conformations was determined (Fig. 3). We propose that it is the spatial isotropy between ligand-protease hydrogen bonds around the active site that influences the non-equilibrium dynamics of the drug-protease interaction. Given that the 'Inner' and 'Top' sub-regions diammetrically oppose each other with respect to the inhibitor, it is the isotropy between the H-bonds of these sub-regions that plays a factor in retaining the inhibitor within the active site of the protease. Such an isotropic arrangement is only exhibited by conformation C1, in which the 'Inner' H-bonds are marginally more frequent than the 'Outer' ones. In all other conformations, the change in the H-bond contributions across all three sub-regions, both directly through the interactions of the P2 subsite and more subtly through the changes in other ligand-protease H-bond interactions, causes a significantly larger association with the 'Top' sub-region as compared with the 'Inner'. Furthermore, whilst no 'Outer' H-bonds are exhibited by C1, there is some contribution in all other conformations. Adoption of conformations such as C2-C4 that exhibit this anisotropy then promotes stronger coupling to the mobile flaps and may result in the drug being pulled away from the catalytic centre over longer timescales. The effects of these anisotropic arrangements of H-bonds on the dynamics of the inhibitor over longer timescales are shown in the following subsection. The frequency of occurrence of each conformation in all systems after further simulation for 2ns is also shown in Fig. 3. From the initial crystallographic position, the drug in the wildtype moves into conformation C1 more frequently than any other conformation, whilst in the G48V single mutant, the preferred conformation is more equally distributed between C1 and C2. This can be explained as a direct effect of the G48V mutation. The H atoms of G148 in the wildtype are free to rotate about the protease backbone. The substituted

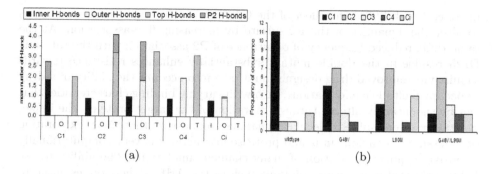

(a)   (b)

Fig. 3. (a) Mean number of instantaneous 'Inner' (I), 'Outer' (O) and 'Top' (T) hydrogen bonds between drug and protease for the G48V/L90M system. (b) Frequency of occurrence of each observed conformation of the P2 subsite from an ensemble size of 15.

valine in the G48V mutant causes increased steric hindrance, preventing complete rotation of the side chain into the flaps and constraining the quinoline moiety of saquinavir through increased hydrophobic interactions. This reduces the available conformational space of the carbonyl oxygen of residue 148, crucial to the formation of C2, and effectively increases the frequency with which C2 occurs. The L90M and G48V/L90M mutants favour C2 with a large reduction in the frequency of occurrence of C1 and also show an increased occurrence of Ci. Furthermore, the L90M system shows marked increase in the persistence of Ci with only one out of five occurrences evolving into another conformation after a further 2 ns of simulation.

4 Conclusions

The change in the frequency of adoption of the two dominant conformations of the P2 subsite of saquinavir, coupled to the change in the ligand-protease H-bond arrangement for each conformation, revealed here by ensemble molecular dynamics, explains the molecular basis for the resistance gained by the G48V and L90M single mutants and the G48V/L90M double mutant of HIV-1 protease against the inhibitor. We show that disruption of the spatially isotropic H-bond arrangements induced by various P2 subsite conformations leads to an increased number of H-bonds, but owing to their anisotropic distribution, also leads to increased dissociation from the catalytic centre. The G48V mutation places conformational restraints on residues 48 and 148 (as the valine side-chain cannot rotate into the flaps), which in turn increases the frequency with which the P2 subsite associates with the flaps of the enzyme. Previous studies using RMSD measurements have shown the flaps to be more flexible than the active site base even in the presence of saquinavir [1]. The L90M mutation substantially decreases the frequency of occurrence of the P2 subsite conformation preferred by the wildtype, in which strong coupling to the catalytic aspartic acid dyad is observed as well as the most spatially uniform H-bond arrangement. When in concert with L90M the effect of the G48V mutation in the double mutant is to alter the dynamics of the P2 subsite by increasing flap-association. At the same time, reduced frequency of occurrence of P2 association with the catalytic D125 residue in the double mutant substantially enhances resistance towards saquinavir. Improved drug design should take into account the ability of a drug to occupy multiple conformations, each with varying binding characteristics that can be significantly altered through mutation of the target protein. The implementation of ensemble molecular dynamics simulations that allow exploration of differential beahviour in ligand-protease interactions is very computationally intensive, requiring execution of many chained simulations. The ability to use lightweight grid middleware solutions such as the AHE is therefore essential to effectively manage such simulations in a seamless manner across multiple grid resources.

Acknowledgements

We thank Dr. Shunzhou Wan for insightful discusisons regarding molecular dynamics simulations and Dr. Simon Clifford for helpful advice regarding *ab initio* methods. We are grateful to EPSRC for funding much of this research through RealityGrid grant GR/R67699, which provided access to the UK national supercomputing resources of CSAR in Manchester and HPCx in Daresbury. The Ph.D. studentship of SKS is also funded by EPSRC. Our work was partially supported by the National Science Foundation under NRAC grant MCA04N014, utilising the TeraGrid cluster at the National Computational Science Alliance in the USA.

References

1. Wittayanarakul, K., Aruksakunwong, O., Saen-oon, S., Chantratita, W., Parasuk, V., Sompornpisut, P., Hannongbua, S.: Insights into saquinavir resistance in the G48V HIV-1 protease: Quantum calculations and molecular dynamic simulations. Biophys. J. **88** (2005) 867–879
2. Perryman, A.L., Lin, J., McCammon, J.A.: HIV-1 protease molecular dynamics of a wild-type and of the V82F/I84V mutant: Possible contributions to drug resistance and a potential new target site for drugs. Protein Sci. **13** (2004) 1108–1123
3. Wlodawer, A., Erickson, J.W.: Structure-based inhibitors of HIV-1 Protease. Annu. Rev. Biochem. **62** (1993) 543–585
4. Wlodawer, A., Vondrasek, J.: Inhibitors of HIV-1 Protease: A Major Success of Structure-Assissted Drug Design. Annu. Rev. Biophys. Biomol. Struct. **27** (1998) 249–284
5. Johnson, V.A., Brun-Vezinet, F., Clotet, B., Conway, B., Kuritzkes, D.R., Pillay, D., Schapiro, J., Telenti, A., Richman, D.: Update of the Drug Resistance Mutations in HIV-1: 2005. Int. AIDS Soc. - USA **13** (2005) 51–57
6. Hoffman, N.G., Schiffer, C.A., Swanstrom, R.: Covariation of amino acid positions in hiv-1 protease. Virology **314** (2003) 536–548
7. Zoete, V., Michielin, O., Karplus, M.: Relation between Sequence and Structure of HIV-1 Protease Inhibitor Complexes: A Model System for the Analysis of Protein Flexibility. J. Mol. Biol. **315** (2002) 21–52

8. Kumar, M., Hosur, M.V.: Adaptability and flexibility of HIV-1 protease. Eur. J. Biochem. **270** (2003) 1231–1239
9. Ermolieff, J., Lin, X., Tang, J.: Kinetic Properties of Saquinavir-Resistant Mutants of Human Immunodeficiency Virus Type 1 Protease and Their Implications in Drug Resistance in Vivo. Biochemistry **36** (1997) 12364–12370
10. Graham, S., Karmarkar, A., Mischkinsky, J., Robinson, I., Sedukin, I.: Web Services Resource Framework. Technical report, OASIS Technical Report (2006) http://docs.oasis-open.org/wsrf/wsrf-ws_resource-1.2-spec-os.pdf.
11. Meagher, K.L., Carlson, H.A.: Solvation Influences Flap Collapse in HIV-1 Protease. Proteins: Struct. Funct. Bioinf. **58** (2005) 119–125
12. Schuettelkopf, A.W., van Aalten, D.M.F.: PRODRG - a tool for high-throughput crystallography of protein-ligand complexes. Acta Crystallogr. **D60** (2004) 1355–1363
13. Frisch, M.J., Trucks, G.W., Schlegel, H.B., Scuseria, G.E., Robe, M.A., Cheeseman, J.R., Zakrzewski, V.G., Montgomery, J.A., Stratman, J., Burant, J.C., et al.: Gaussian 98 (2002) Pittsburgh, PA: Gaussian Inc.
14. Wang, J., Wolf, R.M., Case, D.A., Kollman, P.A.: Development and Testing of a General AMBER Force Field (GAFF). J. Comp. Chem. **25** (2004) 1157–1174
15. Lepsik, M., Kriz, Z., Havlas, Z.: Efficiency of a Second-Generation HIV-1 Protease Inhibitor Studied by Molecular Dynamics and Absolute Binding Free Energy Calculations. Proteins: Struct. Funct. Bioinf. **57** (2004) 279–293
16. Humphrey, W., Dalke, A., Schulten, K.: VMD - Visual Molecular Dynamics. J. Mol. Graph. **14** (1996) 33–38
17. Wang, J.M., Cieplak, P., Kollman, P.A.: How well does a restrained electrostatic potential (RESP) model perform in calculating conformational energies of organic and biological molecules? J. Comp. Chem. **21** (2000) 1049–1074
18. Schafmeister, C.E.A.F., Ross, W.S., Romanovski, V.: LEaP (1995) University of California, San Francisco, CA.
19. Case, D.A., Pearlman, J.C.D., III, T.C., Wang, J., Ross, W., Simmerling, C., Darden, T., Merz, T., Stanton, R., Cheng, A., et al.: AMBER7 (2002) San Francisco, CA: University of California.
20. Jorgensen, W.L., Chandrasekhar, J., Madura, J.D., Impey, R.W., Klein, M.L.: Comparison of simple potential functions for simulating liquid water. J. Chem. Phys. **79** (1983) 926–935
21. Kale, L., Skeel, R., Bhandarkar, M., Brunner, R., Gursoy, A., Krawetz, N., Phillips, J., Shinozaki, A., Varadarajan, K., Schulten, K.: NAMD2: Greater scalability for parallel molecular dynamics. J. Comp. Phys. **151** (1999) 283–312
22. Essmann, U., Perera, L., Berkowitz, M.L., Darden, T.: A smooth particle mesh Ewald method. J. Chem. Phys. **103** (1995) 8577–9593
23. Ryckaert, J.P., Ciccotti, G., Berendsen, H.J.C.: Numerical integration of the Cartesian equations of motion of a system with constraints: Molecular dynamics of n-alkanes. J. Comp. Phys. **23** (1977) 327–341

A Structure-Based Analysis of Single Molecule Force Spectroscopy (SMFS) Data for Bacteriorhodopsin and Four Mutants

Annalisa Marsico, K. Tanuj Sapra, Daniel J. Muller,
Michael Schroeder, and Dirk Labudde

Biotec, Dresden University of Technology, Germany
firstname.lastname@biotec.tu-dresden.de

Abstract. Misfolding of membrane proteins plays an important role in many human diseases such as retinitis pigmentosa, hereditary deafness, and diabetes insipidus. Little is known about membrane proteins as there are only a very few high-resolution structures. Single molecule force spectroscopy is a novel technique, which measures the force necessary to pull a protein out of a membrane. Such force curves contain valuable information on the protein's structure, conformation, and inter- and intra-molecular forces. High-throughput force spectroscopy experiments generate hundreds of force curves including spurious ones and good curves, which correspond to different unfolding pathways. As it is not known what is the origin of the interactions that estabilish unfolding barriers, in the present work we analyse the unfolding patterns coming from experiments of unfolding of bacteriorhodopsin and four mutants (P50A, P91A, P186A and M56). We correlate the postition, magnitude and probability of occurrence of force peaks with the results of a bioinformatics analysis of residue conservations, structural alignments and residue-residue contact area in the wild type and in the mutants, in order to gain insights about the interaction pattern stabilizing bacteriorhodopsin structure. From residue-residue contact area calculations we show that the analysed point mutations do not affect the stability of the protein in a significant way. We conclude that, even if the arrangement of intra-moleular interactions locally change in the mutated structures, the overall structural stability is not affected.

1 Introduction

Integral membrane proteins play essential roles in cellular processes, including photosynthesis, transport of ions and small molecules, signal transduction and light harvesting. Despite the central importance of transmembrane proteins, the number of high-resolution structures remains small due to the practical difficulties in crystallising them [1]. Many human disease-linked point mutations occur in transmembrane proteins. These mutations cause structural instabilities in a transmembrane protein, leading it to unfold or to fold in an alternative conformation [4,11]. Protein folding is described by multidimensional energy landscapes or folding funnels and this is the result of complex inter- and intra-molecular interactions [17]. Atomic Force Microscopy (AFM) is mostly known for its imaging capabilities, but it also provides a novel tool for detecting and locating forces on a single molecule level, like the inter- and intra-molecular interactions

M.R. Berthold, R. Glen, and I. Fischer (Eds.): CompLife 2006, LNBI 4216, pp. 162–172, 2006.
© Springer-Verlag Berlin Heidelberg 2006

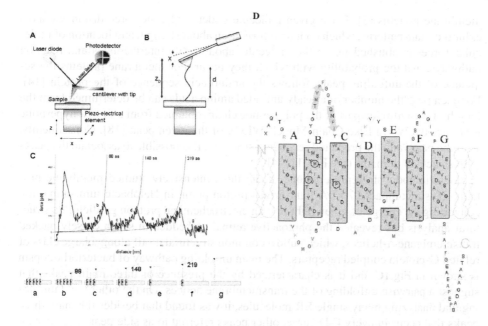

Fig. 1. A) Schematic representation of atomic force microscopy. The sample is mounted on a piezo-electrical element and scanned under a sharp tip attached to the cantilever. The voltage difference of the photodetector is proportional to the deflection of the cantilever. **B)** Unfolding of a transmembrane protein. A single molecule is kept between the tip and the sample while the tip-sample separation is continuously increased. **C)** Typical spectrum obtained from an experiment of unfolding of bacteriorhodopsin with the main peaks fitted by a hyperbolic function (worm like chain model) and correlated to the unfolding of secondary structure elements. **D)** Topology of Bacteriorhodopsin (PDB-ID 1brr) with the in circles mutations.

that stabilise protein structures [7]. Single molecule force spectroscopy experiments allow measuring the stability of membrane proteins and also probing the energy landscapes [6]. In Fig. 1A we show a schematic representation of the force spectroscopy instrumentation. Molecules with complex three-dimensional structures, such as proteins, can be unfolded in a controlled way.

Titin and bacteriorhodopsin are examples of proteins that have been intensively studied [16,6,20]. When transmembrane proteins are unfolded in force spectroscopy experiments, during continuous stretching of the molecule, the applied forces are measured by the deflection of the cantilever and plotted against extension (tip-sample separation), yielding a characteristic Force-Distance curve (F-D curve) for the specific molecule under investigation (see Fig. 1). The F-D curve is the result of subsequent events of molecular interactions [24,16].

From the analysis of single molecule force spectra it is possible to associate the peaks to single potential barriers stabilising segments within membrane proteins. These segments can be represented by transmembrane helices, polypeptide loops or fragments [8]. It is not yet clear how these interactions are established. Currently it is assumed that single amino acids as well as grouped amino acids can stabilise or destabilise membrane proteins [3]. For a given molecule under study, the force-distance curves exhibit certain patterns, which contain information about strength and location of molecular forces established inside the molecule, about stable intermediates and reaction pathways, and the probability with which they occur. For membrane proteins the sequence of the unfolding peaks follows the amino acid sequence of the protein [14]. For each peak the number of already unfolded amino acids can be determined from the length of the unfolded part of the polypeptide chain, obtained from a fit to a hyperbolic function, the Worm Like Chain Model (WLC), of the given peak [18]. Consequently, with the peaks and the predicted secondary structure, it is possible to associate the peaks with structural domains (see Fig. 1) [14].

Bacteriorhodopsin represents one of the most intensively studied membrane proteins. It is a compact 27 kDA light-driven proton pump in Halobacterium salinarum, converting the energy of green light into an electrochemical proteon gradient. Its structural analysis has revealed the photoactive retinal embedded in seven closely packed transmembrane α-helices, which build a common structural motif along a large class of related G-protein coupled receptors. The main unfolding pathway of bacteriorhodopsin is shown in Fig.1C and it is characterized by the presence of three main peaks that suggest a pairwise unfolding of the transmembrane helices. On the other hand unfolding and analyzing many single BR molecules, it was found that besides the main three peaks that occur in every F-D curve, other peaks referred to as side peaks, occur with diffferent probabilities besides the main peaks, indicating that BR exhibit sometimes different unfolding intermediates [14]. The course of a F-D curve represents conformational changes in the protein during the process of unfolding. In order to understand if specific point mutations affect the mechanical stability of Bacteriorhodopsin, we semi-automatically analyze five data sets corresponding to unfolding experiments of Bacteriorhodopsin and four mutants (P50A, P91A, P186A, M56A). We explain briefly the reason for analyzing these specific mutants. The vast majority of transmembrane helices contain significant distortion from ideal helix geometry. Helix distortions are one of the mechanisms for creating structural diversity from the relatively simple building blocks in helix bundle membrane proteins [23]. Most transmembrane helix deformations (ca. 60%) occurr at proline residues. Proline residues, by not contributing the normal hydrogen bond, cause kinks in the folded structure and are widely recognized as playing a significant role in structure, folding and unfolding of proteins [12]. Bacteriorhodopsin contains 11 proline residues, three of which are membrane embedded: Pro-50 (helix B), Pro-91 (helix C), and Pro-186 (helix F) (see Fig. 1D). Prolines substitutions with Alanine are expected to alter the mechanical stability of the protein. The mutation M56A is analyzed because it is known from biophysical studies to be the most stabilizing mutation for Bacteriorhodopsin (positive variation in the free energy) [3]. Here, we analyse force-distance curves relative to Bacteriorhodopsin and its mutants in a semi-automated way, aligning the curves with a dynamic programming approach and detecting and classifying peaks in the force-distance traces using the WLC model. We use different bioinformatics tools (see materials and methods) for deriving conservation scores of the residues, structural alignments of the wild type (WT) with the mutants and residue-residue contact maps analysis. Furthermore, we calculate the differences in residue-residue contact areas between the WT and the mutants and we correlate them with the unfolding barrier classified in the spectra analysis procedure.

2   Materials and Methods

Experimental Setup. Wild-type purple membrane (PM) was extracted from H. salinarum and purified as described in [15]. BR mutants P50A, P91A, P186A, and M56A were a kind gift of Prof. James Bowie (UCLA, USA) and were prepared as described elsewhere [3]. BR was attached nonspecifically to silicon nitride cantilever by applying a contact force of 1 nN between the AFM stylus and the membrane surface. Single-molecule AFM imaging and force spectroscopy was performed as described earlier [14,16]. First, membrane patches were imaged at using contact mode AFM [13]. For force measurements, the AFM stylus was approached to the membrane protein surface while applying a constant force of <1 nN. After a contact time of 500 ms − 1s, the stylus was retracted from membrane surface at a set of different constant velocities.

Semi-Automated Analysis of the Force-Distance (FD) Curves. Our procedure for data analysis is made up of three steps: data preparation and noise reduction; curve alignment with dynamic programming; peak detection and fitting. In the first step we automatically identify and remove curves that do not contain unfolding signals, curves that exhibit an overall length indicative of partial or multiple unfolding events, curves showing peaks due to unspecific interactions (corrupted curves). According to a classical analysis, we determine the zero-force baseline, the contact point and the standard deviation of the noise [9]. In a typical force spectroscopy experiment, the standard deviation in the final part of the spectrum is usually between 10 and 40 pN. In order to even out the noise, in the second step we apply a moving average for smoothing the spectra, combining data points from equally spaced intervals of 0.1 nm. This allows us to reduce the standard deviation of the noise from 40 pN to 14 pN. In the second stage of our method the curves are aligned using global multiple sequence alignment with dynamic programming (see e.g. [2]). For the detailed description of the overall spectra alignment procedure see [10].

The key reason for using a sequence alignment technique is the meaningful definition of matches/mismatches, insertions, and deletions. Matches and mismatches reward/penalise more or less fitting parts of the force curves. Insertions and deletions are important, as peaks in the curves may vary by up to six residues and as peaks may be missing completely between two curves.

In the third stage of our procedure we fit every peak of every aligned force extension curve with the worm like chain model with a persistence length of 0.4 nm and a monomer length of 0.36 nm [18]. The number of extended amino acids at each peak was then calculated using the contour length obtained from the WLC fitting. This allows to assign unfolding events to structural segments as described before [18]. To measure the unfolding force and probability of unfolding for each individual structural segment, every event of each curve was analyzed.

Bioinformatics Tools. *Identification of highly conserved residues: the HSSP Database [19].* We use the HSSP database for the identification of highly conserved residues

(residues with conservation score greater than 80%). In the HSSP database each amino acid position in a protein is associated with a variation entropy score or Shannon entropy measuring the sequence variability. The Shannon entropy is calculated on the basis of the amino acid frequencies at each position within the homologous proteins; variable positions have high entropy and conserved positions have low entropy [19]. Using a color code, we project these pre-calculated variation scores onto Bacteriorhodopsin's structure (pdb_id 1brr) in Pymol. We correlate the position of highly conserved residues to the location of the detected unfolding barriers in the structure.

Structural alignments of bacteriorhodopsin and each mutant: the DaliLite program. We perform structural alignments between bacteriorhodopsin wild type (bR WT) and each mutant (pdb_id 1pxr, 1q5j, 1q5i and 1pxs for P50A, P91A, P186A and M56A) using the DaliLite program for pairwise structure comparison and structure database search [5]. The program output reports Z-scores, which are similarity scores between two PDB structures normalized to the structure size. We determine Z-scores for all the four structural alignments of the WT with the mutants.

Calculation of residue-residue contact area differences between bR WT and mutants: The Contact Map Analysis (CMA) server. The CMA server is a tool for construction and analysis of protein contact maps in order to retrieve residue-residue contacts between two chains or within a single chain [21]. Residue-residue contact area calculations are based on the sum of atom-atom contact surface areas, as implemented in the LPC/CSU software [22]. We apply the method to a single chain of bR WT and the four mutants in order to catch differences in residue-residue contact areas between them. For each protein (bR WT and mutants) and for each residue we evaluate the total number of contacts and we calculate the differences in residue-residue contact areas between the WT and each mutant in the unfolding barriers in the following way:

Let B be the set of residues in a structure segment establishing an unfolding barrier (as detected from the WLC fitting of a given peak in a set of analysed FD curves);

let P be the whole set of residues in the folded portion of the protein structure.

$$\Delta Area_{wt,mutant} = (\sum_{i\in B}\sum_{j\in P} Area(i,j))_{wt} - (\sum_{i\in B}\sum_{j\in P} Area(i,j))_{mutant}$$

where, for each residue i in a detected barrier B we calculate the differences in contact area between the wild type and the mutant, considering in the calculation all the contacts between a given residue i in the barrier B with any other residue j in the protein P (excluded the portion of the protein already pulled out). In the calculation of the contact area differences all the missing and additional contacts of the mutant compared to the wild type are included.

3   Results and Discussion

Unfolding Patterns of Bacteriorhodopsin and Mutants. We show the automated superimposition of 40 spectra for bR WT and the mutants (Fig. 2). The results show that bR WT and its mutants exhibit the same unfolding patterns (peaks at the same positions), inside the experimental errors (±5 amino acids). We can see in Fig. 3 that the

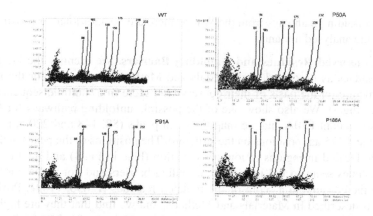

Fig. 2. Automated superimposition of 40 spectra for WT and the Proline mutants and fitting of the detected peaks with the WLC model

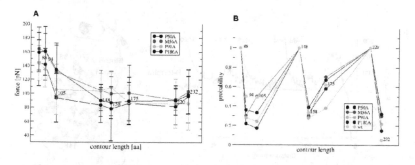

CL	WT			P50A			M56A			P91A			P186A		
	F[pN]	stdev	prob	F[pN]	stdev	prob	F[pN]	stdev	prob	F[pN]	stdev	prob	F[pN]	stdev	prob
88	150	44	1.0	143	33	1.0	163	31	1.0	143	25	1.0	158	28	1.0
94	153	53	0.5	141	23	0.22	159	26	0.3	134	26	0.28	161	28	0.36
105	120	59	0.49	93	35	0.17	130	38	0.5	105	36	0.24	134	38	0.33
148	107	34	1.0	83	25	1.0	104	29	1.0	103	25	1.0	89	23	1.0
158	108	47	0.33	78	46	0.33	100	32	0.3	109	22	0.28	86	23	0.39
175	84	38	0.62	86	30	0.58	100	40	0.7	86	25	0.38	89	34	0.67
220	101	24	1.0	81	15	1.0	91	15	1.0	85	35	1.0	90	21	1.0
232	103	17	0.05	96	38	0.15	102	32	0.3	86	38	0.22	98	27	0.32

Fig. 3. A) For bR WT and for each mutant we show the plot of average unfolding forces relative to the detected peaks; **B)** plot of the probability of occurrence of each peak for wild type and mutants. For bR WT and each mutant we also show a table reporting the total numer of already unfolded amino acids retrieved from the WLC fit of each peak (column CL), the average unfolding force (column F), the standard deviation on the average force (column st-dev) and the probability of occurrence of each peak (column prob).

unfolding forces are also the same within the experimental errors. Furthermore, we can observe that, while the probability of occurrence of the main peaks does not change from the WT to the mutants, the probability of occurrence of the side peaks slightly varies. However, this variation is not so strong to support the idea that mutants exhibit different unfolding pathways from the WT, as there are no additional or missing peaks in any of the analysed mutants.

Highly Conserved Residues and Unfolding Barriers. We identify the highly conserved residues as described in (Materials and Methods), and we color the residues in the structure on the basis of their conservation score in a multiple sequence alignment (see. Fig.4). We also show one of the possible unfolding pathways for Bacteriorhodopsin mutant P50A and we map the main peaks (88, 148 and 220 aa) and side peaks (94 and 175 aa) positions on the structure. The positions of the peaks in the force curves are located in regions of low conservation (blue in Fig.4) and the highly conserved residues seem to fall inside the unfolding barriers and not at the borders. We calculate the conservation score of the mutated residues (Pro50, Pro91, Pro186 and Met56), as described in Materials and Methods, and we find that they are highly conserved, with a conservation score of 0.64 for Pro50, 0.95 for Pro91, 0.97 for Pro186 and 0.60 for Met56[1].

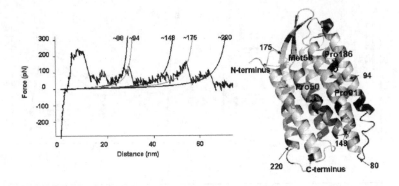

Fig. 4. Example of a spectrum showing the mapping of main peaks (88, 148 and 220 aa) and side peaks (94 and 175 aa) positions in the structure of the bR P50A mutant. The residues in the structure are colored on the basis of their conservation score in a multiple sequence alignment. The color scale ranges from red (highly conserved residues) to blue (weakly conserved residues). The highly conserved residues fall inside two detected unfolding barriers.

Structural Alignments. We ran pairwise structural alignments of the WT and mutant structure. The RMSD of each pairwise alignment are all around 0.7, indicating a good superimposition between the WT and each mutant. The main backbone structure is conserved in the mutants with minimal structural changes.

Contact Area Differences and Unfolding Patterns of Bacteriorhodopsin and Mutants. As the contact area between interacting residues can be thought of as a good descriptor of the pattern of interaction forces stabilizing a protein structure, we calculate the differences in contact area between the WT and each mutant (as described in the section Materials and Methods), in order to see if we can observe significant

---

[1] The conservation score is calculated as $conservation\_score = 1 - vonNeumannEntropy$.

mutation	add-contacts	miss-contacts	Δ Area($\mathring{A}^2$)	P-value
P50/A50	0	0	44.4	< 0.1
P91/A91	1	1	40.6	< 0.05
P186/A186	0	0	43.4	< 0.1
M56/A56	0	2	129.9	< 0.01

Fig. 5. Total differences in contact area for each mutated residue compared to the WT. We show also the number of additional (column add-contacts) and missing contacts (column miss-contacts) for the mutated residue compared to the original in the WT.

changes in the interaction patterns of the mutants. First, we show the differences in contact area only for the residues that are mutated. The results are shown in Fig.5. For instance, for Pro50 we take into account, in the calculation of the contact area, all the residues that are interacting with Pro50, setting the minimum required distance for a contact to 4 Å. We do the same for the mutated residue (Ala50) in the mutant and we compute the differences between the contact areas. We repeat the calculation for the other mutants. As shown in Fig.5, the calculated P-values indicate that the differences in contact area are significant when we limit the calculation just to the single mutated residue. The positive values in the contact area differences for all the four mutants indicate that the total contact area of each mutated residue is decreased when compared to the wild type. However, the pairwise structural alignements of bR wild type with each of the mutants and the unfolding pathways from the AFM experiments indicate that, overall, the structural stability of the mutants is not affected by any of the single point mutations. In order to understand the reason for that, we calculate the total residue-residue contact areas in each sequence segment estabilished from the detected unfolding barriers. We find that in most of the cases these differences are not significant (P-value > 0.5). When these differences are significant, they probably compensate each other, as we can note that in some cases the residue-residue differences in contact areas are positive (indicating that the total contact area in a given structural segment is minor in the mutant than in the wild type) and in other cases the reverse is true. The results are shown and plotted in Fig.7.

Discussion. We show that a semi-automated analysis of FD curves coming from experiments of unfolding of bR WT and four mutants is able to detect unfolding energy barriers inside the structures. We find that the WT and mutants exhibit the same unfolding pattern (same intermediates) inside the experimental errors. Furthermore, we show that there is a possible correlation between the position of the energy barriers and the location of highly conserved residues (e.g. Pro50, Pro91, Pro186 and Met56 are embedded in the structural segments that estabilish the barriers). From the analysis of residue-residue intra-molecular contacts and contact surface area differences between the WT and the mutants we observe that, locally, proline residues and methionine alter the arrangement of the residue-residue interactions (Fig.5). Looking at the different structural segments estabilished from the energy barriers, we find sometimes a different arrangement of the interaction patterns, due to missing or additional residue-residue contacts and significant changes in contact area, between WT and mutants. However, it seems that the local structural alterations compensate each other all over the structure, leaving the three-dimensional structure of the mutants unchanged. This result supports

CL	sequence-segment	P50A		M56A		P91A		P186A	
		ΔArea($\mathring{A}^2$)	P-value	ΔArea($\mathring{A}^2$)	P-value	ΔArea($\mathring{A}^2$)	P-value	ΔArea($\mathring{A}^2$)	P-value
88	(159-230)	-277.6	< 0.1	-321.3	< 0.05	< +392.1	< 0.05	-260.4	< 0.05
94	(149-158)	-51.5	< 0.1	-34.5	n.s.	44.4	< 0.1	+30	n.s.
105	(132-148)	+21.5	n.s.	+6.9	n.s.	+16.8	n.s.	+9.6	n.s.
148	(97-131)	+7.5	n.s.	+68.4	n.s.	-79.5	< 0.1	+139.2	< 0.05
158	(80-96)	+43.4	n.s.	+64.5	n.s.	-71.4	< 0.1	+37.5	n.s.
175	(62-79)	+92.5	< 0.1	-108.5	< 0.05	+61.2	n.s.	+104.4	< 0.05
220	(5-61)	-162.5	< 0.1	-66.7	n.s.	-162.30	< 0.05	-141.4	< 0.05
232	(6-24)	+162.5	< 0.05	+20.0	n.s.	+116.7	< 0.1	+124.2	< 0.1

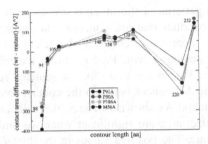

Fig. 6. Differences in residue-residue contact areas for all the residues in each unfolding barrier. In the top table, in column CL we show the number of already unfolded amino acids corresponding to each unfolding barrier. We also show the sequence segment corresponding to the extension of each barrie (e.g. 88 amino acids pulled from the C-terminus of the protein correspond to the sequence/structure portion from amino acid 159 to amino acid 230). Besides the calculation of the total residue-residue contact area difference (column ΔArea(A)) we show the P-value associated to this calculation which indicates the probability of the observed result occurring by chance alone (when the P-value is greater than a threshold of 0.1 we consider the variation not significant (n.s.). We plot the differences in contact areas for all the intermediates (unfolding barriers) and for all the mutants.

CL	P50A		M56A		P91A		P186A	
	missing	additional	missing	additional	missing	additional	missing	additional
88	66	54	63	59	61	54	60	64
94	7	7	7	4	7	4	6	1
105	4	9	6	8	6	9	4	9
148	9	7	9	8	7	6	7	7
158	5	5	6	6	5	4	6	6
175	21	11	18	10	18	8	18	10
219	14	16	14	19	11	18	12	19
232	13	8	14	9	15	8	14	7
total seq.	177	139	177	142	172	131	170	143

Fig. 7. We show for each mutant the number of additional (column additional) and missing residue-residue contacts (column missing) inside each intermetiate or unfolding barrier. We show also, in the last row of the table, the total number of missing and additional contacts calculated over all the sequence of each mutant.

the hypothesis that the collective behaviour of several amino acids is responsible for structural stability of membrane proteins and energy barriers against unfolding rather than single amino acids.

Acknowledgements. Thanks to Andreas Henschel for conservation coloring in PyMol and Jana Clement for spectra alignments pictures. Funding through the EFRE project CODI and FoldUnfold is kindly acknowledged.

References

1. James U. Bowie. Solving the membrane protein folding problem. *Nature*, 438(7068):581–589, 2005.
2. S. R. Eddy. What is dynamic programming? *Nature Biotechnology*, 22(7):909–910, 2004.
3. Salem Faham, Duan Yang, Emiko Bare, Sarah Yohannan, Julian P Whitelegge, and James U Bowie. Side-chain contributions to membrane protein structure and stability. *J Mol Biol*, 335(1):297–305, 2004.
4. S. Filipek, D. C. Teller, K. Palczewski, and R. Stenkamp. The crystallographic model of rhodopsin and its use in studies of other g protein-coupled receptors. *Annu Rev Biophys Biomol Struct*, 32:375–397, 2003.
5. L Holm and Park J. Dalilite workbench for protein structure comparison. *Bioinformatics*, 16(6):566–567, 2000.
6. H. Janovjak, J. Struckmeier, M. Hubain, A. Kedrov, M. Kessler, and D. J. Muller. Probing the energy landscape of the membrane protein br. *Structure*, 12(5):871–879, 2004.
7. A. Janshoff, M. Neitzert, Y. Oberdorfer, and H. Fuchs. Force spectroscopy of molecular systems-single molecule spectroscopy of polymers and biomolecules. *Angew Chem Int Ed Engl*, 39(18):3212–3237, 2000.
8. M. Kessler, K. E. Gottschalk, H. Janovjak, D. J. Muller, and H. E. Gaub. Bacteriorhodopsin folds into the membrane against an external force. *J Mol Biol*, 357(2):644–654, 2006.
9. M. Kuhn, H. Janovjak, M. Hubain, and D. J. Muller. Automated alignment and pattern recognition of single-molecule force spectroscopy data. *J Microsc*, 218(Pt 2):125–132, 2005.
10. A Marsico, K.T Sapra, D Muller, D Labudde, and M Schroeder. A novel pattern recognition algorithm to classify membrane protein unfolding pathways with high-throughput single molecule force spectroscopy. acccepted in J. Bioinformatics.
11. T. Mirzadegan, G. Benko, S. Filipek, and K. Palczewski. Sequence analyses of g-protein coupled receptors: similarities to rhodopsin. *Biochemistry*, 42(10):2759–2767, 2003.
12. T Mogi, L J Stern, B H Chao, and H G Khorana. Structure-function studies on bacteriorhodopsin. viii. substitutions of the membrane-embedded prolines 50, 91, and 186: the effects are determined by the substituting amino acids. *J Biol Chem*, 264(24):14192–14196, 1989.
13. D Muller, H Sass, S Muller, G Buldt, and A Engel. Surface structures of native bacteriorhodopsin depend on the molecular packing arrangement in the membrane. *J Mol. Biol.*, 33(285):1903–1909, 1999.
14. D. J. Muller, M. Kessler, F. Oesterhelt, C. Moller, D. Oesterhelt, and H. Gaub. Stability of bacteriorhodopsin alpha-helices and loops analyzed by single-molecule force spectroscopy. *Biophys J*, 83(6):3578–3588, 2002.
15. D Oesterhelt and W Stoeckenius. Isolation of the cell membrane of halobacterium halobium and its fraction into red and purple membrane. *Methods Enzymol*, (31):667–678, 1974.
16. F. Oesterhelt, D. Oesterhelt, M. Pfeiffer, A. Engel, H. Gaub, and D. J. Muller. Unfolding pathways of individual bacteriorhodopsins. *Science*, 288(5463):143–146, 2000.
17. J. N. Onuchic and P. G. Wolynes. Theory of protein folding. *Current Opinion in Structural Biology*, (14):70–75, 2004.
18. M. Rief, M. Gautel, F. Oesterhelt, J. M. Fernandez, and H. E. Gaub. Reversible unfolding of individual titin immunoglobulin domains by afm. *Science*, 276(5315):1109–1112, 1997.

19. C Sander and R Schneider. Database of homlogy-derived protein structures and structural meaning of sequence alignment. *Proteins*, 101(9):56–68, 1991.
20. K. T. Sapra, H. Besir, D. Oesterhelt, and D. J. Muller. Characterizing molecular interactions in different bacteriorhodopsin assemblies by single-molecule force spectroscopy. *J Mol Biol*, 355(4):640–650, 2006.
21. V Sobolev, E Eyal, S Gerzon, V Potapov, M Babor, J Prilusky, and M Edelman. Space: a suite of tools for protein structure prediction and analysis based on complementarity and environment. *Nucleic Acids Research*, 33(4):39–43, 2005.
22. V Sobolev, A Sorokine, E Prilusky, and M Edelman. Automated analysis of interatomic contacts in proteins. *Bioinformatics*, 15(4):321–332, 1999.
23. S Yohannan, S Faham, D Yang, P Whitelegge, and J Bowie. The evolution of transmembrane helix kinks and the structural diverstity of g protein-coupled receptors. *PNAs*, 101(4):959–963, 2003.
24. X. Zhuang and M. Rief. Single-molecule folding. *Curr Opin Struct Biol*, 13(1):88–97, 2003.

Classifying the World Anti-Doping Agency's 2005 Prohibited List Using the Chemistry Development Kit Fingerprint

Edward O. Cannon and John B.O. Mitchell*

Unilever Centre for Molecular Science Informatics, Department of Chemistry,
University of Cambridge, Lensfield Road, Cambridge CB2 1EW, United Kingdom
* Tel.: +44-1223-762983; Fax: +44-1223-763076
jbom1@cam.ac.uk

Abstract. We used the freely available Chemistry Development Kit (CDK) fingerprint to classify 5235 representative molecules taken from ten banned classes in the 2005 World Anti-Doping Agency's (WADA) prohibited list, including molecules taken from the corresponding activity classes in the MDL Drug Data Report (MDDR). We used both Random Forest and k-Nearest Neighbours (kNN) algorithms to generate classifiers. The kNN classifiers with $k = 1$ gave a very slightly better Matthews Correlation Coefficient than the Random Forest classifiers; the latter, however, predicted fewer false positives. The performance of kNN classifiers tended to decline with increasing k. The performance of the CDK fingerprint is essentially equivalent to that of Unity 2D. Our results suggest that it will be possible to use freely available chemoinformatics tools to aid the fight against drugs in sport, while minimising the risk of wrongfully penalising innocent athletes.

1 Introduction

Doping comes from the Dutch word "doop", meaning a thick liquid or sauce and originally a South African drink, drunk to help make an individual work harder. Here, we discuss illegal doping in sport, the objective of which is to enhance athletic performance, with little thought as to either the consequences for athletes' health or the integrity of competition. The issue of doping in sport is further complicated by a minefield of legal, political, and ethical questions. The urgency and importance of the battle against drugs in sport was underlined when several of the world's leading cyclists were forcibly withdrawn on the eve of the 2006 Tour de France, following an investigation by Spanish police.

The WADA [1] prohibited list contains 11 different classes of substance: one of these, alcohol (P1), has just one member and is not considered further. Anabolic agents (S1) are artificial synthetic analogues of the male sex hormone testosterone. They are used to promote growth of the skeletal muscles and red blood cells; particularly useful in events such as weightlifting or the 100m sprint, whereby these substances increase muscle size and strength allowing the athlete to train harder. Hormones and related substances (S2) include: erythropoietin, growth hormones, gonadotrophins, insulin and

M.R. Berthold, R. Glen, and I. Fischer (Eds.): CompLife 2006, LNBI 4216, pp. 173–182, 2006.
© Springer-Verlag Berlin Heidelberg 2006

corticotrophins. These substances are taken by athletes to stimulate cell growth and red blood cell production and to increase sugar levels in the blood to avoid fatigue.

The primary medical use of beta-2 agonists (S3) is to treat asthmatic patients during an asthma attack. The drugs are used to open up the airways in the lungs which become restricted following an asthma attack. They are now being used in sport because if injected into the bloodstream they have a powerful anabolic effect that can cause muscle mass to increase and body fat to drop. Anti-estrogenic agents (S4) are substances that prevent the full expression of estrogen. Examples of anti-estrogens include tamoxifen and clomiphene.

Diuretics (S5), normally used to treat heart failure or high blood pressure, have been abused in sport for weight loss and elimination of drugs from the system. Diuretics work by increasing urine production in the kidneys. Sports where diuretics might be abused for promoting weight loss include boxing and lightweight rowing, and indeed any sports where competitors are required to reduce their body weight to below a specified level. Diuretics have been abused as masking agents to dilute the concentration of substances in the urine and avoid detection of other performance-enhancing drugs. Stimulants (S6) increase the activity of the sympathetic nervous system. Examples of stimulants include cocaine, amphetamine and modafinil; caffeine has recently been removed from the WADA prohibited list. These substances make the user feel more alert, energetic and able to concentrate. Narcotics (S7) enhance performance in sport by acting as pain killers. Narcotics allow an injured athlete to continue to train and compete by relieving pain. Examples of narcotics banned in sport include heroin, morphine and fentanyl.

Cannabinoids (S8) have been used to treat pain, migraine, insomnia, nausea and high blood pressure. They are used in sport to relax an athlete before competition. Glucocorticosteroids (S9) are now used as anti-inflammatory agents to treat arthritis and dermatitis. Examples of glucocorticosteroids include hydrocortisone and fludrocortisone acetate. Beta blockers (P2) act as performance-enhancing drugs by lowering the human heart rate and blood pressure, particularly useful in Olympic sports such as archery or shooting where the beta blockers provide more time for the athlete to aim in between heart beats. Examples of beta blockers include acebutolol, alprenolol, nadolol and atenolol.

The repertoire of substances used as doping agents in sport is continually evolving. This leads to an "arms race" between cheats and testers. The former are engaged in the design and synthesis of novel drugs, exemplified by "designer steroids" [2,3] such as tetrahydrogestrinone (THG), which has recently gained notoriety in track and field athletics. The WADA list of prohibited substances uses the phrase "and other substances with a similar chemical structure or similar biological effect(s)" to prohibit analogues of known performance-enhancing molecules. This is a very delicate area legally and ethically, since the authorities run the risk of criminalising athletes who ingest substances which are in some way "similar", without any hard evidence of bioactivity.

Prior to our work, interest in chemoinformatics approaches to drugs in sport appears to have been limited to the single study of Kontaxakis and Christodoulou [4], devoted to the prediction of chromatographic retention times of prohibited substances using an artificial neural network. Nonetheless, chemoinformatics may have an important role to play, since much of the discipline is built around, firstly, quantifying chemical similarity and, secondly, predicting bioactivity – exactly the two issues that are most relevant in the present context. In recent work [5], we have built classifiers which can be used to predict whether a given molecule is likely to exhibit the bioactivity specific to any particular class of prohibited substances.

Our approach has a number of advantages, not least of which is putting the definition of chemical similarity on a quantitative (algorithmic) footing, which should be less vulnerable to legal challenge than a purely qualitative definition. It can also identify molecules unlikely to be bioactive and hence reduce the likelihood of athletes being unjustifiably penalised. We anticipate that in practice such classifiers would be used to complement, rather than replace, experimental methods such as assays [3]. Experimental methods would allow confirmation of the bioactivities suggested by chemoinformatics. The use of classifiers such as ours on large databases or libraries of molecules can help the authorities predict where in chemical space their opponents are likely to be sourcing the next (or even current) generation of designer drugs. This would be highly beneficial, since it seems almost certain that much drug abuse in sport involves bioactive substances that are not currently known to, and hence not specifically looked for by, the drug testing regime.

In this paper, we will demonstrate that the freely available CDK Fingerprint [6] can be used to generate excellent classifiers. This is part of the Chemistry Development Kit, described as "a freely available open-source Java library for Structural Chemo- and Bioinformatics" [7]. This decouples the classifiers from the commercial fingerprints such as Unity 2D [8] and MACCS [9], which had been the basis of the successful classifiers in our previous work [5]. We will show that Random Forest is particularly suitable for minimising false positives. For kNN classifiers, we will find that $k = 1$ is most successful. We will also consider the class-specific predictive abilities of our classifiers, which exhibit a fairly consistent pattern. We believe that our work facilitates the use of chemoinformatics in the fight against doping in sport.

2 Methods

2.1 Datasets

All methods were applied to a dataset of 5235 molecules, some derived directly from the prohibited list and others taken from activity classes in the MDDR database (Version 2003.1) [9] corresponding to each WADA prohibited class of substance. The use of MDDR molecules of the corresponding bioactivities was necessary since the number of explicitly named molecules in the WADA list is relatively low, and justified by the "similar chemical structure or similar biological effect(s)" criterion. Our dataset contained: 47 anabolic agents (S1), 272 hormones and related substances (S2), 367 beta-2 agonists (S3), 928 anti-estrogenic agents (S4), 995 diuretics and masking agents (S5), 804 stimulants (S6), 195 narcotics (S7), 995 cannabinoids (S8), 26 glucocorticosteroids (S9), 239 beta-blockers (P2) and 367 explicitly allowed substances.

2.2 Fingerprints

This work considers two fingerprints, the Chemical Development Kit (CDK) fingerprint and the Unity 2D fingerprint. The CDK fingerprint used in this work is modelled on the Daylight [10] fingerprint. It operates by running a breadth-first search starting at each atom in the molecule and produces a string representation of paths up to six atoms in length. The software is written in Java and uses the Java hashing function in combination with a pseudorandom number generator with a default range of 0-1023. The number indicates a position in a fingerprint of length 1024 bits that is set to 1, based on the paths computed for the molecule.

The Unity 2D fingerprint is composed of 992 feature bits. It is also similar to the Daylight fingerprint, the difference being that the Unity fingerprint segregates different path lengths into different regions of the fingerprint [11]. Unity was the best performing fingerprint in our recent work [5], hence Unity provides an important benchmark.

This work is underpinned by the "Similar Property Principle", that molecules close together in the chemical space defined by our descriptors are likely to share similar properties (in this case bioactivities).

2.3 Classification

The two machine learning algorithms used in this work are k-Nearest Neighbours and Random Forest. These algorithms were run using R software [12]. In all cases the classification was performed in a binary fashion, such that a query molecule was either predicted to be part of a prohibited class under question or was not.

In our k-Nearest Neighbour (kNN) classifiers, the class of a query molecule is determined by the majority vote of the class labels (member or non-member) of its k nearest neighbours, according to Euclidean distance in descriptor space, with tied votes resolved randomly.

Random Forest [13] generates a forest of decision trees. At each node of each tree, a descriptor is chosen for branch splitting; this is not selected from the full set of available descriptors, but from a random subset of candidates. The parameter *mtry* indicates how many descriptors will be randomly selected as candidates at each node in the tree. Its default value was used in this work, defined as the square root of the number of bits in the fingerprint (rounded down to an integer). Hence for the 1024 bit CDK fingerprint *mtry* is taken as 32, and for the 992 bit Unity 2D fingerprint the default *mtry* is 31. For each tree, branches continue to be subdivided while the minimum number of observations in each leaf is no less than a pre-determined *nodesize* value. Branches are not pruned back. The Random Forest algorithm produces one output per molecule per tree. Each output classifies the molecule into either the category of member or that of non-member of a particular prohibited class. The outputs of the trees are aggregated using majority voting. We used 500 trees per Random Forest (*ntree* = 500).

We used fivefold cross-validation everywhere. This means that results for the Random Forest classifiers are based on five runs, each using a different 20% of the dataset as an independent test set, with the results being aggregated. Each molecule thus appears in exactly one of the five test sets (and exactly four of the five training sets). A similar procedure was used in the kNN work, with 20% of the molecules being predicted based on their nearest neighbours in the remaining 80%. This nearest neighbour prediction test was repeated five times on mutually exclusive test sets. Thus each molecule was predicted once, and the results aggregated.

2.4 Performance Measures

For each of the classifiers operating on each of the 10 prohibited classes, a 2×2 confusion matrix was generated, giving the numbers of:

- True positives (t_p), correctly classified members of the class;
- True negatives (t_n), correctly classified non-members;
- False positives (f_p), non-members misclassified as members;
- False negatives (f_n), members misclassified as non-members.

Since each classifier was run separately against each of the 10 WADA classes, a false positive could arise in two different ways. One is that a molecule from an incorrect class is predicted as positive, for instance a member of S1 being labelled as a member of P2. The other is that an explicitly allowed substance is predicted as a member of the WADA class under test. A given test molecule could be classified by our methods as belonging to any combination of the 10 classes (or none). The "correct" labels of our 5235 molecules are, however, unique with each molecule being assigned membership of either zero or one WADA class.

Using the numbers of true and false positives and negatives, we calculated a version of the Matthews Correlation Coefficient with a slight modification, which we introduced in recent work [5]:

$$MCC^* = \frac{t_p t_n - f_p f_n}{MAX[1, \sqrt{(t_p + f_p)(t_p + f_n)(t_n + f_p)(t_n + f_n)}]} .$$

The modification involves the MAX function in the denominator and ensures that MCC^* is defined even if one of the four sums inside the square root is zero, a situation which may occur if no positives are predicted (and thus $MCC^* = 0$). Baldi *et al.* [14] have shown that the limiting value of the unmodified MCC as ($t_p + f_p$) tends to zero is, as expected, zero. This may be considered by some a more mathematically elegant way of ensuring that the coefficient is defined; nonetheless our introduction of MCC^* provides a pragmatic solution. The possible range of MCC^* values is from -1 (perfect anticorrelation), through 0 (random performance) to +1 (perfect correlation).

3 Results and Discussion

The Random Forest results are illustrated in Fig. 1. The levels of performance obtained with CDK and Unity are almost identical, both overall and across the individual classes. Unity does a little better on S1, but conversely CDK is superior in classifying S9. Comparison of the first two lines of Table 1 shows that the overall MCC^*s, with t_p, t_n, f_p and f_n aggregated over the ten classes, of the two Random Forest classifiers are virtually identical (MCC^* is 0.8143 for Unity and 0.8136 for CDK). The principal purpose of Unity's inclusion here is comparison with the new results for the freely available CDK fingerprint; the results for Unity are naturally very close to those we obtained in previous work [5] on an extremely similar dataset. In that work, Unity was shown to perform better than four other fingerprint definitions in classifying prohibited substances.

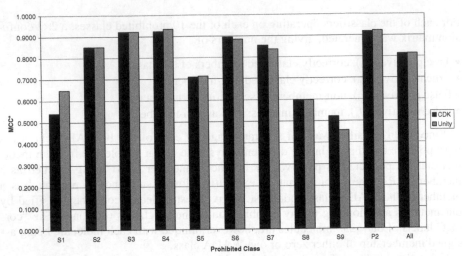

Fig. 1. MCC^* values obtained for each prohibited class using Random Forest classifiers based on the CDK (left hand side of each pair of bars) and Unity (right) fingerprints

The Random Forest classifiers have the very useful property of predicting very few false positives. For Unity, only 147 false positives are predicted; this amounts to only 0.3% of the individual class assignments that ought correctly to be negatives (0.35% for CDK). Even considering that there are ten possible banned classes that a molecule could be assigned to, these figures suggest that the overall probability of an inactive molecule being wrongly classified as a positive by this Random Forest classifier is approximately 3% for Unity (approximately 3.5% for CDK). The false positive rate would be further reduced by combining the chemoinformatics approach with suitable

Table 1. Performance of the classifiers aggregated across all ten prohibited classes

FP	Method	t_p	t_n	f_p	f_n	MCC^*
CDK	RF	3493	47318	164	1375	0.8136
Unity	RF	3482	47335	147	1386	0.8143
CDK	1NN	4091	46763	719	777	0.8297
Unity	1NN	4124	46766	716	744	0.8342
CDK	3NN	3813	46792	690	1055	0.7962
Unity	3NN	3833	46808	674	1035	0.8005
CDK	5NN	3592	46799	683	1276	0.7673
Unity	5NN	3605	46792	690	1263	0.7683
CDK	10NN	3098	46876	606	1770	0.7063
Unity	10NN	3193	46783	699	1675	0.7098
CDK	20NN	2665	46972	510	2203	0.6530
Unity	20NN	2688	46895	587	2180	0.6474

assays [3]. Minimising false positives is important for legal reasons, and for the credibility and integrity of the anti-doping process; wrongful disqualification of athletes is to be avoided so far as possible.

Fig. 2 shows the performance of the kNN classifiers with $k = 1$, which we shall call 1NN classifiers (those with $k = 3$ are called 3NN classifiers *etc.*). The performance of CDK is very similar to that of Unity, except that it fares less well on class S1. The overall MCC^* values, shown in Table 1, are very similar for 1NN and Random Forest classifiers. In fact, 1NN achieves a slightly higher value than Random Forest in each case. Unity does very marginally better than CDK. An important difference is that, despite the very similar MCC^* values, the 1NN classifiers predict many more false positives, but fewer false negatives, than Random Forest (Table 1). As a consequence, 1NN gives a higher recall but lower precision for positives. This is true for both CDK and Unity fingerprints.

This illustrates the point that kNN generates models which are local in nature, with the class membership of a test molecule being predicted based on a very small number of its neighbours. This is especially true for the 1NN models. We believe that this makes the kNN method especially suitable for identifying members of those classes which correspond to several different clusters in chemical space. This is likely to occur when interaction with any one of a plurality of receptors can give rise to the specified bioactivity.

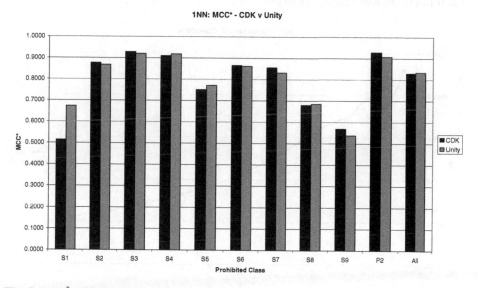

Fig. 2. MCC^* values obtained for each prohibited class using 1NN classifiers based on the CDK (left hand side of each pair of bars) and Unity (right) fingerprints

The four classifiers generated from Unity and CDK, Random Forest and 1NN (illustrated in Fig. 1 and Fig. 2) are in excellent agreement about the relative degrees of difficulty of predicting the ten prohibited classes. The six independent correlation coefficients between their sets of class-specific MCC^* values are all in the range r = 0.9154 (Unity-RF *vs* CDK-1NN) to r = 0.9906 (Unity-RF *vs* Unity-1NN). Although the smallest classes, S1 and S9, are amongst the hardest to predict, there is only a weak relationship between class size and MCC^*. The overall consensus ranking of the classes, in decreasing order of prediction quality, is:

$$S3 \approx S4 \approx P2 > S2 \approx S6 \approx S7 > S5 > S1 \approx S8 \approx S9.$$

We have also evaluated (Fig. 3 and Table 1) the performance of the kNN classifiers for higher values of k, for both the CDK and Unity fingerprints. There are two salient features of these results. Firstly, the MCC^* values tend to deteriorate for higher values of k. Secondly, the differences in performance between the two fingerprints are tiny. The fall-off with increasing k reinforces the local nature of the successful kNN models. For these data, at least, inclusion of additional neighbours generally reduces the MCC^* obtained. This indicates that the potential benefit of having information from more molecules is outweighed by the fact that these extra molecules are further away from that being classified. Fig. 3 contains some information additional to that in Table 1, in particular the inclusion of $k = 2$, $k = 4$ and $k = 15$. The slight recovery between $k = 2$ and $k = 3$ may be related to the random resolution of ties in the $k = 2$ case. This mirrors the observation, in a rather different field, by Lam and Suen [15] that augmenting an odd number of classifiers by an additional one can have a deleterious effect on overall prediction quality. Having an odd number of voters for a binary classification problem is an obvious way of avoiding problems with tied votes. This may be important when the number of voters is small, which is the case for our kNN, but not for our Random Forest, classifiers.

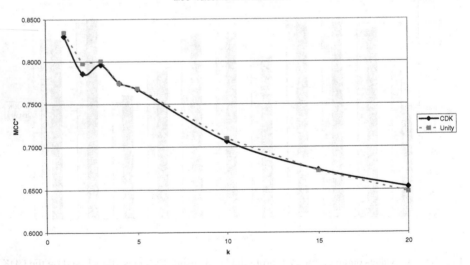

Fig. 3. MCC^* values obtained by kNN classifiers as a function of k for CDK (solid line) and Unity (broken line) fingerprints; based on results aggregated over all ten classes

4   Conclusions

We have successfully categorised molecules into WADA prohibited classes using both Random Forest and k-Nearest Neighbours algorithms, with Matthews Correlation

Coefficients above 0.8. In addition, we have shown that the freely available CDK fingerprint performs almost exactly as well as Unity 2D, which we previously demonstrated to be the best of five commercial fingerprints for this purpose. Although the 1NN algorithm (kNN with $k = 1$) gives the slightly higher MCC^*, the Random Forest classifier produces fewer false positives. Our best Random Forest models have a false positive rate, aggregated over all classes, of around 3%. The relative prediction accuracies of the different prohibited classes are very similar for the four different classifiers comprising Random Forest and 1NN algorithms with Unity and CDK fingerprints.

We find that 1NN is clearly the best kNN model for both fingerprints. The use of 2NN models is problematic due to the occurrence of tied votes, which are then resolved randomly. We favour the use of odd numbers of votes in classification problems of this kind. We also argue that the highly local nature of our 1NN models makes them particularly suitable for assigning molecules to classes of prohibited substances which comprise more than one cluster in chemical space.

These results suggest that it will be possible to create chemoinformatics-based classifiers, using freely available software, to determine whether novel molecules should be assigned to WADA prohibited classes. This will be especially powerful in combination with complementary experimental methods. Such tools will aid the fight against drug abuse in sport, while protecting competitors against unjustified sanctions.

Acknowledgements

We thank Unilever plc and the EPSRC for funding.

References

1. World Anti-Doping Agency (WADA), Stock Exchange Tower, 800 Place Victoria, (Suite 1700), P.O. Box 120, Montreal, Quebec, H4Z 1B7, Canada; http://www.wada-ama.org/
2. Handelsman, D.J., Designer Androgens in Sport: When too Much is Never Enough. Sci. STKE, Issue **244** (2004) pe41
3. Death, A.K., McGrath, K.C.Y., Kazlauskas, R., Handelsman, D.J., Tetrahydrogestrinone is a Potent Androgen and Progestin. J. Clin. Endocrinol. Metab. **89** (2004) 2498-2500
4. Kontaxakis, S.G., Christodoulou, M.A., A Neural Network System for Doping Detection in Athletes. Proceedings 4th International Conference on Technology and Automation, Thessaloniki, Greece, October 2002
5. Cannon, E.O., Bender, A., Palmer, D.S., Mitchell, J.B.O., Chemoinformatics-based Classification of Prohibited Substances Employed for Doping in Sport. J. Chem. Inf. Model., **submitted**
6. http://cdk.sourceforge.net/api/
7. Steinbeck, C., Han, Y., Kuhn, S., Horlacher, O., Luttmann, E., Willighagen, E., The Chemistry Development Kit (CDK): An Open-Source Java Library for Chemo- and Bioinformatics. J. Chem. Inf. Comput. Sci., **43** (2003) 493-500
8. Tripos Inc., 1699 South Hanley Road, St. Louis, MO 63144-2319, USA; http://www.tripos.com
9. Elsevier MDL, 2440 Camino Ramon, San Ramon, CA 94583, USA; http://www.mdli.com
10. Daylight Chemical Information Systems, Inc. 120 Vantis - Suite 550 - Aliso Viejo, CA 92656, USA; http://www.daylight.com/

11. Wild, D., Blankley, C.J., Comparison of 2D Fingerprint Types and Hierarchy Level Selection Methods for Structural Grouping Using Ward's Clustering. J. Chem. Inf. Comput. Sci., **40** (2000) 155-162

12. R Development Core Team (2005). R: A Language and Environment for Statistical Computing. R Foundation for Statistical Computing, Vienna, Austria; ISBN 3-900051-07-0; http://www.R-project.org.

13. Breiman, L., Random Forests. Machine Learning, **45** (2001) 5-32

14. Baldi, P, Brunak, S., Chauvin, Y., Andersen, C.A.F., Nielsen, H., Assessing the Accuracy of Prediction Algorithms for Classification: An Overview. Bioinformatics **16** (2000) 412-424

15. Lam, L., Suen, C.Y., Application of Majority Voting to Pattern Recognition: An Analysis of its Behavior and Performance. IEEE Trans Systems, Man and Cybernetics **27** (1997) 553-567

A Point-Matching Based Algorithm for 3D Surface Alignment of Drug-Sized Molecules

Daniel Baum and Hans-Christian Hege

Zuse Institute Berlin (ZIB), Germany
{baum, hege}@zib.de

Abstract. Molecular shapes play an important role in molecular interactions, e.g., between a protein and a ligand. The 'outer' shape of a
molecule can be approximated by its solvent excluded surface (SES). In
this article we present a new approach to molecular surface alignment
which is capable of identifying partial similarities. The approach utilizes
an iterative point matching scheme which is applied to the points representing the SES. Our algorithm belongs to the multi-start methods. We
first generate a number of initial alignments that are locally optimized
by an iterative surface point matching algorithm which tries to maximize
the number of matched points while minimizing the distance between the
matched points. The algorithm identifies similar surface regions given by
the matched surface points. This makes it well suited for multiple alignment of molecular surfaces. The subalgorithm proposed for distributing
points uniformly across a surface might be of general interest for the
comparison of molecular surfaces.

1 Introduction

Molecular alignment is a widely used tool in pharmaceutical drug design. An
alignment of molecular structures allows one to identify commonalities and differences among the molecular structures. Molecular alignment is thus a prerequisite
for pharmacophore identification using Structure Activity Relationship (SAR)
approaches.

There exist many molecular alignment algorithms (see, e.g., [1]). Most of
these algorithms consider the molecular 'skeleton'. But there also exist approaches considering the 'outer' shape of the molecule, i.e., its molecular surface,
e.g. [2,3,4,5,6]. Ritchie et al. [2] represent the molecular surface using spherical
harmonics. While this approach is fast, it is restricted to globular molecules and
is not suitable for partial surface matching. Goldman et al. [3] use quadratic
shape descriptors. Single points on the surface together with the principal directions of curvature are used to align molecular surfaces. Cosgrove et al. [4] and
Hofbauer [6] identify circular regions of approximately constant curvature into
which they position points. They then apply clique detection on those points
to align the molecular surfaces. Exner et al. [7] patchify the molecular surface
into overlapping surface patches of similar properties employing fuzzy sets. This
patchification is later used in a docking scheme [5] whereby they utilize geometric

M.R. Berthold, R. Glen, and I. Fischer (Eds.): CompLife 2006, LNBI 4216, pp. 183–193, 2006.

hashing for the identification of complementary surface regions. Their approach could also be applied to the alignment of molecular surface regions.

Molecular surfaces are also used in virtual screening for computing molecular descriptors, e.g. [8,9,10], which are used to screen data bases of potential drugs. Zauhar et al. [8] generate *shape signatures* from the molecular surface by propagating a random ray inside the molecular surface. Shape signatures represent a probability distribution of the measured distances within the molecular surface. This one-dimensional signature is used to compare the shapes of molecules. The approaches by Stiefl et al. [9] and Bender et al. [10] both use point distributions on the surfaces gained either directly from the molecular surface triangulation [10], or by approximating the molecular surface by points on a 3D uniform grid [9]. While in the first approach the points are not uniformly distributed across the molecular surface, the second approach positions the points not directly onto the surfaces and the point distribution is not transformation invariant.

In this paper we introduce a new transformation invariant algorithm for distributing points uniformly across a molecular surface. We also propose a new molecular alignment algorithm based on uniformly distributed points on the surface.

2   Methods

In this section we describe the whole process of aligning two molecular surfaces based on a partial surface matching. At the core of the algorithm lies the representation of the solvent excluded surface (SES) by point sets. We have developed a new method to uniformly distribute a rather small number of points on a surface, given as a triangle mesh. The alignment procedure is a multi-start approach. We first generate a number of initial transformations of the surface point sets based on the identification of small common substructures of the molecular skeletons. To each initial transformation we apply an iterative point matching scheme [11,12,13] which was modified to meet the needs of surface alignment, including the handling of physico-chemical properties.

2.1   Surface Point Representation

To efficiently compute molecular surface alignments based on an iterative point matching scheme, we need to uniformly distribute points on the molecular surface. We compute high-resolution triangle meshes of the SES based on the algorithm of Totrov et al. [14], implemented in the molecular visualization and analysis tool *amiraMol* [15]. These triangle meshes are partitioned into equally sized surface patches using the graph-partitioning software library METIS [16], version 4.0. Into each surface patch we position a single point, representing the patch. Although this initial point positioning scheme works rather well, we need to relax the points to obtain a uniform point distribution. We have developed an approximate Voronoi diagram computation method on 2-manifold triangle meshes based on the work of Deussen et al. [17], who use Voronoi diagrams to distribute points regularly in 2D. We also investigated the use of the remeshing approach proposed by Surazhsky et al. [18] which worked well for most surface meshes, but was unable to move close points in the presence of creases. Apart from this problem, the point distribution algorithm proposed here is much faster while not running into problems with creases.

Initial Point Positioning. The input to our point distribution algorithm is a 2-manifold triangle mesh $\mathcal{M}$. Let $V = \{v_1, \ldots, v_m\}$ be the set of vertices of $\mathcal{M}$ and $T = \{t_1, \ldots, t_n\}$ the set of triangles of $\mathcal{M}$. For $t \in T$ we denote by $v(t, i)$ the i'th vertex of t, $i \in \{1, 2, 3\}$. Furthermore, we denote by $\mathbf{x}(v)$ the coordinates of vertex $v \in V$, by $\mathbf{x}(t)$ the coordinates of the barycenter of $t \in T$, by $\mathbf{n}(t)$ the normal vector of t, and by $A(t)$ the area of t. Let $N(v)$ denote the triangle-neighbors of vertex $v \in V$, i.e. all triangles incident to v, and let $N(t)$ denote the triangle-neighbors of triangle $t \in T$, i.e. the triangles that share a common edge with t. Let $S(\mathcal{M})$ be a connected subpatch of $\mathcal{M}$, i.e. for each two triangles $t', t'' \in S(\mathcal{M}), t' \neq t''$, there exists a $k > 0$ and a path $(t^0, t^1, \ldots, t^{k-1}, t^k)$ with $t^0 = t'$, $t^k = t''$, such that $t_i \in N(t_{i-1}) \cap S(\mathcal{M}), \forall i = \{1, \ldots, k\}$.

From the surface area $A(T) := \sum_{t \in T} A(t)$ and the specified point density ρ we compute the number of points $N = A(T) \cdot \rho$ to be distributed on $\mathcal{M}$. For partitioning $\mathcal{M}$ into N patches, we use the METIS method `PartMeshDual`, which needs as input the connectivity of the triangle mesh $\mathcal{M}$, i.e. T and for each $t \in T$ its vertices $v(t, i), i \in \{1, 2, 3\}$, and N. `PartMeshDual` first converts the mesh-graph into its dual by representing each triangle t by a node which is connected to the nodes representing the triangles $t^i \in N(t), i \in \{1, 2, 3\}$. The `PartMeshDual` algorithm aims at minimizing the number of edge cuts of the dual graph while partitioning the dual graph into N equally sized connected node-partitions. The node-partitioning is then assigned to the original mesh $\mathcal{M}$, giving a triangle-partitioning $\mathcal{M} = \bigcup_{i=1}^{N} S_i(\mathcal{M})$, with $S_i(\mathcal{M}) \cap S_j(\mathcal{M}) = \emptyset, \forall i \neq j$.

To obtain the initial point positioning, for each subpatch $S_i(\mathcal{M})$ we compute its barycenter $\mathbf{x}(S_i(\mathcal{M})) := \sum_{t \in S_i(\mathcal{M})} A(t)\mathbf{x}(t) / \sum_{t \in S_i(\mathcal{M})} A(t)$ and project it onto $\mathcal{M}$ using its normal vector $\mathbf{n}(S_i(\mathcal{M})) := \sum_{t \in S_i(\mathcal{M})} A(t)\mathbf{n}(t) / \sum_{t \in S_i(\mathcal{M})} A(t)$ or the vector $-\mathbf{n}(S_i(\mathcal{M}))$, depending on whether the barycenter is inside or outside of $\mathcal{M}$, respectively. To do this efficiently, we use an octree in which we store all triangles of $\mathcal{M}$. This allows us to quickly identify the triangles that are intersected by a ray from $\mathbf{x}(S_i(\mathcal{M}))$ with direction $\mathbf{n}(S_i(\mathcal{M}))$ or $-\mathbf{n}(S_i(\mathcal{M}))$. We denote the points representing $S_i(\mathcal{M})$ by p_i and its coordinates by $\mathbf{x}(p_i)$.

Point Positioning Optimization. Let $P := \{p_1, \ldots, p_N\}$ be the set of points distributed on $\mathcal{M}$. Let $t(p), p \subset P$, be the triangle in which $\mathbf{x}(p)$ lies. The optimization algorithm works as follows. First, compute the Voronoi diagram for $X(P) := \{\mathbf{x}(p_1), \ldots, \mathbf{x}(p_N)\}$ on $\mathcal{M}$. Second, compute the center of each Voronoi cell, V_i, and move $\mathbf{x}(p_i)$ to the center of V_i. We shortly recall the definition of the Voronoi diagram. For some space Ω, the Voronoi diagram of a finite set of points $X := \{x_1, \ldots, x_n\} \subseteq \Omega$ partitions Ω into Voronoi cells V_i, such that for some distance metric d_Ω it is true that $d_\Omega(y, x_i) < d_\Omega(y, x_j), \forall y \in V_i, \forall i \neq j$. In our case, where $\Omega = \mathcal{M}$, the appropriate distance metric would be the geodesic distance. However, the exact computation of the Voronoi diagram on a triangle mesh is very expensive to compute. We have therefore developed the following approximation method.

Let $E = \{e_1, \ldots, e_s\}$ be the set of edges of $\mathcal{M}$, and for some edge $e \in E$, let $t(e, i)$, $i \in \{1, 2\}$, be the triangles adjacent to e. We now consider the dual mesh $\tilde{\mathcal{M}}$ of $\mathcal{M}$, where each triangle $t \in T$ is replaced by a vertex, $\tilde{v}(t)$, with coordinates $\mathbf{x}(\tilde{v}(t)) := \mathbf{x}(t)$, and each edge e is replaced by its dual edge $\tilde{e} = (\tilde{v}(t(e, 1)), \tilde{v}(t(e, 2)))$, connecting the dual vertices of $t(e, 1)$ and $t(e, 2)$. We denote the set of vertices of $\tilde{\mathcal{M}}$ by $\tilde{V}$, and the set of edges of $\tilde{\mathcal{M}}$ by $\tilde{E}$. For each $t \in T$ we now extend $\tilde{E}$ by edges from $\tilde{v}(t)$ to the dual vertices of the triangles adjacent to t's vertices $v(t, i)$, $i \in \{1, 2, 3\}$. This means, we define the extended dual edge set by $\tilde{E}' := \bigcup_{t \in T} \bigcup_{i \in \{1,2,3\}} \bigcup_{t' \in N(v(t,i))} (\tilde{v}(t), \tilde{v}(t'))$ (cf. Fig. 1).

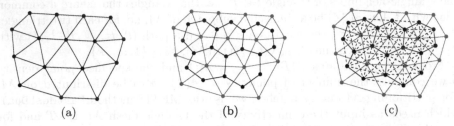

Fig. 1. (a) Mesh $\mathcal{M}$. (b) Mesh $\mathcal{M}$ (grey) and dual mesh $\tilde{\mathcal{M}}$ (black). (c) Same as (b) plus extended edges (dashed).

Instead of computing the exact Voronoi diagram for $X(P)$ on $\mathcal{M}$, we compute an approximation by collecting for each $p \in P$ all triangles whose barycenters are closer to $\mathbf{x}(t(p))$ than to any other $\mathbf{x}(t(p')), p' \in P$. The geodesic distances from $\mathbf{x}(t(p))$ to the barycenters of the triangles of $\mathcal{M}$ are approximated by the length of the shortest paths on the extended dual edge set $\tilde{E}'$, denoted by $d_{\tilde{E}'}(\cdot, \cdot)$, whereby we weight each edge by the Euclidean distance between the barycenters connected by the edge. The shortest paths on the weighted graph $G = (\tilde{V}, \tilde{E}')$ are computed using a modified version of Dijkstra's algorithm starting from all $\tilde{v}(t(p))$ simultaneously. For each vertex $\tilde{v} \in \tilde{V}$, we not only store the current shortest distance, but also the point $p \in P$ it is currently assigned to.

The Voronoi cell approximation leads to a new partitioning $\mathcal{M} = \bigcup_{i=1}^{N} S_i'(\mathcal{M})$, with $S_i'(\mathcal{M}) := \{t \mid d_{\tilde{E}'}(\tilde{v}(t(p_i)), \tilde{v}(t)) < d_{\tilde{E}'}(\tilde{v}(t(p_j)), \tilde{v}(t)), \forall j \neq i\}$. With this partitioning we proceed as with the initial partitioning, resulting in new coordinates $\mathbf{x}(p_i)$ for all points $p_i \in P$. We repeat this procedure until no points are moved anymore or until a maximum number of iterations has been performed.

The reason for computing the shortest paths on $G = (\tilde{V}, \tilde{E}')$ instead of on $G = (\tilde{V}, \tilde{E})$ lies in the much better approximation of the geodesic distances than when using $\tilde{E}$. In order to produce good results, $\mathcal{M}$ needs to have a rather fine resolution. We have obtained good results for a point density of 20 points per $\mathring{A}^2$ for $\mathcal{M}$ (cf. Fig. 2).

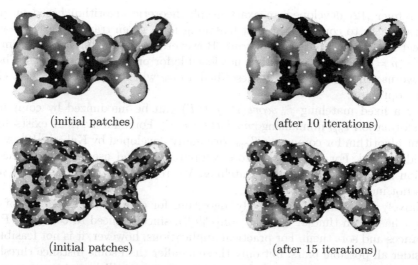

(initial patches)   (after 10 iterations)

(initial patches)   (after 15 iterations)

Fig. 2. Patchification and point distribution on the SES of Tipranavir for different resolutions and at different times of iteration. Top row: 139 points with an approximate point distance of 2.0 Å. Bottom row: 559 points with an approximate point distance of 1.0 Å.

2.2   Surface Point Matching

Let P and Q be two surface point sets with corresponding coordinate sets $X(P)$ and $X(Q)$ generated using the point distribution algorithm described in Sec. 2.1. We define a *matching* on P and Q as a bijection $f : P' \to Q'$, with $P' \subseteq P$ and $Q' \subseteq Q$. A matching f can be written as a set f^* of pairs from $P \times Q$ with $(p, q) \in f^* \Leftrightarrow f(p) = q$. The number of pairs $|f^*|$ will be referred to as the *size* of f. We define the root mean square distance (rmsd) of the two point sets with respect to a matching f and a rigid body transformation T as

$$rmsd(P, Q; f; T) := \sqrt{\frac{\sum_{(p,q) \in f^*} \|\mathbf{x}(p) - T(\mathbf{x}(q))\|^2}{|f^*|}}. \tag{1}$$

We further define the matching score w.r.t. T as

$$score(P, Q; f; T) := \frac{|f^*|}{\min(|P|, |Q|)} \cdot e^{-\alpha \cdot rmsd(P,Q;f;T)}, \tag{2}$$

with $\alpha \in \mathbb{R}^+$. The parameter α allows us to weight the importance of the rmsd. For a given transformation T we compute matchings of sizes 1 to $\min(|P|, |Q|)$ using a greedy strategy [12]. Set $f_0^* = \emptyset$, $P_0 = P$, and $Q_0 = Q$. From f_i^* we compute f_{i+1}^* by adding $(p', q') \in P_i \times Q_i$ with $d(p', q') <= d(p, q), \forall (p, q) \in P_i \times Q_i$. We then set $P_{i+1} = P_i \setminus \{p'\}$ and $Q_{i+1} = Q_i \setminus \{q'\}$. We compute the score for each of these matchings and select the one with the highest score, which we denote by f_{opt}.

Kirchner [12] developed an exact graph-theoretic algorithm based on augmented paths to compute the best matching for a given transformation T with respect to the objective function *score*. However, this algorithm has a runtime of $\mathcal{O}(n^3)$, $n = \min(|P|, |Q|)$, which is not feasible for practical applications. Therefore we use the greedy strategy described above which in most cases gets close to the optimum.

For a fixed matching f, $score(P, Q; f; T)$ can be maximized by computing transformation T_{opt} minimizing $rmsd(P, Q; f; T)$. Fortunately, there exists an efficient algorithm for computing T_{opt}, originally developed by Kabsch [19], which is widely used. For f_{opt} we therefore compute T_{opt} and use it as new transformation for computing a better matching. We repeat this procedure until *score* does not increase anymore.

Naively programmed the greedy algorithm for computing matchings of size 1 to n has a run time of $\mathcal{O}(|P||Q| \cdot \ln|P||Q|)$, since we need to compute $|P||Q|$ distances and sort them. For practical applications, however, it is not feasible to consider all possible pairs, but only those smaller than some distance threshold δ. This reduces the time needed for sorting, but we still have to compute the distances of all possible matching pairs. We can further improve on that by using a regular grid, such that we store for each grid cell all points of P having a smaller distance to the grid cell than δ. For each grid cell we can sort the points stored according to the distances to the grid cell. Then, instead of computing the distances of some point $q' \in Q$ to all points $p \in P$, in a first step we consider only those points that are closer to q' than some threshold $\tilde{\delta} <= \delta$. If q' is matched to a point p' with $d(p', q') =< \tilde{\delta}$, we do not have to compute the distances to $p'' \in P$ with $d(p'', q') > \tilde{\delta}$. If we run out of feasible matching pairs with distances smaller than $\tilde{\delta}$ we increase $\tilde{\delta}$ and search for new feasible pairs, but only for those points $q \in Q$ that have not yet been matched.

We can further reduce the number of surface points that can be matched by forbidding points to get matched if the angle between their normals is larger than some angle β. In general, we set β to $60°$. δ is generally set to 2.0Å.

Using all these improvements, the run time of the greedy algorithm can be drastically reduced. Run times for the algorithm are given in Sec. 3.2.

2.3   Handling of Physico-Chemical Properties

If we want to integrate physico-chemical properties into the proposed point matching scheme, we have two options.

The first option, proposed by Baum [13], uses weighted distances according to the properties of the matched points. While this keeps the number of points small, it does not always allow to distinguish between matchings in which the properties of the matched points are equal and in which they differ. Consider two matchings of equal size and $rmsd(P, Q; f; T)$ equal to 0. Since the weights scale the distances, if the distance is 0 there is nothing to be scaled. While this is an artificial scenario, it nevertheless reveals the problem with this approach.

The second option, which will be described here, uses additional points on the surface. These points will not be distributed across the whole surface, but will only be placed in regions where a certain property is present. The importance of properties will not be described by weights influencing the distance measure, but by the density of the point distribution. If we want to consider m properties, the point sets P and Q each decompose into m subsets, i.e., $P = \bigcup_{i=1}^{m} P_i$, with $P_i \cap P_j = \emptyset, \forall i \neq j$, and $Q = \bigcup_{i=1}^{m} Q_i$, with $Q_i \cap Q_j = \emptyset, \forall i \neq j$. We only allow two points p and q to match, if they are of the same property, i.e., $p \in P_i$ and $q \in Q_i$. Due to this we need to modify our scoring function. The new scoring function is now defined as

$$score^*(P, Q; f; T) := \frac{|f^*|}{\sum_{i=1}^{m} \min(|P_i|, |Q_i|)} \cdot e^{-\alpha \cdot rmsd(P,Q;f;T)}, \qquad (3)$$

The algorithm presented in Sec. 2.2 only needs to be modified in that the points of a feasible matching pair must be of the same property. We maintain a separate grid for each property to facilitate fast searching.

2.4   Generating Initial Transformations

For the generation of initial transformations used as input to the surface point matching algorithm described in Sec. 2.2, we apply a heuristic, which is based on the observation that for a good surface matching at least a few atoms of both molecules should be close to each other. We therefore use the same method as in [13], which identifies small common substructures of the molecules and applies least-squares fitting to these substructures. In all tests we performed, this heuristic yielded good results. Alternatively, we could use clique detection on the surface points together with shape indices [3] to reduce the number of edges in the correspondence graph.

3   Results

Times given in this paper refer to a 3GHz Intel Xeon processor.

3.1   Surface Point Representation

As Fig. 3 shows, optimization of the point distribution converges very quickly. The largest improvement is gained in the first step and after about 10 iterations, only a small further improvement might be gained. The mean distances and standard deviations have been averaged over 150 conformers of Tipranavir for a point distance of 2Å (a), and over 63 conformers of Amprenavir for a point distance of 1Å (b). The improvement can be visually inspected in Fig. 2. The algorithm took about half a second to optimize the point distribution on the molecular surface of Tipranavir with a point distance of 2Å and 10 iterations. For a point distance of 1Å it took slightly more than 1 second.

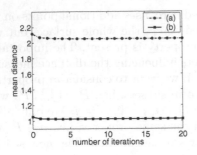 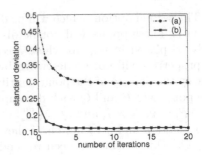

Fig. 3. Optimization of point distribution. Left: Evolution of mean geodesic distances. Right: Evolution of standard deviation of geodesic distances. (a) For a point distance of 2.0 Å. (b) For a point distance of 1.0 Å.

3.2 Surface Point Matching

In this section we present the results of the new surface alignment approach applied to two data sets. For both data sets we used the active conformers extracted from crystallized complexes gained from the PDB [20]. The active conformers were parametrized with the Merck Molecular Force Field (MMFF) [21]. The parametrization was needed for assigning donor, acceptor, aromatic, and hydrophobic properties to the atoms.

For each conformer we computed a point representation of the molecular surface containing so called *geometry* points, *donor* points and *acceptor* points. The geometry points were equally distributed across the whole molecular surface, representing the geometry of the surface, hence they are called geometry points. Donor and acceptors points were distributed in those surface areas assigned to atoms with donor and acceptor properties, respectively. All points were distributed with a geodesic distance to its neighbors of 2Å.

We computed pairwise alignments for each conformer of each set with all other conformers of the same set. We generated the initial transformations by identifying small common substructures of atoms, which were used for the pre-alignment. For each initial transformation we applied the algorithm described in Sec. 2.2. We computed the score (see Eq. 3) of each matching and sorted the matchings according to their scores. For each pairwise comparison we selected the first matching with an rmsd of the transformed atom positions to the experimental atom positions smaller than 1.0Å. If such a matching did not exist, we took the next matching with rmsd close to 1.0Å.

Thermolysin Inhibitors. As first data set we used Thermolysin inhibitors which were also used to evaluate the surface alignment approaches by Cosgrove et al. [4] and Hofbauer [6]. We extracted seven inhibitors from the crystallized complexes (PDB entries) 1TLP, 1TMN, 3TMN, 4TMN, 5TLN, 5TMN, and 6TMN.

For computing the initial transformations we used common substructures (subsets of atoms) of size at least 4, resulting in up to a thousand initial transformations for each pair of molecules. The results of the pairwise surface alignments are shown in Table 1.

Table 1. Pairwise surface alignment of Thermolysin inhibitors with a surface point distance of 2 Å. The times in the second column give the overall time needed for comparing the molecule in the first column with the other molecules. In the other columns, the first number gives the rmsd to the position within the complex, the second number gives the ranking due to the *score** value.

		5TLN	1TLP	1TMN	3TMN	4TMN	5TMN	6TMN
5TLN	6.4s	−/ −	0.76/ 16	1.01/ 2	0.87/ 8	0.99/ 4	0.95/ 1	0.86/ 9
1TLP	19.7s	0.74/ 16	−/ −	0.72/ 1	0.64/ 1	0.97/ 10	0.91/ 1	1.01/ 1
1TMN	15.8s	0.91/ 1	0.74/ 1	−/ −	0.86/ 1	0.58/ 1	0.96/ 9	0.59/ 1
3TMN	5.3s	0.86/ 8	0.87/ 1	0.71/ 2	−/ −	1.18/ 53	0.92/ 2	0.67/ 3
4TMN	21.8s	0.72/ 1	0.82/ 2	0.51/ 1	0.50/ 53	−/ −	0.46/ 3	0.5/ 11
5TMN	22.9s	0.61/ 1	0.80/ 1	0.88/ 5	0.95/ 1	0.49/ 3	−/ −	0.39/ 1
6TMN	24.9s	0.66/ 1	0.86/ 1	0.63/ 1	0.62/ 3	0.56/ 11	0.38/ 1	−/ −

Table 2. Pairwise surface alignment of HIV-protease inhibitors with a surface point distance of 2 Å. See Table 1 for further explanations.

		APV	IDV	LPV	NFV	RTV	SQV	TPV
APV	4.3s	−/ −	0.76/ 9	0.90/ 1	0.67/ 1	0.95/ 2	0.84/ 2	1.30/ 14
IDV	13.1s	0.99/ 1	−/ −	0.85/ 1	0.91/ 1	0.76/ 1	0.77/ 2	1.08/ 14
LPV	14.9s	0.36/ 2	0.81/ 1	−/ −	0.76/ 5	0.92/ 1	1.00/ 1	0.95/ 18
NFV	19.7s	0.18/ 4	0.91/ 1	0.89/ 5	−/ −	0.77/ 7	0.86/ 6	0.55/ 7
RTV	26.5s	0.88/ 4	0.69/ 5	0.88/ 1	0.76/ 24	−/ −	1.35/ 6	0.68/ 8
SQV	10.5s	0.87/ 1	0.75/ 2	0.93/ 2	0.98/ 1	0.89/ 7	−/ −	0.73/ 2
TPV	9.4s	1.63/ 18	1.24/ 18	1.95/ 18	0.54/ 7	0.73/ 8	0.95/ 9	−/ −

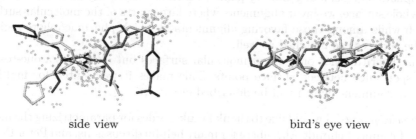

side view   bird's eye view

Fig. 4. Best alignment of Amprenavir (light grey) and Tipranavir (dark grey). The spheres denote the matched surface points. Where the points appear denser, the molecules have similar hydrogen bonding properties.

HIV Protease Inhibitors. As second data set we used HIV-protease inhibitors: Amprenavir (APV), Indinavir (IDV), Lopinavir (LPV), Nelfinavir (NFV), Ritonavir (RTV), Saquinavir (SQV), and Tipranavir (TPV). We extracted the inhibitors from crystallized complexes and compared the alignments with the experimental positions in the following complexes (PDB entries): 1HPV for APV; 1C6Y, 1HSG, 1HSH, 1K6C, 1K6P, 1K6T, 1K6V, 1SDT, 1SDU, 1SDV, and 1SGU for IDV; 1MUI and 1RV7 for LPV; 1OHR for NFV; 1HXW, 1N49, and 1SH9 for RTV; 1C6X, 1C6Y, 1C6Z, 1C70, 1FB7 and 1MTB for SQV; 1D4Y for TPV.

For computing the initial transformations we used common substructures of size at least 5, resulting in up to a few hundred initial transformations for each pair of molecules. The results of the pairwise surface alignments are shown in Table 2. An alignment of Amprenavir and Tipranavir is shown in Fig. 4.

4   Conclusion

The contribution of this paper is twofold. First, we have presented a new algorithm for uniformly distributing points on molecular surfaces. It applies a simple optimization scheme based on the approximation of Voronoi diagrams on triangle meshes. The point distribution algorithm works very fast and could be of interest for other algorithms beside the one presented in this paper, such as MaP [9] and MOLPRINT(3D) [10]. The second contribution is a new molecular surface alignment algorithm based on uniformly distributed surface points. It allows to easily integrate several molecular properties, such as atom type (donor, acceptor, aromatic, hydrophobic), charge, etc.

As was shown in Sec. 3.2, the algorithm is very well capable of producing feasible alignments. It is faster than other surface algorithms, such as those proposed by Cosgrove et al. [4] and Hofbauer [6]. The run times of both algorithms are much longer, both for the preprocessing step which is in the order of one minute and the alignment which might take several minutes. For our algorithm, the preprocessing time is in the order of a few seconds and the alignment only takes several seconds, depending on the size of the molecule, of course.

By taking into account pharmacophore points, which are positioned in pharmacophore regions, and geometry points, which are uniformly distributed across the whole surface, we favor alignments where large parts of the molecular surface match while among those favoring alignments in which pharmacophore regions on the surfaces are matched as well.

Our algorithm not only aligns molecular surfaces but also gives a one-to-one correspondence between surface points. This makes it well suited for multiple surface alignment, which will be described elsewhere.

Acknowledgments. I would like to thank Frank Cordes for parametrizing the molecules, Johannes Schmidt-Ehrenberg for many helpful discussions, and Peter Deuflhard for his continuous support. The work has been supported by the German Federal Ministry of education and research (grant no. 031U109A/031U209A, Berlin Center for Genome Based Bioinformatics).

References

1. Lemmen, C., Lengauer, T.: Computational methods for the structural alignment of molecules. J. Comput. Aid. Mol. Des. **14** (2000) 215–232
2. Ritchie, D.W., Kemp, G.J.L.: Fast computation, rotation, and comparison of low resolution spherical harmonic molecular surfaces. J. Comp. Chem. **20**(4) (1999) 383–395

3. Goldman, B.B., Wipke, W.T.: Quadratic shape descriptors. 1. rapid superposition of dissimilar molecules using geometrically invariant surface descriptors. J. Chem. Inf. Comp. Sci. **40**(3) (2000) 644–658

4. Cosgrove, D., Bayada, D.M., Johnson, A.P.: A novel method of aligning molecules by local surface shape similarity. J. Comput. Aid. Mol. Des. **14** (2000) 573–591

5. Exner, T.E., Keil, M., Brickmann, J.: Pattern recognition strategies for molecular surfaces. II. Surface complementarity. J. Comp. Chem. **23** (2002) 1188–1197

6. Hofbauer, C.: Molecular Surface Comparison. A Versatile Drug Discovery Tool. PhD thesis, Technische Universität Wien (2004)

7. Exner, T.E., Keil, M., Brickmann, J.: Pattern recognition strategies for molecular surfaces. I. Pattern generation using fuzzy set theory. J. Comp. Chem. **23** (2002) 1176–1187

8. Zauhar, R.J., Moyna, G., Tian, L., Li, Z., Welsh, W.J.: Shape signatures: a new approach to computer-aided ligand- and receptor-based drug design. J. Med. Chem. **46**(26) (2003) 5674–5690

9. Stiefl, N., Baumann, K.: Mapping property distributions of molecular surfaces: algorithm and evaluation of a novel 3D quantitative structure-activity relationship technique. J. Med. Chem. **46**(8) (2003) 1390–1407

10. Bender, A., Mussa, H.Y., Gill, G.S., Glen, R.C.: Molecular surface point environments for virtual screening and the elucidation of binding patterns (MOLPRINT 3D). J. Med. Chem. **47** (2004) 6569–6583

11. Akutsu, T.: Protein structure alignment using a graph matching technique. Genome Informatics **6** (1995) 1–8

12. Kirchner, S.: Ein Approximationsalgorithmus zur Berechnung der Ähnlichkeit dreidimensionaler Punktmengen. Diploma Thesis, Department of Computer Science, Humboldt University Berlin (2003)

13. Baum, D.: Multiple Semi-flexible 3D Superposition of Drug-sized Molecules. In: Computational Life Sciences: First International Symposium, CompLife 2005. Volume 3695 of Lecture Notes on Computer Science., Konstanz, Germany, Springer (2005) 198–207

14. Totrov, M., Abagyan, R.: The contour-buildup algorithm to calculate the analytical molecular surface. J. Struct. Biol. **116** (1995) 138–143

15. AmiraMol, AmiraDeconv - Extensions for Amira 3.1. Zuse Institute Berlin (ZIB) and Mercury Computer Systems - TGS Group, http://amira.zib.de/Amira31-MolDeconv-manual.pdf (2003)

16. Karypis, G., Kumar, V.: METIS, a Software Package for Partitioning Unstructured Graphs and Computing Fill-Reduced Orderings of Sparse Matrices. University of Minnesota, Department of Computer Science (1998)

17. Deussen, O., Hiller, S., van Overveld, C.W.A.M., Strothotte, T.: Floating points: A method for computing stipple drawings. Comput. Graph. Forum **19**(3) (2000)

18. Surazhsky, V., Gotsman, C.: Explicit surface remeshing. In: Symposium on Geometry Processing. (2003) 20–30

19. Kabsch, W.: A discussion of the solution for the best rotation to relate two sets of vectors. Acta Crystallographica A **34** (1978) 827–828

20. Protein Data Bank (PBD). (http://www.rcsb.org/pdb/)

21. Halgren, T.A.: Merck molecular force field. I-V. J. Comp. Chem. **17**(5&6) (1996) 490–641

Adaptive Approach for Modelling Variability in Pharmacokinetics

Andrea Y. Weiße[1,2], Illia Horenko[2], and Wilhelm Huisinga[1,2]

[1] DFG Research Center MATHEON, and Department of Mathematics
[2] Informatics, Free University Berlin, Arnimallee 6, 14195 Berlin, Germany

Abstract. We present an improved adaptive approach for studying systems of ODEs affected by parameter variability and state space uncertainty. Our approach is based on a reformulation of the ODE problem as a transport problem of a probability density describing the evolution of the ensemble of systems in time. The resulting multidimensional problem is solved by representing the probability density w.r.t. an adaptively chosen Galerkin ansatz space of Gaussian densities. Due to our improvements in adaptivity control, we substantially improved the overall performance of the original algorithm and moreover inherited to the numerical scheme the theoretical property that the number of Gaussian distributions remains constant for linear ODEs. We illustrate the approach in application to dynamical systems describing the pharmacokinetics of drugs and xenobiotics, where variability in physiological parameters is important to be considered.

1 Introduction

The medical benefits of a drug depend not only on its biological effect at the target protein, but also on its "life cycle" within the organism - from its absorption into the blood, distribution to tissue and its eventual breakdown or excretion by the liver and kidneys. Pharmacokinetics is the study of the drug-organism interaction, in particular the investigation of absorption, distribution, metabolism, and excretion (ADME) processes [9]. Studying ADME profiles is widely used in drug discovery to understand the properties necessary to convert leads into good medicaments [16,3]. Physiologically based pharmacokinetic (PBPK) models aim at describing pharmacokinetic processes on a mechanistic basis. They model the body as a network of organ or tissue compartments that are interconnected by blood flow (see Fig. 1). From a mathematical point of view, a PBPK model comprises a system of coupled ordinary differential equations (ODEs). These equations involve physiological and physicochemical parameters, each of which is typically affected with uncertainty and some degree of variability due to inter- and intra-individual variations.

There exist theoretical and numerical tools for investigating ODEs with initial values and/or parameter uncertainty distributions. One class of approaches is represented by Monte Carlo methods based on a sampling of the initial distribution and subsequent solution of the underlying ODE for each of the sampling

M.R. Berthold, R. Glen, and I. Fischer (Eds.): CompLife 2006, LNBI 4216, pp. 194–204, 2006.
© Springer-Verlag Berlin Heidelberg 2006

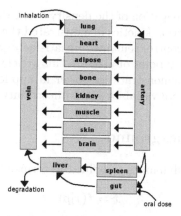

Fig. 1. A typical PBPK model. We distinguish between physiological and physicochemical parameters. Physiological parameters comprise organ volumes and blood flows; physicochemical parameters such as partition coefficients or solubility describe properties of the compound.

points (e.g. [15]). While this is the method of choice for problems with many parameters and degrees of freedom in order to avoid the "curse of dimensionality", the questions of numerical accuracy, reliability, and adaptivity still remain partially unclear. A second class of methods are sparse grid techniques [17,4] or particle methods [13,11]. In contrast to frequently used conventional grid methods, they both scale reasonably well for medium dimensional problems. Sparse grids work best for smooth *anisotropic* densities with the grid being aligned to the propagated objects. Adaptivity is used to generate an optimal sparse grid in order to minimize the approximation error beneath some predefined threshold. However, for *isotropic* problems, e.g., for classical Liouville or Fokker–Planck equations with Gaussian initial density, the adaptive sparse grid methods end up by full grids, i.e., they are practically not applicable for high–dimensional isotropic problems. A third class of approaches is known as the stochastic finite elements (SFEMs) approach [7,8,12]. This method represents the overall statistical response of the system by a linear combination of orthogonal basis functions. However, in the available form, this approach cannot be applied to higher dimensional problems with different time and length scales as it is typical for reaction kinetics and pharmacokinetic models.

In this article, we present a theoretical framework and an improved *adaptive* numerical approach for systems of ODEs affected by parameter variability and uncertainty distributions. The problem is reformulated in terms of the well-established Frobenius-Perron theory, involving the semigroup of Frobenius-Perron operators. In order to approximate the semi-group numerically, we adopt and substantially extend the adaptive Gaussian-based particle method TRAIL [6] that has originally been developed in the context of molecular dynamics [6] and recently be transferred to reaction systems [5]. The approach is based on two ingredients, (i) a time-dependent Galerkin ansatz space of Gaussian basis functions , and (ii) a propagation of the density w.r.t. the Galerkin ansatz space. First applications to reaction kinetics demonstrated the potential of the method, however the adaptivity control remained unsatisfactory from both the theoretical as well as the efficiency point of view. In this article we propose an improved adaptivity control and demonstrate its power in application to typical systems in pharmacokinetics, where variability and uncertainty play an important role.

2   Theoretical Background

Let us assume that the dynamics is defined in terms of some ODE

$$\dot{x} = f(x|p) \tag{1}$$

with some continuously differentiable right hand side $f(\cdot|p) : \mathbf{X} \to \mathbb{R}^d$ parameterized by $p \in \mathbf{Y} \subset \mathbb{R}^m$, the vector of fixed parameters. In the setting of pharmacokinetics, $\mathbf{X} \subset \mathbb{R}^d$ is the space of concentrations in d compartments, and p represents physiological and physicochemical parameters. Given the initial state $x(0) = x_0$, the solution of the initial value problem (1), $x(t|p) : \mathbb{R} \to \mathbf{X}$ describes the concentration-time behavior in d different tissues and can formally be written in terms of the flow $\Phi^t(\cdot|p)$

$$x(t|p) = \Phi^t(x_0|p) \tag{2}$$

that is known to be invertible. Now assume that the parameters are specified in terms of some statistical distribution in contrast to a fixed numerical value. Rather than solving (1) for some initial value x_0 and a single set of parameters, we are interested in capturing the effects of distributed parameters on the evolution of the dynamics, i.e, the initial value. Then, eq. (1) becomes an ODE with random parameters and possibly random initial conditions that can be interpreted as the evolution of an entire population. We can easily extend equation (1) to account for influences of distributed parameters by extending the state space to $\mathbf{Z} := \mathbf{X} \times \mathbf{Y}$ and setting $\dot{p}$ to zero

$$\begin{pmatrix} \dot{x} \\ \dot{p} \end{pmatrix} = F(x,p) = \begin{pmatrix} f(x|p) \\ 0 \end{pmatrix}, \tag{3}$$

since parameters are assumed to be constant in time.

Denote by $\mathcal{L}^2(dxdp)$ the space of square integrable functions. The semigroup of Frobenius-Perron operators $P_t : \mathcal{L}^2(dxdp) \to \mathcal{L}^2(dxdp)$ associated with (3) describes the evolution of a given density $u_0 \in \mathcal{L}^2(dxdp)$ in time according to $P_t u_0 = u_t$. Since the flow Φ^t is invertible and differentiable, P_t is explicitly defined by

$$P_t u_0(x,p) = u_0\left(\Phi^{-t}(x,p)\right) \cdot \left| \frac{d\Phi^{-t}(x,p)}{dxdp} \right| \tag{4}$$

where the last term denotes the determinant of the Jacobian of Φ^{-t} [10]. In broad terms, definition (4) can be interpreted as follows: the value of density u_t at (x,p)

is given by the value of u_0 at the pre-image of (x, p) corrected according to the dynamics (expanding or contracting directions). The infinitesimal generator $\mathcal{A}$ of the semigroup of Frobenius-Perron operators is defined by

$$\mathcal{A}u = -\operatorname{div}(F \cdot u) = -\sum_{i=1}^{n} \frac{\partial}{\partial z_i}(F \cdot u), \tag{5}$$

where $z \in \mathbf{Z}$ and $n = d + m$ is the dimension of $\mathbf{Z}$.

3   Adaptive Density Propagation

For the density propagation of Liouville type problems, Horenko and Weiser developed a multidimensional, adaptive particle method to describe the propagation of distributions in non-linear dynamical systems, called TRAIL (trapezoidal rule for adaptive integration of Liouville dynamics) [6]. The adaptive discretization scheme is based on a semi-discretization in time and subsequent approximation of the stationary spatial problem. The key idea is to approximate the distribution u w.r.t. a time-adapted Galerkin basis of Gaussian ansatz functions.

Semi-discretization in time. Consider a probability distribution u_0 characterizing variability and uncertainty of parameters and state variables. Then, at any later time $t > 0$ the distribution u_t is given by $u_t = P_t u_0$ involving the semi-group of Frobenius-Perron operators. The basic idea of temporal semi-discretization is to approximate $\{P_t\}_{t \geq 0}$ by a simpler and numerically treatable semi-group $\{R_t\}_{t \geq 0}$. Loosely speaking, since $P_t = \exp(t\mathcal{A})$, we define $R_t = r(t\mathcal{A})$ based on some possible rational approximation $r(\cdot)$ of the exponential function $\exp(\cdot)$. Relevant in our context are $R_t = (\operatorname{Id} - t\mathcal{A}/2)^{-1}(\operatorname{Id} + t\mathcal{A}/2)$ denoting the trapezoidal rule and $R_t = \operatorname{Id} + t\mathcal{A}$ denoting the explicit Euler scheme. Finally, the numerical scheme exploits the semi-group property to approximate P_t for some large $t > 0$ according to $P_t = P_{\tau_n} \circ \cdots \circ P_{\tau_n} \approx R_{\tau_n} \circ \cdots \circ R_{\tau_n}$ for some adaptively chosen sequence of time steps $t = \tau_n + \cdots + \tau_1$.

Spatial discretization of stationary problem. At the very beginning, the initial distribution u_0 is approximated by a finite sum of Gaussian distributions, i.e.,

$$u_0 = \sum_{j=1}^{N_{t_0}} \omega_j(t_0) B_j(\cdot\,; t_0) + \delta_{t_0} \tag{6}$$

such that $\|\delta_{t_0}\| < \text{TOL}$, and for $j = 1, \ldots, N_t$ and $t \geq 0$

$$B_j(z; t) := \exp\left\{ \frac{1}{2}(z - \mu_j(t))^T G_j(t)(z - \mu_j(t)) + a_j(t) \right\}. \tag{7}$$

The parameters $\mu_j(t)$, $G_j(t)$, and $a_j(t)$ with $j = 1, \ldots, N_t$ denote the corresponding means, inverses of the covariance matrices and normalization constants[1],

---

[1] The constants are chosen in such a way that $B_j(\cdot\,; t)$ is normalized to one, resulting in $a_j(t) = \ln(\det(2\pi\Sigma_j(t))^{-\frac{1}{2}})$.

respectively. For details, see [6,5]. The initial approximation also defines the Galerkin basis $\{B_j(\cdot; t_0) : j = 1, \ldots, N_{t_0}\}$ at time t_0.

In each step, the scheme comprises two steps: (i) adaptation of the Galerkin basis w.r.t. the underlying dynamics; (ii) optimal representation of the time-propagated density w.r.t. the adapted Galerkin basis. The propagation of the Galerkin basis is performed w.r.t. to the locally (around each mean μ_j) linearized dynamics guaranteeing that the Gaussian distributions remain Gaussian in time. The parameters evolve according to the ODEs:

$$\dot{a}_j = -\text{trace}(DF(\mu_j)) \tag{8}$$

$$\dot{\mu}_j = F(\mu_j) \tag{9}$$

$$\dot{G}_j = -DF(\mu_j)^T G_j - G_j\, DF(\mu_j), \tag{10}$$

where $DF(z)$ denotes the Jacobian of F at $z \in \mathbf{Z}$. The representation of the time-propagated density w.r.t. to the new Galerkin basis is realized via the trapezoidal rule. The new coefficients $\{\omega_j(t + \tau) : j = 1, \ldots, N_t\}$ are optimized according to

$$\|\delta_{t+\tau}\| = \left\|\left(1 - \frac{\tau}{2}\mathcal{A}\right) u(\cdot, t + \tau) - \left(1 + \frac{\tau}{2}\mathcal{A}\right) u(\cdot, t)\right\| = \min \tag{11}$$

with

$$u(\cdot, t + \tau) = \sum_{j=1}^{N_t} \omega_j(t + \tau) B_j(\cdot; t + \tau) \,.$$

This can efficiently be performed by approximating the involved norm by a Monte Carlo sampling, reformulating eq. (11) as a least squares problem to be solved by means of a qr-algorithm [6], and noting that the action of the infinitesimal generator $\mathcal{A}$ on a Gaussian basis function can be computed exactly at any given state $z \in \mathbf{Z}$.

Adaptivity in time and space. The crucial ingredient of the TRAIL scheme is the adaptive choice of the next time step and the adaptation of the Galerkin basis (increasing or decreasing the number of basis functions) to keep the local error below a user-defined local tolerance TOL. In the original TRAIL scheme, temporal adaptivity is realized by a step size control based on a comparison of the trapezoidal rule and the Euler scheme, while spatial adaptivity is realized by exploiting properties of the qr-algorithm in the optimization step combined with some "accuracy matching" (splitting of the local tolerance into some temporal and spatial local tolerance), for details, see [6].

4   Improved Adaptivity

It is well-known that for linear ODEs an initial Gaussian distribution stays Gaussian in time. In terms of the TRAIL scheme this means that the number of Gaussian basis functions should stay constant for linear ODEs. However, application to simple linear ODEs revealed that the number of Gaussians basis functions does in general not stay constant unless an a-priori unknown small maximal time step is introduced. Since the numerical effort of the TRAIL scheme scales cubically with the number of Gaussian basis functions, this result is unsatisfactory from both a numerical (adaptivity and efficiency) point of view as well as from a theoretical point of view. Rather one would like to design an adaptivity control that allows for efficiency and inherits the theoretical properties for linear ODEs to the numerical scheme. A thorough analysis of the performance for linear problems revealed that new ansatz functions are added due to a too coarse time discretization resulting in an overestimation of the spatial error. At the same time, conservation of (probability) mass is poor. On the other hand, bounding the time-step from above by some maximal time step (a-priori unknown and in general depending on the ODE and the initial distribution) enforced a constant number of ansatz functions, but slowed down the integration drastically.

The improved adaptivity control is based on the key idea to control the time-step not only based on a comparison of two different numerical schemes (in the case of TRAIL the trapezoidal rule and the Euler scheme), but also based on the spatial discretization error (estimator): For linear ODEs the time-step is rejected and subsequently decreased whenever the spatial error estimator exceeds the spatial tolerance. In addition, we replace the Euler method by a second order Runge-Kutta methods with $R_t = \mathrm{Id} + t\mathcal{A} + t^2\mathcal{A}^2/2$ for a more efficient performance. For non-linear ODEs, the same is done based on a (local) linearization of the ODE. Hence in broad terms, the time step is controlled in such a way that a change in the number of basis functions is only due to non-linear effects of the underlying dynamics. In the next section we demonstrate the improved adaptivity control in application to two typical pharamcokinetics models.

5   Numerical Examples

This section illustrates the numerical scheme in application to two model problems in pharmacokinetics. For a comparison of our method with the Monte Carlo and the finite element approach in application to a small model from reaction kinetics see [5].

The first model is a very simple empirical two-compartment model that is frequently used in population pharmacokinetics to analyze large data sets resulting from clinical studies. It comprises two compartments, a central and a peripheral one. Typically, the central compartment is thought of as blood compartment, while the peripheral compartment is empirically chosen. In contrast to physiologically based models this type of model is empirical. A compound is transferred from the central to the peripheral compartment with some transfer rate k_t, where it is eliminated with some elimination rate k_e. The resulting ODEs are:

$$\begin{pmatrix} \dot{C}_c \\ \dot{C}_p \end{pmatrix} = \begin{pmatrix} -k_t & 0 \\ k_t & -k_e \end{pmatrix} \begin{pmatrix} C_c \\ C_p \end{pmatrix}, \tag{12}$$

where C_c and C_p denote the concentrations of the compound in the central and peripheral compartment. The evolution of the system for initially distributed concentrations is depicted in Fig. 2. In the left column, fixed parameter values are chosen: $k_t = 0.5$ and $k_e = 0.6$, while in the right column parameter variability according to $k_t \sim \mathcal{N}(0.5, 0.01)$ and $k_e \sim \mathcal{N}(0.6, 0.01)$ is taken into account.

Though variances of k_t and k_e are comparatively small, we observe considerable effects on the joint distribution of central and peripheral concentrations. By extending the state space, the ODE has become non-linear. Still, the number of ansatz functions remains constant for the simulation time performed. This might indicate that non-linear effects have not been dominating. Applying the original TRAIL scheme without the presented improvements results in a large number of Gaussian ansatz functions (> 100 depending on the user prescribed tolerance) for both the linear model (only concentrations) as well as the non-linear model (concentration and parameters).

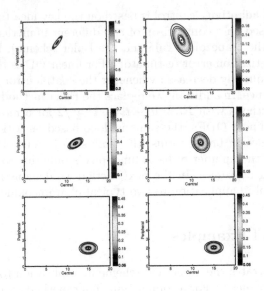

Fig. 2. Simulations of system (12) with initial concentrations $C_c(0) \sim \mathcal{N}(15, 1)$ and $C_p(0) \sim \mathcal{N}(2, 0.1)$. The left column shows the evolution of the initial density *without* parameter variation, whereas in the right column the state space has been extended by k_t and k_e. In both cases joint distributions of central and peripheral concentrations are shown for time points $t = 0, 0.4, 1.2\,$h (bottom to top).

Next, we consider the evolution of styrene concentrations after inhalation in human body. We use a physiologically-based pharmacokinetic model that has been developed in the context of toxicological risk assessment [14,1,2]. The PBPK model comprises the organs/tissues liver, adipose, muscle, and vessel rich tissue that are interconnected by the blood flow. Based on the law of mass action a system of coupled ODEs describing the time-evolution of styrene concentrations in the above organs and tissues is established.

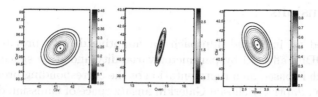

Fig. 3. Simulations of the ten dimensional extended styrene system. Two dimensional projections of the high dimensional joint distribution are shown for concentrations in liver & vessel rich tissue (left), venous blood & liver (center), and V_{max} with liver concentration (right) are shown.

For org $\in$ {liver, adipose, muscle, and vessel rich tissue} we get

$$V_{\mathrm{org}} \frac{\mathrm{d}C_{\mathrm{org}}}{\mathrm{d}t} = Q_{\mathrm{org}} \cdot (C_{art} - \frac{C_{\mathrm{org}}}{P_{\mathrm{org}}}), \qquad (13)$$

where V_{org}, Q_{org} and P_{org} denote the volume, blood flow and the so-called tissue partition coefficient, respectively. Saturable metabolization of styrene in liver is assumed to be represented by an additional non-linear Michaelis-Menten term in the liver compartment. The equations for venous and arterial blood are given by

$$V_{ven} \frac{\mathrm{d}C_{ven}}{\mathrm{d}t} = \sum_{i \in \text{ tissues}} Q_i \cdot \frac{C_i}{P_i} - C_{ven} \cdot Q_{tot}$$

$$V_{art} \frac{\mathrm{d}C_{art}}{\mathrm{d}t} = Q_{tot} \cdot (C_{ven} - C_{art}) + Q_{alv} \cdot (C_{inh} - \frac{C_{art}}{P_{air}}) ,$$

where C_{inh} denotes styrene concentration in inhaled air, Q_{alv} the alveolar flow, and Q_{tot} the total blood flow. In [2], it was found that the system is sensitive to parameter uncertainty w.r.t. V_{max} and K_m in liver tissue, the blood:air partition coefficient P_{air}, and the partition coefficient in adipose tissue P_{fat}. Hence, in addition to ODEs for the concentrations we account for uncertainty by extending the state space by these four parameters.

As a simulation output we obtain the evolution of a ten dimensional probability distribution. Projections of this distribution onto one dimensional subspaces representing a compartment (and not one of the four parameters) monitor the concentration distribution in the respective compartment. Projections onto two dimensions may reveal correlations of concentrations in different compartments or of concentrations with parameter values as shown in Fig. 3. The initially uncorrelated distributions develop correlations in time, both between different tissues (Fig. 3, left and center) and between liver concentration and V_{max} (Fig. 3, right). As we would expect, liver concentration is negatively correlated to V_{max}. Also we would expect a positive correlation for liver and venous concentrations, since both compartments are directly coupled by the liver blood flow. The positive correlation of vessel rich and liver tissue on the other hand is not obvious on first sight. Knowledge about the effect of variability and uncertainty is important information for risk assessment studies, as in [1].

6   Conclusions

We presented an improved approach for adaptive density propagation in the context of ODEs affected by parameter variability and state space uncertainty. Our approach is based on a representation of the corresponding probability density w.r.t. an adaptively chosen Galerkin ansatz space of Gaussian distributions. Due to our improvements in adaptivity control, the theoretical property that the number of Gaussian distribution stays constant for linear ODEs is now inherited to the numerical scheme. Since the numerical efforts scale with the third power of the number of Gaussian basis functions, we managed to substantially improve the overall performance. In addition the conservation of (probability) mass improved.

Most often, Monte Carlo (MC) approaches are applied to study dynamical systems with distributed parameters and states. The MC methods generate an ensemble of sampling points that approximate the statistical distribution. However, in contrast to molecular dynamics, where the underlying Hamiltonian structure implies conservation of phase volume and probability density along trajectories, this properties does rarely hold in reaction kinetics and pharmacokinetics. As a consequence, a single sampling point is of limited use and only in form of expectation values relevant information can be extracted. Moreover, the control of the approximation error still remains partially unclear. The adaptive approach to density propagation presented herein generates a continuous approximation of the density in time. Since the density is approximated in terms of Gaussian distributions, the approach is expected to be efficient whenever the underlying dynamics results in densities that are "sufficiently smooth", as it seems to be the case for pharmacokinetics problems. Due to the improved adaptivity control it might also become possible to selectively study the influence of non-linear effects on the overall dynamics by monitoring the dimension of the Galerkin basis, i.e., the number of basis functions, since only the non-linear part of the dynamics is able to increase or decrease the number of Gaussians. This would allow to extract a completely different and very interesting type of information and is currently under investigation.

The results in application to pharmacokinetic models demonstrate the advantages of the approach presented. As a result of the simulation studies, detailed information on the distribution in state space and, e.g, the correlation between different parameters is available. These are important data for toxicological risk assessments [1], or to study the variability of a drug exposure in an entire population (an information becoming more and more important).

Acknowledgment

A.Y.W. has been supported by the International Max Planck Research School for Computational Biology and Scientific Computing (IMPRS-CBSC), Berlin, and by the Konrad-Zuse-Zentrum für Informationstechnik Berlin (ZIB). I.H.

acknowledges financial support by SFB 450. W.H. acknowledges financial support by the DFG Research Center MATHEON "Mathematics for key technologies", Berlin.

References

1. K. Abraham, H. Mielke, W. Huisinga, and U. Gundert-Remy. Elevated internal exposure of children in simulated acute inhalation of volatile organic compounds: effects of concentration and duration. *Arch Toxicol*, 79(2):63–73, Feb 2005.
2. K. Abraham, H. Mielke, W. Huisinga, and U. Gundert-Remy. Internal exposure of children by simulated acute inhalation of volatile organic compounds: the influence of chemical properties on the child/adult concentration ratio. *Basic Clin Pharmacol Toxicol*, 96(3):242–243, Mar 2005.
3. A.P. Beresford, H.E. Selick, and Michael H. Tarbit. The emerging importance of predictive adme simulation in drug discovery. *DDT*, 7:109–116, 2002.
4. M. Griebel and G. W. Zumbusch. Adaptive sparse grids for hyperbolic conservation laws. In M. Fey and R. Jeltsch, editors, *Hyperbolic Problems: Theory, Numerics, Applications. 7th International Conference in Zürich, February 1998*, International Series of Numerical Mathematics 129, pages 411–422, Basel, 1999. Birkhäuser. http://wissrech.ins.uni-bonn.de/research/pub/zumbusch/hyp7.pdf.
5. I. Horenko, S. Lorenz, Ch. Schütte, and W. Huisinga. Adaptive approach for nonlinear sensitivity analysis of reaction kinetics. *J. Comp. Chem.*, 26(9):941–948, 2005.
6. Illia Horenko and Martin Weiser. Adaptive integration of molecular dynamics. *Journal of Computational Chemistry*, 24(15):1921–1929, 2003.
7. Andreas Keese. A review of recent developments in the numerical solution of stochastic partial differential equations (stochastic finite elements). Informationbericht 2003-6, Department of Computer Science, Technical University Braunschweig, Brunswick, Germany, Institute of Scientific Computing, TU Braunschweig, 2002.
8. M. Kleiber and T. D. Hien. *The stochastic finite element method, Basic perturbation technique and computer implementation.* J. Wiley and Sons, 1992.
9. Younggil Kwon. *Handbook of Essential Pharmacokinetics, Pharmacodynamics, and Drug metabolism for Industrial Scientists.* Kluwer Academic/Plemun Publishers, 2001.
10. A. Lasota and M. C. Mackey. *Chaos, Fractals, and Noise: Stochastic Aspects of Dynamics, 2nd Ed.* Springer-Verlag, Berlin, 1995.
11. Marc A. Schweitzer M. Griebel. *Meshfree Methods for Partial Differential Equations.* Lecture Notes in Computational Science and Engineering. Springer, Berlin, Heidelberg, New York, 2003.
12. Hermann G. Matthies and Marcus Meyer. Nonlinear galerkin methods for the model reduction of nonlinear dynamical systems. Informationsbericht 2002-3, Department of Computer Science, TU Braunschweig, Germany, March 2002.
13. H. Neunzert, A. Klar, and J. Struckmeier. Particle methods: Theory and applications. In *ICIAM 95: proceedings of the Third International Congress on Industrial and Applied Mathematics held in Hamburg, Germany*, 1995.

14. J.C. Ramsey and M.E. Andersen. A physiologically based description of the inhalation pharmacokinetics of styrene in rats and humans. *Toxicol. Appl. Pharmacol.*, 73:159–175, 1984.
15. D. Talay. Probabilistic numerical methods for partial differential equations: elements of analysis. In D. Talay and L. Tubaro, editors, *Probabilistic Models for Nonlinear Partial Differential Equations*, Lecture Notes in Mathematics 1627, pages 48–196. Springer, 1996.
16. H. van Waterbeemd and E. Gifford. Admet in silico modelling: towards prediction paradise? *Nature*, 2:192–204, 2003.
17. Christoph Zenger. Sparse grids. In Wolfgang Hackbusch, editor, *Parallel Algorithms for Partial Differential Equations*, volume 31 of *Notes on Numerical Fluid Mechanics*, pages 241–251. Vieweg, 1991.

A New Approach to Flux Coupling Analysis of Metabolic Networks

Abdelhalim Larhlimi and Alexander Bockmayr

DFG-Research Center Matheon, FB Mathematik und Informatik, Freie Universität
Berlin, Arnimallee, 3, 14195 Berlin, Germany
larhlimi@mi.fu-berlin.de, bockmayr@mi.fu-berlin.de

Abstract. Flux coupling analysis is a method to identify blocked and
coupled reactions in a metabolic network at steady state. We present
a new approach to flux coupling analysis, which uses a minimum set
of generators of the steady state flux cone. Our method does not require to reconfigure the network by splitting reversible reactions into
forward and backward reactions. By distinguishing different types of reactions (irreversible, pseudo-irreversible, fully reversible), we show that
reaction coupling relationships can only hold between certain reaction
types. Based on this mathematical analysis, we propose a new algorithm
for flux coupling analysis.

1 Introduction

Constraint-based methods for network-based metabolic pathway analysis have
attracted growing interest in recent years [6]. The constraints that have to hold
in a metabolic network at steady state include stoichiometry and thermodynamic
irreversibility. These two classes of constraints not only determine all the possible
flux distributions over the metabolic network at steady state. They also induce
different dependencies between the reactions. For example, some reactions, which
are called *blocked reactions* [1] or *strictly detailed balanced reactions* [9], may be
incapable of carrying flux under steady state conditions. Another possibility are
coupled reactions [1], where a zero flux through one reaction implies a zero flux
through other reactions. Since a zero flux through some reaction corresponds to
the deletion of the corresponding gene, such dependencies also link the metabolic
to the gene regulatory network.

The elucidation of blocked and coupled reactions helps to better understand
metabolic interactions within cellular networks. In this context, the *Flux Coupling Finder (FCF)* framework [1] has been developed to identify blocked and
coupled reactions in metabolic networks. Maximizing the flux through a reaction under stoichiometric and thermodynamic constraints allows one to decide
whether this reaction is blocked. Similarly, comparing flux ratios is the basis to
determine whether two reactions are coupled.

The FCF framework [1] requires a reconfiguration of the metabolic network.
All reversible reactions are split into a forward and a backward reaction, which

M.R. Berthold, R. Glen, and I. Fischer (Eds.): CompLife 2006, LNBI 4216, pp. 205–215, 2006.
© Springer-Verlag Berlin Heidelberg 2006

both are constrained to be non-negative. This implies that the number of variables and constraints increases. Since FCF uses linear fractional programming to identify the maximum and minimum flux ratios for every pair of metabolic fluxes, a very big number of linear optimization problems has to be solved. Therefore, FCF may not scale well for genome-scale models of complex microorganisms which involve a large number of reactions. Furthermore, as a consequence of the network reconfiguration, FCF cannot compute directly coupling relationships for reversible reactions. Since all reversible reactions are split into a forward and a backward reaction, FCF only computes interactions between reaction directions (see Fig. 1). A post-processing step is needed to deduce reaction couplings involving reversible reactions in the original network.

The goal of this work is to develop a new method to identify blocked and coupled reactions in a metabolic network at steady state. The organization of the paper is as follows. We start in Sect. 2 with some basic facts from polyhedral theory and the definition of the steady state flux cone of a metabolic network. This leads to a classification of reactions according to their reversibility type. In Sect. 3, we show mathematically that coupling relationships depend on the reversibility type of the reactions. Our main result is that reaction couplings can only hold between certain reaction types and that they can be computed using a minimum set of generating vectors of the flux cone. In Sect. 4, we propose a new algorithm to identify blocked and coupled reactions. Finally, some computational results are given to compare the new method with the original FCF framework.

2   The Steady State Flux Cone of a Metabolic Network

2.1   Polyhedral Cones

We start with some basic facts about polyhedral cones (see e.g. [8]). A set $C \subseteq \mathbb{R}^n$ is called a *(convex) cone* if $\alpha x + \beta y \in C$, whenever $x, y \in C$ and $\alpha, \beta \geq 0$. A cone C is *polyhedral*, if $C = \{x \in \mathbb{R}^n \mid Ax \geq 0\}$, for some real matrix $A \in \mathbb{R}^{m \times n}$. In this case, $\text{lin.space}(C) = \{x \in \mathbb{R}^n \mid Ax = 0\}$ is called the *lineality space* of C. A cone C is *finitely generated* if there exist $x^1, \ldots, x^s \in \mathbb{R}^n$ such that $C = \text{cone}\{x^1, \ldots, x^s\} \stackrel{\text{def}}{=} \{\alpha_1 x^1 + \ldots + \alpha_s x^s \mid \alpha_1, \ldots, \alpha_s \geq 0\}$. A fundamental theorem of Farkas-Minkowski-Weyl asserts that a convex cone is polyhedral if and only if it is finitely generated. In the sequel, we will consider only polyhedral cones.

An inequality $a^T x \geq 0, a \in \mathbb{R}^n \setminus \{0\}$, is *valid* for C if $C \subseteq \{x \in \mathbb{R}^n \mid a^T x \geq 0\}$. The set $F = C \cap \{x \in \mathbb{R}^n \mid a^T x = 0\}$ is then called a *face* of C. Let t be the dimension of the lineality space of C. A *minimal proper face* of C is a face of dimension $t + 1$.

If we select for each minimal proper face $G^k, k = 1, \ldots, s$, a vector $g^k \in G^k \setminus \text{lin.space}(C)$, and a vector basis $b^1, \ldots, b^t$ of $\text{lin.space}(C)$, then for all $v \in C$, there exist $\alpha_k, \beta_l \in \mathbb{R}, \alpha_k \geq 0$ such that

$$v = \sum_{k=1}^{s} \alpha_k g^k + \sum_{l=1}^{t} \beta_l b^l. \tag{1}$$

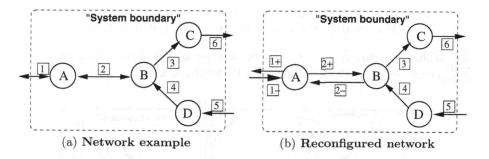

(a) **Network example**   (b) **Reconfigured network**

Fig. 1. The reversible reaction 2 in Fig. 1(a) is split into a forward and backward reaction 2^+ and 2^-. According to FCF, a zero flux through reaction 3 (resp. 4) implies a zero flux through reaction 2^+ (resp. 2^-), i.e., a negative (resp. positive) flux through the reversible reaction 2. However, neither a zero flux through reaction 3 nor a zero flux through reaction 4 implies a zero flux through reaction 2.

2.2   The Steady State Flux Cone

Mathematically, the stoichiometric and thermodynamic constraints that have to hold in a metabolic network have the form [5]

$$Sv = 0, \ v_i \geq 0, \text{ for } i \in Irr. \tag{2}$$

Here, S is the $m \times n$ stoichiometric matrix of the network, with m metabolites (rows) and n reactions (columns), and $v \in \mathbb{R}^n$ is the *flux vector*. The set $Irr \subseteq \{1, \ldots, n\}$ defines the *irreversible* reactions in the network, while $Rev = \{1, \ldots, n\} \setminus Irr$ is the set of *reversible* reactions. The set of all solutions of the constraint system (2), which corresponds to the set of all possible flux distributions through the network at steady state, defines a polyhedral cone,

$$C = \{v \in \mathbb{R}^n \mid Sv = 0, \ v_i \geq 0, i \in Irr\} \tag{3}$$

which is called the *steady state flux cone*.

Example 1. Consider the hypothetical network ILLUSNET depicted in Fig. 2. It consists of thirteen metabolites $(A, \ldots, O)$, and nineteen reactions $(1, \ldots, 19)$. The steady state flux cone is defined by $C = \{v \in \mathbb{R}^{19} \mid Sv = 0, v_i \geq 0 \text{ for all } i \in Irr\}$, with the stoichiometric matrix S and the set of irreversible reactions $Irr = \{1, 2, 3, 4, 5, 6, 7, 8\}$. Fig. 2 shows four pathways

$$\begin{aligned}
g^1 &= (2, 2, 1, 0, 0, 0, 0, 0, \quad 2, \quad 2, \quad 1, \quad 1, 0, 0, 0, 0, 0, 0, 0), \\
g^2 &= (0, 0, 1, 2, 0, 0, 0, 0, \quad 0, \quad 0, -1, -1, 2, 0, 0, 0, 0, 0, 0), \\
g^3 &= (0, 0, 0, 0, 1, 1, 1, 0, -1, -1, \quad 0, \quad 0, 0, 0, 0, 0, 0, 0, 0), \\
g^4 &= (0, 0, 0, 0, 1, 1, 0, 1, -1, -1, \quad 0, \quad 0, 0, 0, 1, 1, 0, 0, 0).
\end{aligned}$$

representing the four minimal proper faces G^k, $k = 1, 2, 3, 4$ of the network. The lineality space lin.space$(C) = \{v \in C \mid v_i = 0, i \in Irr\}$ has dimension 2. It can be generated by the pathways

$$b^1 = (0, 0, 0, 0, 0, 0, 0, 0, 0, 0, 0, 0, 0, -1, 1, 1, 1, 0, 0, 0),$$
$$b^2 = (0, 0, 0, 0, 0, 0, 0, 0, 0, 0, 0, 0, 0, -1, 1, 0, 0, 1, 1, 1, 0).$$

An arbitrary pathway $v \in C$ can be written in the form combination $v = \sum_{k=1}^{4} \alpha_k g^k + \sum_{l=1}^{2} \beta_l b^l$, for some $\alpha_k \geq 0$ and $\beta_1, \beta_2 \in \mathbb{R}$.

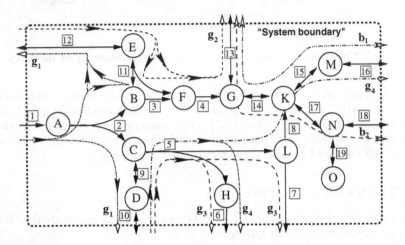

Fig. 2. Network example (ILLUSNET) with representative pathways

2.3   Reaction Classification

A reversible reaction $j \in Rev$ is called *pseudo-irreversible* [4] if $v_j = 0$, for all $v \in \text{lin.space}(C)$. A reversible reaction that is not pseudo-irreversible is called *fully reversible*.

Inside each minimal proper face, the irreversible and the pseudo-irreversible reactions take a unique direction. More precisely, we have the following properties.

Theorem 1 ([4]). *Let G be a minimal proper face of the flux cone C and let $j \in \{1, \ldots, n\}$ be a reaction.*

- *If $j \in Irr$ is irreversible, then $v_j > 0$, for all $v \in G \setminus \text{lin.space}(C)$, or $v_j = 0$, for all $v \in G$. Furthermore, $v_j = 0$, for all $v \in \text{lin.space}(C)$.*
- *If $j \in Rev$ is pseudo-irreversible, then the flux v_j through j has a unique sign in $G \setminus \text{lin.space}(C)$, i.e., either $v_j > 0$, for all $v \in G \setminus \text{lin.space}(C)$, or $v_j = 0$, for all $v \in G \setminus \text{lin.space}(C)$, or $v_j < 0$, for all $v \in G \setminus \text{lin.space}(C)$. For all $v \in \text{lin.space}(C)$, we have again $v_j = 0$.*
- *If $j \in Rev$ is fully reversible, there exists $v \in \text{lin.space}(C)$ such that $v_j \neq 0$. We can then find pathways $v^+, v^-, v^0 \in G \setminus \text{lin.space}(C)$ with $v_j^+ > 0$, $v_j^- < 0$ and $v_j^0 = 0$.*

Example 2. In the ILLUSNET network, the reactions $9, 10, 11$ and 12 are pseudo-irreversible, while reactions $15, 16, 17, 18$ are fully reversible. In the context of the minimal proper face G^1, all the pseudo-irreversible reactions become positive, i.e., $v_i > 0$ for all $v \in G^1 \setminus \text{lin.space}(C)$, while the pseudo-irreversible reactions 9 and 10 become negative in the context of the faces G^3 and G^4. The reaction 19 has zero flux in all the minimal proper faces and in the lineality space of the flux cone.

3 Flux Coupling Analysis Based on the Reversibility Type of Reactions

For the rest of the paper, we assume that $G^k, k = 1, \ldots, s$, are the minimal proper faces of the steady state flux cone C, represented by vectors $g^k \in G^k \setminus \text{lin.space}(C)$, and that $b^l, l = 1, \ldots, t$, is a vector basis of $\text{lin.space}(C)$. Based on Theor. 1, we define the following decomposition of the reaction set $\{1, \ldots, n\}$, which reflects that pseudo-irreversible reactions taking the same direction in all minimal proper faces behave like irreversible reactions.

- $Irev = Irr \cup \{i \mid i \text{ is pseudo-irreversible and } v_i \geq 0, \text{ for all } v \in C \text{ or } v_i \leq 0, \text{ for all } v \in C\}$.
- $Prev = \{i \mid i \text{ is pseudo-irreversible and there exist } v^+, v^- \in C \text{ such that } v_i^+ > 0, v_i^- < 0\}$,
- $Frev = \{i \mid i \text{ is fully reversible}\}$.

Definition 1. *A reaction $i \in \{1, \ldots, n\}$ is blocked if $v_i = 0$, for all $v \in C$. Otherwise, the reaction i is unblocked.*

First, we characterize blocked reactions using generators of the cone. The following proposition follows directly from (1).

Proposition 1. *For any reaction $i \in \{1, \ldots, n\}$, the following are equivalent:*

1. *The reaction i is blocked.*
2. *$g_i^k = 0$, for all $k = 1, \ldots, s$, and $b_i^l = 0$, for all $l = 1, \ldots, t$.*

In our analysis, we will first compute the blocked reactions using Prop. 1. Afterwards, we will identify coupled reactions based on the subsequent results. Here, it is enough to consider only unblocked reactions. The proofs of these properties can be found in Appendix A.

Definition 2 ([1]). *Let i, j be two unblocked reactions. The coupling relationships $\overset{=0}{\rightarrow}, \overset{=0}{\leftrightarrow}, \rightsquigarrow^\lambda$ are defined in the following way:*

- *$i \overset{=0}{\rightarrow} j$ if for all $v \in C$, $v_i = 0$ implies $v_j = 0$.*
- *$i \overset{=0}{\leftrightarrow} j$ if for all $v \in C$, $v_i = 0$ is equivalent to $v_j = 0$.*
- *$i \rightsquigarrow^\lambda j$ if there exists $\lambda \in \mathbb{R}$ such that for all $v \in C, v_j = \lambda v_i$.*

Obviously, $i \rightsquigarrow^\lambda j$ implies $i \overset{=0}{\leftrightarrow} j$. Moreover, $i \overset{=0}{\leftrightarrow} j$ is equivalent to ($i \overset{=0}{\rightarrow} j$ and $j \overset{=0}{\rightarrow} i$). The next results shows that the relations $i \overset{=0}{\rightarrow} j$, $i \overset{=0}{\leftrightarrow} j$, $i \rightsquigarrow^\lambda j$ cannot hold for arbitrary pairs of reactions.

Theorem 2. *Let i, j be two unblocked reactions such that at least one of the relations $i \overset{=0}{\to} j$, $i \overset{=0}{\leftrightarrow} j$ or $i \curvearrowright^\lambda j$ is satisfied. Then either (a) or (b) holds:*

(a) *i and j are both (pseudo-)irreversible: $i, j \in Irev \cup Prev$.*
(b) *i and j are both fully reversible: $i, j \in Frev$.*

In the following, we study the coupling relationships for the different types of reactions. We first consider the case $i \in Prev$.

Proposition 2. *Suppose i, j are unblocked, $i \in Prev$ and $j \in Irev \cup Prev$. Then the following are equivalent:*

1. $i \overset{=0}{\to} j$
2. $i \overset{=0}{\leftrightarrow} j$
3. $i \curvearrowright^\lambda j$
4. $g_j^k = \lambda g_i^k$, for all $k = 1, \ldots, s$.

In each of these cases, $j \in Prev$.

Next, we characterize the case $i \in Frev$.

Proposition 3. *Suppose i, j are unblocked and $i \in Frev$ is fully reversible. Then the following are equivalent:*

1. $i \overset{=0}{\to} j$
2. $i \overset{=0}{\leftrightarrow} j$
3. $i \curvearrowright^\lambda j$
4. $g_j^k = \lambda g_i^k$, for all $k = 1, \ldots, s$, and $b_j^l = \lambda b_i^l$, for all $l = 1, \ldots, t$.

In each of these cases, $j \in Frev$.

Finally, we have to consider $i \in Irev$.

Proposition 4. *Suppose i, j are unblocked, $i \in Irev$ and $j \in Irev \cup Prev$. Then the following are equivalent:*

1. $i \overset{=0}{\to} j$ *holds in the flux cone C.*
2. $i \overset{=0}{\to} j$ *holds in all minimal proper faces $G^k, k = 1, \ldots, s$.*
3. $g_i^k = 0$ *implies $g_j^k = 0$, for all $k = 1, \ldots, s$.*

If also $j \overset{=0}{\leftrightarrow} i$ or $i \curvearrowright^\lambda j$, then $j \in Irev$.

Corollary 1. *Suppose i, j are unblocked and $i, j \in Irev$. Then we have:*

a) *$i \overset{=0}{\leftrightarrow} j$ iff $g_i^k = 0$ is equivalent to $g_j^k = 0$, for all $k = 1, \ldots, s$.*
b) *$i \curvearrowright^\lambda j$ iff $g_j^k = \lambda g_i^k$, for all $k = 1, \ldots, s$.*

Table 1 summarizes the different possible coupling relationships. Note that $i \overset{=0}{\to} j$, $i \overset{=0}{\leftrightarrow} j$ and $i \curvearrowright^\lambda j$ are equivalent for $i, j \in Prev$ or $i, j \in Frev$.

Example 3. Fig. 3 shows all coupled reactions in the network from Fig. 2. We see that many reactions depend on reaction 3. A zero flux for this reaction implies a zero flux for the reactions $1, 2, 4, 11$ and 12. Thus, reaction 3 plays a crucial role in the network. Reaction 19 is blocked, because it is involved neither in the definition of the minimal proper faces nor in the definition of the lineality space.

Table 1. Reaction coupling cases

i/j	Irev $\overset{=0}{\rightarrow}$	Irev $\overset{=0}{\leftrightarrow}$	Irev $\leadsto^\lambda$	Prev $\overset{=0}{\rightarrow}$	Prev $\overset{=0}{\leftrightarrow}$	Prev $\leadsto^\lambda$	Frev $\overset{=0}{\rightarrow}$	Frev $\overset{=0}{\leftrightarrow}$	Frev $\leadsto^\lambda$
Irev	Prop.4	Cor.1	Cor.1	Prop.4					
Prev				Prop.2	Prop.2	Prop.2			
Frev							Prop.3	Prop.3	Prop.3

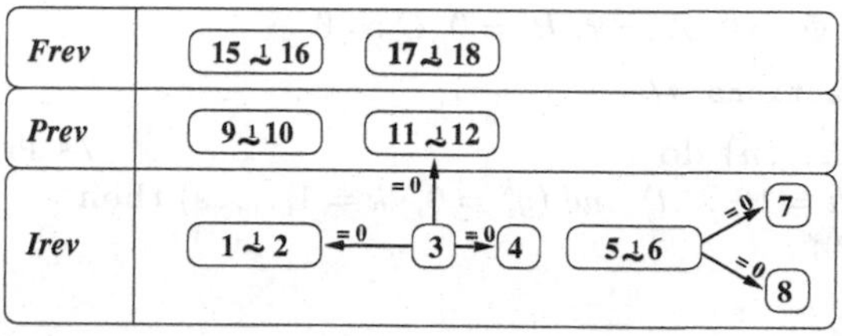

Fig. 3. Coupled reactions in ILLUSNET

4 Identifying Blocked and Coupled Reactions

4.1 Improving the FCF Prodecure

It follows from Sect. 3 that coupling relationships depend on the reversibility type of the reactions. Irreversible and pseudo-irreversible reactions cannot be coupled with fully reversible reactions. According to Table 1, to detect coupling relationships, we do not have to explore exhaustively all possible reaction pairs. We can improve the FCF procedure significantly by applying linear-fractional programming only in those cases where coupling relationships can occur. All the possible cases are given in Table 1. An empty entry indicates that the corresponding coupling relationship is not possible. Note that the reversibility type of a reaction can be identified without computing a set of generating vectors of the flux cone. However, this improved version of FCF still requires a network reconfiguration, which leads to a large number of linear optimization problems that have to be solved.

4.2 A New Algorithm

The results in Sect. 3 also suggest a new algorithm to identify blocked and coupled reactions. This method does not require any reconfiguration of the metabolic network. It is only based on the reversibility type of the reactions and a minimum set of generators of the flux cone. The basic steps of this new algorithm are as follows. First, we compute a set of generators of the flux cone C using existing software for polyhedral computations. Second, we classify the reactions according to their reversibility type. This classification allows us to determine whether a coupling between two reactions is possible. Finally, we apply the results from Sect. 3 to identify blocked and coupled reactions. For a more detailed description, see Algorithm 1.

Input : Sets $Irev, Prev, Frev \subseteq \{1, \ldots, n\}$;
for each minimal proper face $G^k, k = 1, \ldots, s$, a generating vector
$g^k \in G^k \setminus \text{lin.space}(C)$;
a vector basis $b^1, \ldots, b^t$ of lin.space(C).

Output: Blocked reactions: $\Phi = \{i \mid i \text{ is blocked}\}$;
Coupled reactions: $A = \{(i,j) \mid i \leftrightsquigarrow^\lambda j, 1 \le i < j \le n\}$,
$B = \{(i,j) \mid i \overset{=0}{\leftrightarrow} j, 1 \le i < j \le n, (i,j) \notin A\}$,
$C = \{(i,j) \mid i \overset{=0}{\rightarrow} j, (i,j) \notin A \cup B, (j,i) \notin A \cup B\}$.

Initialization: $\Phi := \emptyset$, $A := \emptyset$, $B := \emptyset$, $C := \emptyset$.

```
/* Blocked reactions */
```

foreach $i \in \{1, \ldots, n\}$ **do**   /* Proposition 1 */
    if $(b_i^l = 0, \forall l = 1, \ldots, t)$ *and* $(g_i^k = 0, \forall k = 1, \ldots, s)$ **then**
        add(i, Φ);
    end
end

```
/* Coupled reactions */
```

$Irev := Irev \setminus \Phi$, $Prev := Prev \setminus \Phi$;

foreach $i, j \in Prev$ *with* $i < j$ **do**   /* Proposition 2 */
    if $\exists \lambda \in \mathbb{R}$ *such that* $g_j^k = \lambda g_i^k, \forall k = 1, \ldots, s$ **then** add$((i,j), A)$;
end
foreach $i, j \in Frev$ *with* $i < j$ **do**   /* Proposition 3 */
    if $\exists \lambda \in \mathbb{R}$ *such that* $b_j^l = \lambda b_i^l, \forall l = 1, \ldots, t$, *and* $g_j^k = \lambda g_i^k, \forall k = 1, \ldots, s$ **then**
    add$((i,j), A)$;
end
foreach $i \in Irev, j \in Irev \cup Prev$ **do**   /* Proposition 4 */
    if $g_i^k \neq 0$ *or* $g_j^k = 0, \forall k = 1, \ldots, s$ **then** add$((i,j), C)$;
end
foreach $(i,j) \in C$ *with* $i, j \in Irev$ *and* $i < j$ **do**   /* Corollary 1 */
    if $(j, i) \in C$ **then**
        remove$((i,j), C)$, remove$((j,i), C))$;
        if $\exists \lambda \in \mathbb{R}$ *such that* $g_j^k = \lambda g_i^k, \forall k = 1, \ldots, s$ **then** add$((i,j), A)$;
        else add$((i,j), B)$;
    end
end

Algorithm 1. Blocked and coupled reactions finder

4.3 Computational Results

Both our new algorithm and the FCF algorithm have been implemented in the
Java language. The FCF procedure was realized using CPLEX 9.0 (a state-of-
the-art solver for linear and integer programming problems) accessed via Java.
To compute a set of generating vectors of the steady state flux cone, our algo-
rithm uses the software *cdd* [2], which is a C++ implementation of the Double
Description method of Motzkin et al. for general convex polyhedra in $\mathbb{R}^n$.

To compare the two approaches, we computed blocked and coupled reactions for some genome-scale networks. The computations were performed on a Linux server with a AMD Athlon Processor 1.6 GHz and 2 GB RAM. We present computation times for models of the human red blood cell [13], the human cardiac mitochondria [12], the central carbon metabolism of *E. coli* [3,10], the *E. coli K-12* (iJR904 GSM/GPR) [7], and the *H. pylori* (iIT341 GSM/GPR) [11]. We refer to [1] for a discussion of the biological aspects of flux coupling analysis.

Tab. 2 summarizes our computational results. It shows that flux coupling analysis can be done extremely fast if a set of generators of the flux cone is available. Computing such a set is the most time-consuming part in our algorithm. However, it should be noted that this step has an interest in its own. We obtain similar information as by computing the elementary flux modes or extreme pathways of the network, see [4] for more information. The overall running time of the new algorithm is still significantly faster than the original FCF method.

Table 2. Metabolic systems, with the number of blocked reactions (*Blk*), the size of the sets *Irev, Prev, Frev*, the running time (in seconds) of computing a set of generators (*MMB*), reaction coupling using this set (*FCMMB*), and reaction coupling using the FCF procedure (*FCF*)

Metabolic network	Blk	Irev	Prev	Frev	MMB	FCMMB	FCF
Red Blood Cell	0	31	14	6	2.32	0.26	110.65
Central metabolism of *E. coli*	0	92	18	0	214.49	2.55	477.14
Human cardiac mitochondria	121	83	3	9	1262.65	0.34	13426.91
Helicobacter pylori	346	128	15	39	13551.44	0.43	318374.15
E. coli K-12	435	480	49	110	261306.15	5.32	≥ 1 week

5   Conclusion

In this paper, we have introduced a new method for flux coupling analysis based on a refined analysis of the steady state flux cone. By distinguishing three types of reactions (irreversible, pseudo-irreversible, fully reversible), we study mathematical dependencies between coupling relationships and the reversibility type of the reactions. The results that have been obtained allow improving the flux coupling finder (FCF) and lead to a new algorithm for identifying blocked and coupled reactions in a metabolic network.

References

1. A.P. Burgard, E.V. Nickolaev, C.H. Schilling, and C.D. Maranas. Flux coupling analysis of genome-scale metabolic network reconstructions. *Genome Res.*, 14:301–312, 2004.

2. K. Fukuda and A. Prodon. Double description method revisited. In *Combinatorics and Computer Science*, pages 91–111. Springer, LNCS 1120, 1995. Software available: http://www.ifor.math.ethz.ch/~fukuda/cdd_home/.
3. S. Klamt and J. Stelling. Combinatorial complexity of pathway analysis in metabolic networks. *Mol. Bio. Rep.*, 29:233–236, 2002.
4. A. Larhlimi and A. Bockmayr. Minimal metabolic behaviors and the reversible metabolic space. Preprint 299, DFG Research Center Matheon, December 2005. http://page.mi.fu-berlin.de/~bockmayr/MMB.pdf.
5. J.A. Papin, J. Stelling, N. D. Price, S. Klamt, S. Schuster, and B. O. Palsson. Comparison of network-based pathway analysis methods. *Trends Biotechnol.*, 22(8):400–405, 2004.
6. N. D. Price, J. L. Reed, and B. O. Palsson. Genome-scale models of microbial cells: evaluating the consequences of constraints. *Nat Rev Microbiol.*, 2(11):886–97, 2004.
7. J.L. Reed, T.D. Vo, C.H. Schilling, and B.Ø. Palsson. An expanded genome-scale model of Escherichia coli K-12 (iJR904 GSM/GPR). *Genome Biol.*, 4(9):R54.1–R54.12, 2003.
8. A. Schrijver. *Theory of Linear and Integer Programming*. Wiley, 1986.
9. S. Schuster and R. Schuster. Detecting strictly detailed balanced sub-networks in open chemical reaction networks. *J. Math. Chem.*, 6:17–40, 1991.
10. J. Stelling, S. Klamt, K. Bettenbrock, S. Schuster, and E.D. Gilles. Metabolic network structure determines key aspects of functionality and regulation. *Nature*, 420:190–193, 2002.
11. I. Thiele, T.D Vo, N.D. Price, and B.Ø. Palsson. An expanded metabolic reconstruction of Helicobacter pylori (iIT341 GSM/GPR). *J. Bacteriol.*, 187(16):5818–5830, 2005.
12. T.D Vo, H.J. Greenberg, and B.Ø. Palsson. Reconstruction and functional characterization of the human mitrochondrial metabolic network based on proteomic and biochemical data. *J. Biol. Chem.*, 279(38):39532–39540, 2004.
13. S.J. Wiback and B.O. Palsson. Extreme pathway analysis of human red blood cell metabolism. *Biophys. J.*, 83:808–818, 2002.

Appendix A

Proof of Theorem 2: First suppose $i \in Irev \cup Prev$ and $j \in Frev$. Since $j \in Frev$, there exists $v \in \text{lin.space}(C)$ such that $v_j \neq 0$. Since $i \in Irev \cup Prev$, we have $v_i = 0$, so $i \overset{=0}{\nrightarrow} j$.

Now suppose that $i \in Frev$ and $j \in Irev \cup Prev$. Since j is unblocked, there exists $w \in C$ such that $w_j \neq 0$. Since i is fully reversible, there exists $b \in \text{lin.space}(C)$ such that $b_i \neq 0$. Define $v = w - (w_i/b_i) \cdot b$. It follows $v \in C$, $v_i = 0$ and $v_j = w_j \neq 0$, which implies $i \overset{=0}{\nrightarrow} j$.   □

Proof of Proposition 2: (3) ⇒ (2) ⇒ (1) is immediate.

(1) ⇒ (4): Let $K^+ = \{k \mid g_i^k > 0\}$ and $K^- = \{k \mid g_i^k < 0\}$. Since $i \in Prev$, there exist $v^+, v^- \in C$ with $v_i^+ > 0$ and $v_i^- < 0$. If $K^+ = \emptyset$ (resp. $K^- = \emptyset$), we would have $v_i \leq 0$ (resp. $v_i \geq 0$), for all $v \in C$, which is a contradiction. So both K^+ and K^- must be non-empty. Let $p \in K^+$ and $q \in K^-$. Define $w = g_i^p \cdot g^q - g_i^q \cdot g^p$. Then $w \in C$ and $w_i = 0$. Since $i \overset{=0}{\to} j$, we get $w_j =$

$g_i^p g_j^q - g_i^q g_j^p = 0$, or $g_j^p / g_i^p = g_j^q / g_i^q \stackrel{\text{def}}{=} \lambda$, independently of the choice of p and q. We conclude $g_j^k = \lambda g_i^k$, for all $k \in K^+ \cup K^-$. Since $i \stackrel{=0}{\rightarrow} j$, this holds for all $k = 1, \ldots, s$.

(4) $\Rightarrow$ (3): For all $v \in C$, there exists $b \in \text{lin.space}(C)$ and $\alpha_k \geq 0$ such that $v = \sum_{k=1}^{s} \alpha_k g^k + b$. Since $i \in Prev$ and $j \in Irev \cup Prev$, we have $b_i = b_j = 0$. It follows that $v_j = \sum_{k=1}^{s} \alpha_k g_j^k = \sum_{k=1}^{s} \alpha_k \lambda g_i^k = \lambda v_i$.   □

Proof of Proposition 3: (3) $\Rightarrow$ (2) $\Rightarrow$ (1) and (3) $\Leftrightarrow$ (4) are immediate.

To prove (1) $\Rightarrow$ (3), we suppose $i \stackrel{=0}{\rightarrow} j$. Since i is fully reversible, there exists $b \in \text{lin.space}(C)$, with $b_i \neq 0$. Given $v \in C$, define $w = v - (v_i/b_i) \cdot b$. Then $w \in C$ and $w_i = 0$. Since $i \stackrel{=0}{\rightarrow} j$, we get $w_j = v_j - (v_i/b_i)b_j = v_j - (b_j/b_i)v_i = 0$. Defining $\lambda = b_j/b_i$, this shows $v_j = \lambda v_i$, for all $v \in C$.   □

Proof of Proposition 4: (1) $\Rightarrow$ (2) $\Rightarrow$ (3) is obvious, so we have to prove only (3) $\Rightarrow$ (1). For all $v \in C$, there exist $b \in \text{lin.space}(C)$ and $\alpha_k \geq 0$ such that $v = \sum_{k=1}^{s} \alpha_k g^k + b$. Since $i \in Irev$ and $j \in Irev \cup Prev$, we get $b_i = b_j = 0$. By the definition of $Irev$, either $g_i^k \geq 0$, for all $k = 1, \ldots, s$, or $g_i^k \leq 0$, for all $k = 1, \ldots, s$. Suppose $v_i = \sum_{k=1}^{s} \alpha_k g_i^k = 0$. It follows $g_i^k = 0$, for $k = 1, \ldots, s$. Using (3), we obtain $g_j^k = 0$ for $k = 1, \ldots, s$, and so $v_j = \sum_{k=1}^{s} \alpha_k g_j^k = 0$.

Under the hypotheses of Prop. 4, suppose $i \stackrel{=0}{\leftrightarrow} j$. Clearly, $j \stackrel{=0}{\rightarrow} i$. If $j \in Prev$, then by Prop. 2, $i \in Prev$, which is a contradiction. So $j \in Irev$. Similarly, if $i \rightsquigarrow^\lambda j$, then $i \stackrel{=0}{\leftrightarrow} j$, and again $j \in Irev$.   □

Proof of Corollary 1: a) Suppose $g_j^k = 0$ is equivalent to $g_i^k = 0$, for all $k = 1, \ldots, s$. Then $g_i^k = 0$ implies $g_j^k = 0$, and vice versa, for all $k = 1, \ldots, s$. Since $i, j \in Irev$, we may apply Prop. 4 and get $i \stackrel{=0}{\rightarrow} j$ and $j \stackrel{=0}{\rightarrow} i$. So $i \stackrel{=0}{\leftrightarrow} j$.
b) Suppose $g_j^k = \lambda g_i^k$, for all $k = 1, \ldots, s$. For all $v \in C$, there exist $b \in \text{lin.space}(C)$ and $\alpha_k \geq 0$ such that $v = \sum_{k=1}^{s} \alpha_k g^k + b$. Since $i, j \in Irev$, we have $b_i = b_j = 0$. It follows that $v_j = \sum_{k=1}^{s} \alpha_k g_j^k = \sum_{k=1}^{s} \alpha_k \lambda g_i^k = \lambda v_i$.   □

Software Supported Modelling in Pharmacokinetics

Regina Telgmann[1], Max von Kleist[2], and Wilhelm Huisinga[2]

[1] CiT GmbH,
Oldenburger Str. 200,
D-26180 Rastede, Germany
r.telgmann@cit-wulkow.de
[2] Freie Universität Berlin,
Department of Mathematics and Informatics,
Arnimallee 6,
D-14195 Berlin, Germany
{kleist, huisinga}@math.fu-berlin.de

Abstract. A powerful new software concept to physiologically based pharmacokinetic (PBPK) modelling of drug disposition is presented. It links the inherent modular understanding in pharmacology with orthogonal design principles from software engineering. This concept allows for *flexible* and *user-friendly* design of pharmacokinetic whole body models, data analysis, hypotheses testing or extrapolation. The typical structure of physiologically-based pharmacokinetic models is introduced. The resulting requirements from a modelling and software engineering point of view and its realizations in the software tool MEDICI-PK [9] are described. Finally, an example in the context of drug-drug interaction studies is given that demonstrates the advantage of defining a whole-body pharmacokinetic model in terms of the underlying physiological processes quite impressively: A system of 162 ODEs is automatically compiled based on the specification of 7 local physiological processes only.

1 Introduction

Pharmacokinetics is the study of the time course of drug and metabolite levels in different fluids, tissues, and excreta of the body [12]. This includes the investigation and understanding of the processes of absorption, distribution, metabolism, and excretion (ADME). The pharmacokinetic profile of a drug strongly influences its delivery to biological targets, thereby affecting its efficacy and potential side effects. Following studies in the late 1990s indicating that poor pharmacokinetics and toxicity were important causes of costly late-stage failures in drug development, it has been widely perceived that these areas need to be considered as early as possible in the drug discovery process [1]. Today's combinatorial chemistry and high throughput screening methods enlarged the space of drug candidates significantly, creating actual needs for *in silico* pharmacokinetic analysis to support the drug development pipeline. The pharmacokinetics of a compound are

M.R. Berthold, R. Glen, and I. Fischer (Eds.): CompLife 2006, LNBI 4216, pp. 216–225, 2006.

typically understood, analyzed and interpreted in the context of their underlying ADME processes. However, there is no unique mathematical model for any of these processes; usually a number of different models with different underlying assumptions, parameterization and applicability are concurrent.

To efficiently support *in silico* modelling and simulation in pharmacokinetics, we propose to inherit the inherent modular structure, which is based on the physiological processes, to the software tool. We describe the concepts of modularity and orthogonality as fundamental principles for the design of a virtual lab in pharmacokinetics. The above mentioned design principles have recently been realized successfully in the software tool MEDICI-PK, that is especially designed to fit the needs in pharmacokinetic modelling. Our approach is illustrated by an example from drug-drug interaction studies.

2   Mathematical Modelling in Physiologically Based Pharmacokinetics (PBPK)

A physiologically based pharmacokinetic (PBPK) whole body model is a special type of compartmental model, in which the compartments represent anatomical volumes, such as organs or tissues. The compartments are connected in an anatomically meaningful way, to simulate drug exchange via the blood flow. The conceptional representation of a 15 organ PBPK model is shown in Fig. 1. Each

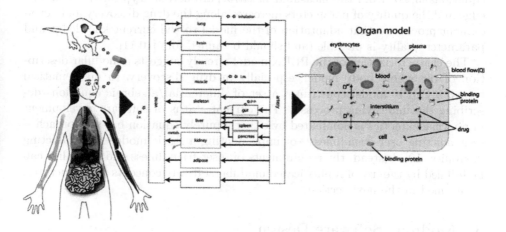

Fig. 1. Organ structure of a physiologically based pharmacokinetic model

compartment is further subdivided into the four phases: erythrocytes, plasma, interstitium and cellular space (see Fig. 1). Many physiological processes in pharmacokinetics are accessible for a mechanistic description at this resolution. Typically, following processes are modelled:

- Convection of drug molecules by the blood flow
- Binding to macromolecules in plasma and interstitial space

- Distribution into tissue
- Diffusion or active transport across the cellular membrane
- Metabolism or interaction with metabolic networks or signalling pathways etc.

There are many levels of mathematical description for a certain biological process (e.g., each of the physiological processes stated in the above list). To give an example, the process of protein binding (complex formation) between a compound and some macromolecule can be explicitly modelled in terms of the corresponding differential equations derived from the law of mass action. However, often it is assumed that the binding process is fast in comparison to other processes and therefore in dynamical equilibrium (quasi-stationarity). This results in some algebraic equation, often still accounting for saturation effects of the binding process. A further simplification finally results in a linear algebraic equation that is not capable of accounting for saturation effects, however it may be directly parameterized in terms of a frequently generated *in vitro* parameter.

Each mathematical model has its range of applicability and typically requires different knowledge about the process and in particular different input parameters. In broad terms, a chosen model will be a compromise between detailed mechanistic description and required "quality" of the input parameters. At early stages of drug discovery, frequently measured *in vitro* parameters are used to parametrize early PBPK models. Either the parameter can directly be used in the model, or relevant model parameters are estimated through mechanistic equations [3,4,5] from the measured *in vitro* parameters. Typically, the knowledge and the quality of parameters increases along the drug discovery and development process so that adaptation of the model to the current knowledge and parameter quality is possible (and should be aimed for) [10,11].

The characterization of the PBPK model already suggests a modular description of the whole body model, especially in drug discovery. In mathematical terms, a PBPK model constitutes a set of differential/algebraic equations describing the underlying processes. The current status of software development in pharmacokinetics is dominated by either a purely equation based approach— contradicting user-friendliness—or implementing a static model—contradicting flexibility [10]. Instead, the requirements on user-friendliness and flexibility can be fulfilled by the use of sophisticated modular software concepts and structures, as outlined in the next section.

3   Modular Software Design

To support the specification of a whole-body pharmacokinetic model, a variety of physiological processes (as mentioned above) and a corresponding collection of different mathematical models have to be regarded. In practice, identical processes in different compartments will be described by identical mathematical models. From a software engineering point of view it is important to encapsulate the mathematical descriptions into modular parts. These modular parts ("models")–collected in a model library–are defined only once and can be reused inside the whole PK model wherever suitable. This prevents redefinition and rewriting and ensures the even treatment of identical processes wherever wanted. The models are specified in terms of concentrations and parameters, but (a prerequisite for this approach) do not rely on any specific parameter values. This allows for the evaluation of identical mathematical models in different contexts (by means of different parameter sets).

A given topology (like the 15 organs example) combined with a selection of models from the library (one for each physiological process) builds a description of the PBPK model which is still independent of specific values. It may be evaluated for any selection of parameter values. We call this description a 'full body template' (see Fig.2). The parameters to which the models refer can be classified as (i) compound dependent, (ii) species/individual dependent, (iii) dependent on the compound and species or (iv) independent (general parameters). This classification suggests the introduction of four corresponding software objects, each building an orthogonal structure of its own, independent of the 'full body template'.

For the full specification of a PBPK model, a 'full body template' and values of the required parameters given in the mentioned (i)-(iv) parameter objects have finally to be linked (see Fig. 2) - this is realized by the 'simulation object'.

The models address the concentration of a compound by fixed terms (e.g. "Comp1", "Comp2"). They are independent of an actual compound selection,

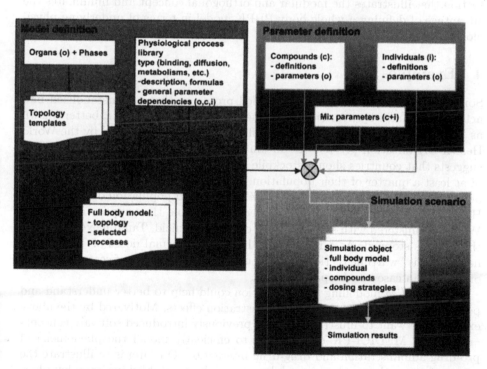

Fig. 2. Schematic illustration of the orthogonal approach to software supported pharmacokinetic modelling

which makes it necessary to assign the actual selection of compounds to these expressions. Models which consider only one compound need no assignment, since this is internally handled, however, multi-compound models (e.g. needed in metabolism models for interactions between compounds) necessitate explicit mapping of the terms addressing the different compound concentrations. For instance, the simple metabolism model

$$V\frac{\mathrm{d}}{\mathrm{d}t}C^{(1)} = k_1 C^{(1)} - k_2 C^{(1)} C^{(2)}$$
$$V\frac{\mathrm{d}}{\mathrm{d}t}C^{(2)} = -k_1 C^{(1)} + k_2 C^{(1)} C^{(2)}$$

will be evaluated for two compounds A and B only if the mapping between compounds (A, B) and concentration terms $(C^{(1)}, C^{(2)})$ is performed, e.g., compound A to $C^{(1)}$ and compound B to $C^{(2)}$. This has to be done inside the 'simulation object' where the 'full body template' and the compound(s) are specified. As soon as all missing mappings are defined, the PBPK model is completed.

When starting the simulation, the resulting differential equation system is automatically generated, including the assignment of all compound-specific parameter values and species-specific physiological parameter values to the respective processes. An example from drug-drug interaction studies is given in the next section that illustrates the modular and orthogonal concept and illuminates the advantage of defining a whole-body PBPK model in terms of underlying physiological processes quite impressively.

4   Examples

Some drugs are administered as so-called pro-drugs that are metabolized into active compounds by liver enzymes. One example is Oseltamivir, better known as Tamiflu [8]. Tamiflu is the main antiflu medicine recommended by the World Health Organization (WHO) [2]. In anticipation of a flu pandemic, the WHO suggests that countries should stockpile enough Tamiflu to allow the treatment of at least a quarter of their population. At present time, however, the supplies of Tamiflu are enough to cover about 2% of the world population only. Recently, Hill et al. [7] highlighted a way to effectively double the supplies of Tamiflu: When administered with a second drug, called probenicid, Tamiflu excretion into the urine is stopped. As a result, only half of the normal doses of Tamiflu are needed. This "wartime tactic" could be used to double power of scarce resources of Tamiflu in case of a flu pandemic [2].

Here, *in silico* modelling and simulation could help to better understand and possibly further optimize the co-administration effects. Motivated by the above example, we want to illustrate how the previously introduced software concepts –realized in MEDICI-PK–can be used to efficiently model the phenomena of pro-drug administration and drug-drug interaction. Our aim is to illustrate the power of our orthogonal and modular approach by establishing complex pharmacokinetic models. It is explicitly not our aim to reproduce experimental data; this is work in progress.

The starting point is the definition of the building blocks in our PBPK model. This is done in terms of the relevant physiological processes like: (a) i.v. absorption, (b) linear protein binding, (c) passive diffusion, (d) tissue distribution (according to [6]), (e) saturable metabolism, (f) renal excretion. Each of the processes (modules) is defined in a 'model basis' by a corresponding mathematical equation. For instance, the processes of saturable metabolism is defined by

$$v_{\text{meta}} = \frac{V_{\text{max}}^{\text{meta}} C_u}{K_{\text{m}}^{\text{meta}} + C_u},$$ (1)

with maximum reaction velocity $V_{\text{max}}^{\text{meta}}$ and Michaelis-Menten constant $K_{\text{m}}^{\text{meta}}$. The concentration of unbound drug is denoted by C_u. The process of excretion is specified by

$$v_{\text{excr}} = \left(Q_{\text{GF}} + \frac{V_{\text{max}}^{\text{ren}}}{K_{\text{m}}^{\text{ren}} + C_u} \right) (1 - F_{\text{re-abs.}}) \cdot C_u$$ (2)

The parameter $V_{\text{max}}^{\text{ren}}$ denotes the maximum velocity of the saturable active tubular excretion process with Michaelis-Menten constant $K_{\text{m}}^{\text{ren}}$. Q_{GF} and $F_{\text{re-abs.}}$ are the glomerular filtration rate and the fraction, that is passively reabsorbed. The renal excretion has been modelled as a function of three processes: (i) passive glomerular filtration (efflux), (ii) active tubular secretion (efflux) and (iii) passive reabsorption. Metabolic clearance in the kidney has been neglected. In total, seven local processes have been defined to model the whole body pharmacokinetics of the three compounds.

The overall PBPK model is then defined by the 'full body template' that links the local physiological process modules on the organ level. For efficiency, it is possibe to define a generic organ structure, which is taken as a default for the initialization of the entire list of organ models. Subsequent individual modifications are possible in order to model organ specific processes, like e.g., excretion by the kidneys. Next, we specify the physiological parameters of the considered species, in our case a 250 g weighting male rat. These values are later needed to fill the model parameters. Finally, we specify the compound-specific data. Motivated by the Tamiflu example, we consider three compounds named A, B, and C; a pro-drug, an active metabolite and a competitive inhibitor for the secretion (of compound B).

At this stage, the three constituents are completely independent. The PBPK model is specified in terms of parameters, however, no actual numerical values are assigned in the model. Only if we map the specific numerical values of the parameters (corresponding to the compound and the species of interest), we obtain a fully specified and ready to simulate so-called 'simulation object'. The advantage of this orthogonal specification and data management is a large flexibility. The same PBPK model can be used for different species and compounds, while the same species database can be used in studies of different models etc. We now demonstrate how to user-friendly and efficiently set up a model for three compounds interacting in a way motivated by the Tamiflu example in three steps. An overview over the necessary modelling steps to be performed is given in Table 1.

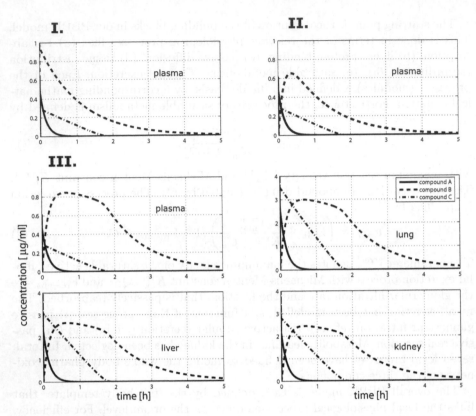

Fig. 3. Simulation results as concentration vs. time profiles in venous plasma for (**I.**) *independent pharmacokinetics* (top left) and (**II.**) *conversion of pro-drug A to B* (top right). The simulation results for (**III.**) *competition for tubular excretion* of compounds B and C are shown in the middle and bottom panels. The middle left and right panels shows the concentration vs. time profiles in the venous plasma and interstitial space of the lung, while the bottom panels show the profiles in the cellular space of the liver (bottom left) and kidney.

Independent pharmacokinetics. To start with, the pharmacokinetics of the three compounds A, B and C are simulated independently. This is easily performed by creating a 'simulation object', which links the full body model, the species data (rat) and the respective compound data. As a consequence, MEDICI-PK automatically generates a set of model equations for each compound. In this example we have identical models for the three compounds. The resulting pharmacokinetic profiles are shown in Fig. 3 (top left panel) for an intravenous administration of 2 mg/(kg bodyweight) of each compound (modelling details in Table 1). Compounds A, B and C show very different pharmacokinetic profiles. This is due to their distinctive distribution characteristics in the various tissues and due to their different elimination characteristics. While compound C is eliminated in an almost constant fashion, compound A and B are eliminated in an exponential fashion. Plasma levels of compound B are substantially higher than plasma levels of compound A and C respectively. This is because compound B is mainly distributed in the plasma, with significantly lower concentrations in the interstitium and cellular space.

Table 1. Brief overview over the performed simulations

description	interactions	dosing
independent pharmacokinetics	-	A: 2[mg/kg body weight] i.v. B: 2[mg/kg body weight] i.v. C: 2[mg/kg body weight] i.v.
conversion of pro-drug A to active metabolite B	A → B	A: 2[mg/kg body weight] i.v. C: 2[mg/kg body weight] i.v.
conversion of pro-drug A to B competition for active tubular excretion between B and C	A → B⊥⊢ C	A: 2[mg/kg body weight] i.v. C: 2[mg/kg body weight] i.v.

Conversion of pro-drug A to B. We next demonstrate how to link the pharmacokinetics of compound A and B. In physiological terms, we want to model the conversion of compound A into compound B in the liver (see Table 1). Given the 'full body template' from the first simulation scenario, this requires only a single change, namely the adaptation of the metabolism model chosen for compound B in the liver. We define the so-called multi-compound metabolism model

$$v_{\text{meta}} = -\frac{V_{\max}^{(1)} C_u^{(1)}}{K_m^{(1)} + C_u^{(1)}} + \frac{V_{\max}^{C_2} C_u^{(2)}}{K_m^{(2)} + C_u^{(2)}}$$

and subsequently map B to $C^{(1)}$ and A to $C^{(2)}$ in the 'simulation object'. Assuming no i.v. administration of compound B, the simulation results are shown in Fig. 3 (top right panel) for intravenous administration of 2 mg/(kg body weight) of compound A and C.

While the simulation of non-interacting compounds based on the same PBPK model can be solved by successive simulation of a single compound at a time, the consideration of (dynamic) interactions requires to establish a joint model for the interacting compounds. In MEDICI-PK, this is automatically generated exploiting the described software concept. This will become even more obvious in the next case.

Competition for active tubular excretion. Finally, we want to include the drug-drug interactions between compound B and C (modelling details in Table 1). In physiological terms, compound C will compete with compound B for active excretion. Given the 'full body template' from the second simulation scenario, this again requires only a single change. We modify eq. (2) to include the competition by

$$v_{\mathrm{CL_{ren}}}^{(1)} = \left(Q_{\mathrm{GF}} + \frac{V_{\mathrm{max}}^{(1)}}{K_{\mathrm{m}}^{(1)}(1 + \frac{C_u^{(2)}}{K_{\mathrm{m}}^{(2)}}) + C_u^{(1)}} \right) \left(1 - \mathrm{F_{re-abs.}}^{(1)} \right) \cdot C_u^{(1)}$$

$$v_{\mathrm{CL_{ren}}}^{(2)} = \left(Q_{\mathrm{GF}} + \frac{V_{\mathrm{max}}^{(2)}}{K_{\mathrm{m}}^{(2)}(1 + \frac{C_u^{(1)}}{K_{\mathrm{m}}^{(1)}}) + C_u^{(2)}} \right) \left(1 - \mathrm{F_{re-abs.}}^{(2)} \right) \cdot C_u^{(2)}$$

As with the case of Oseltamivir-Probenicid competitive inhibition, uni-directed inhibition can be achieved by greatly diverging K_{m} values (factor 10^4) for compounds B and C. After mapping B to $C^{(1)}$ and C to $C^{(2)}$ in the 'simulation object', the simulation is performed; the results are shown in Fig. 3 (middle and bottom panels) for intravenous administration of 2 mg/(kg body weight) of compound A and C. This example nicely illustrates the phenomenon of extended drug exposition of compound B (active metabolite) as a result of a drug-drug interaction.

In our physiological context (Fig. 1), a total of 162 ordinary differential equations is necessary to simulate the pharmacokinetics of the three compounds, including drug-drug interactions; – on the basis of the presented concepts, MEDICI-PK generates these equations from seven user-defined local physiological models only!

5 Conclusion and Outlook

Considerable progress has been made in the development of *in silico* models to predict and understand the pharmacokinetics of new compounds, in particular in early drug discovery. As a result, modelling and simulation is possible prior to any *in vivo* experiments, solely based on *in vitro* data. We present the principles and concepts of a software design that efficiently allows to build up PBPK models in terms of the underlying physiological processes, combining user-friendliness and flexibility. These principles and concepts are the basis of the software tool MEDICI-PK that has been used to illustrate our approach.

We believe that the combination of *in vitro* experiments and *in silico* modeling has the potential to drastically increase the insight and knowledge about relevant physiological and pharmacological processes in drug discovery. In anticipation of modelling not only the distribution of the drug in the body, but also its effect (disease modelling), one future challenge will be the combination of pharmacokinetics and effect related metabolic networks or signalling pathways for a better understanding of the disease dynamics. An example would be the treatment-induced selective pressure on viral dynamics. The presented concepts, realized in MEDICI-PK, are powerful and flexible enough to also support these future tasks.

Acknowledgements

It is a pleasure to thank Michael Wulkow (CiT, Rastede) for fruitful and constructive discussions. M.v.K. and W.H. acknowledge financial support by the

DFG Research Center MATHEON "Mathematics for key technologies: Modelling, simulation, and optimization of real-world processes", Berlin.

References

1. H. van de Waterbeemd and E. Gifford, ADMET in silico modelling: towards prediction paradise? Nature Reviews Drug Discovery (2003) Vol.2:3 192-204
2. D. Butler, Wartime tactic doubles power of scarce bird-flu drug. Nature (2005) Vol.438 6
3. P. Poulin and F.P. Theil, A priori prediction of tissue:plasma partition coefficients of drugs to facilitate the use of physiologically-based pharmacokinetic models in drug discovery. Journal of Pharmaceutical Sciences (2000) Vol.89:1 16-35
4. P. Poulin, K. Schoenlein and F.P. Theil, Prediction of adipose tissue: plasma partition coefficients for structurally unrelated drugs. Journal of Pharmaceutical Sciences (2001) Vol.90:4 436-447
5. P. Poulin and F.P. Theil, Prediction of pharmacokinetics prior to in vivo studies. I.Mechanism-based prediction of volume of distribution. Journal of Pharmaceutical Sciences (2002) Vol.91:1 129-56
6. P. Poulin F.P. and Theil, Prediction of pharmacokinetics prior to in vivo studies. II. Generic physiologically based pharmacokinetic models of drug disposition. Journal of Pharmaceutical Sciences (2002) Vol.91:5 1358-70
7. G. Hill, T. Cihlar, C. OO, E.S. Ho, K. Prior, H. Wiltshire, J. Barrett, B. Liu and P. Ward, The Anti-Influenza Drug Oseltamivir Exhibits Low Potential To Induce Pharmacokinetic Drug Interactions Via Renal Secretion—Correlation Of In Vivo And In Vitro Studies. Drug Metabolism and Disposition (2002) Vol.30:1 13-19
8. G. He, J. Massarella and P. Ward, Clinical Pharmacokinetics of the Prodrug Oseltamivir and its Active Metabolite Ro 64-0802. Clinical Pharmacokinetics (1999) 37:6 471-484
9. W. Huisinga, R. Telgmann and M. Wulkow, The Virtual Lab Approach to Pharmacokinetics: Design Principles and Concept. Drug Discovery Today (2006) (accepted)
10. M. Rowland, L. Balant and C. Peck, Physiologically Based Pharmacokinetics in Drug Development and Regulatory Science: A Workshop Report (Georgetown University, Washington, DC, May 29-30, 2002) AAPS Pharmaceutical Science (2004); 6 (1) 1-12
11. M. von Kleist and W. Huisinga, Hierachical Approach to Physiologically Based Pharmacokinetics: Refining Models with Increasing Knowledge (2006) in preparation
12. M. Gibaldi and D. Perrier, Pharmacokinetics 2nd ed. Marcel Dekker (1982), New York

On the Interpretation of High Throughput MS Based Metabolomics Fingerprints with Random Forest

David P. Enot, Manfred Beckmann, and John Draper

Institute of Biological Sciences, University of Wales, Aberystwyth, SY23 3DA, UK
{dle, meb, jhd}@aber.ac.uk

Abstract. We discuss application of a machine learning method, Random Forest (RF), for the extraction of relevant biological knowledge from metabolomics fingerprinting experiments. The importance of RF margins and variable significance as well as prediction accuracy is discussed to provide insight into model generalisability and explanatory power. A method is described for detection of relevant features while conserving the redundant structure of the fingerprint data. The methodology is illustrated using two datasets from electrospray ionisation mass spectrometry from 27 *Arabidopsis* genotypes and a set of transgenic potato lines.

1 Introduction

The generation, representation and interpretation of metabolomics data plays increasingly important role in many research fields ranging from plant phenotyping, food safety, pharmacology to human nutrition[1]. A true metabolomics approach is *data driven* as the scientific discovery of relevant metabolic changes is usually done without much *a priori* knowledge about the problem. This is contrasted with metabolite profiling that focuses on analysing a small targeted subset of compounds. Metabolite fingerprinting provides a comprehensive view of the metabolome with a potential access to 100's of compounds in a single run[2]. Data analysis plays a central role in the process as it aims to derive meaningful knowledge that could be used to design new experiments or to perform complementary chemical identifications. As any experimental observation is always accompanied with some degree of uncertainty and imprecision due to the randomness and variability of most biological phenomena, the aim of statistical analysis is to evaluate if an observed effect is genuine or has happened by chance. In 'omics' technologies, the number of the training samples is often limited and commonly less than the dimension of the feature space, significant effects are almost always detected. This is even more important in metabolomics because of a combination of factors both associated with the complexity of relations between metabolites in metabolic pathways and natural variability of biological materials. Despite the implementation of controlled experimental environments and the relative standardisation of analytical protocols, extraction of meaningful knowledge remains challenging to a certain

M.R. Berthold, R. Glen, and I. Fischer (Eds.): CompLife 2006, LNBI 4216, pp. 226–235, 2006.
© Springer-Verlag Berlin Heidelberg 2006

extent and is a perilous task. When linking mathematical modelling to any biological problems statistical significance must always be compared to the biological significance of the conclusions. Many statistical and machine learning techniques can be successfully applied to produce accurate models, but this is often at the expense of generating classification models that are opaque to further interpretability especially when dimensionality reduction, feature selection or orthogonal signal correction are utilised. We stress that the goal of a data modelling experiment is not only to discriminate sample classes with high accuracy but also to identify and interpret metabolome signals important to the biological problem. In other words, model robustness must be assessed alongside an understanding of its mechanism and the highlighted features. Here, we discuss the use of Random Forest in order to fulfill both modelling goals and to provide scientists with necessary, comprehensible and safe measures to assess results and interpret their experiments.

2 Materials and Methods

2.1 Random Forest

Random Forest (RF) is a tree classification algorithm that grows an ensemble of trees[4]. In order to keep a low degree of correlation between the trees, three elements differ from the general approach to build a decision tree: 1) each tree of the forest is built on a bootstrap sample of the initial training set (bagging) 2) at each node, the best split is chosen among a subset of the initial pool of variables (random input vector); 3) there is no pruning step, *i.e.* each tree is grown large and probably overfits the bootstrap sample. Designation of the final class membership takes place by determining the winning class from the votes on the overall ensemble of trees (aggregation). Briefly, the choice of RF for the present investigation was dictated by several factors. First, the model does not overfit as the generalisation error will reach a certain value no matter how many trees are built (*Strong law of large numbers*)[4]. In practice, RF produces highly competitive models often outperforming all other approaches including support vector machines. Secondly, RF can deal with highly dimensional and correlated datasets without an initial reduction of dimensionality of the dataset allowing direct high throughput comparison of for example raw fingerprint data containing the whole compositional information. Finally, RF is easy to use *off the shelf* and requires very little parametrisation which makes this algorithm reproducible across sites/experiments and safe to use for scientists even with little machine learning knowledge. In this respect, we are interested here in two properties of RF.

Prediction votes and margins. Generalisability of a model is usually assessed by looking at the number of samples correctly described. However, the class boundary complexity is not represented explicitly and there is no guarantee to get a good class membership for new samples especially when the training set is not big enough. In this paper, we examine a useful by-product of voting methods, *prediction votes* which may be used to assess the confidence in predicting individual observations. Lets consider $\mathbf{X}$ the FI-MS fingerprint matrix of dimension $N \times p$, Y a class vector of each N observations ($Y = (Y_1, Y_i, ..., Y_N), Y_i \in \{1, K\}$), a forest of b trees T and I a variable indicating the class prediction from T_b. The margin $mg(\mathbf{X}, Y)$ is by definition the difference between the score (averaged number of votes) for the true class and the largest score of the rest of the classes:

$$mg(\mathbf{X}, Y) = av_b I(T_b(\mathbf{X}) = Y) - \max_{j \neq Y} av_b I(T_b(\mathbf{X}) = Y)$$

The margin measures the extent to which the average number of votes at X,Y for the right class exceeds the average vote for any other class. The larger the margin, the higher is the confidence that the example belongs to the actual class. A positive margin means that the example is correctly classified.

Variable importance. A perceived drawback of RF compared to classic decision trees is the lack of interpretability of the final model. A partial solution to this issue relies on RF's ability to produce an importance score for each variable of the model. In the implementation used for this study, the importance score is determined by comparing the performance of each tree on the original matrix and that obtained in the samples left out for construction of the tree with each variable randomly permuted. The higher the mis-classification when a variable is masked, the more important the variable for the problem under study.

2.2   Linear Discriminant Analysis

Linear Discriminant Analysis (LDA) is chosen as a reference algorithm in this study due to its popularity among biologists (thanks largely to its simple concept and graphical representation) and also because it has well explored statistical foundations. The aim of LDA is to find a projection P that maximises between-class separability (S_B) while minimising within-class variability (S_W). This is accomplished by solving ($S_W^{-1} S_B$)$P = P\Lambda$ which is equivalent to the eigensystem problem of $S_W^{-1} S_B$. However, in sample size problems (as in metabolomics), S_W is always singular and/or unstable requiring first reduction of the data dimension by techniques such as Principal Components Analysis (PCA). Because the determination of the optimal number of principal components representative for a dataset is not trivial and the number of dimensions varies from one comparison to another introducing a bias to the estimation of the separability measure, we have opted for a 2 step procedure proposed in [5]: the number of principal components to retain is equal to the rank of the covariance matrix $\mathbf{X}$ (usually $N - 1$) and S_W is replaced by a version where the less reliable eigenvalues have been replaced. From Λ, canonical correlation R is obtained by the following transformation: $R = \sqrt{\Lambda/(\Lambda + 1)}$. R defines the proportion of explained variance in the projection P and takes values between 0 and 1. According to our experience, this *flavour* of LDA tends to outperform the usual approach encountered in metabolomics literature (often described as Discriminant function Analysis) especially when the ratio between data points and variables is very small.

2.3   Implementation

All calculations were implemented in R on a dual processor PowerPC G5 (1.8 GHz) with 2GB SDRAM. Two additional R packages *randomForest* and *ROCR* were used to perform RF and receiver operating characteristic (ROC) analysis. Prediction votes of the *out of bag* training samples are used to compute the average margin $\overline{mg}(\mathbf{X}, Y)$. Test set margins are determined from the prediction votes of the test samples across all the trees of the RF model.

2.4   Biological Material

For this study we have used data from flow infusion electrospray ionisation mass spectrometry (FIE-MS). The advantage of this fingerprinting technique is that acquired signals (ion mass to charge ratio: m/z) can be linked directly to candidate metabolites by virtue of atomic mass[3,2]. To illustrate our approach, two sources of data are presented here:

Potato tubers - Only a subset from the data published in[3] is used in this study. It comprises the two near isogenic potato lines Désirée 1 and 2 (De1 and De2) and three transgenic lines developed in the cultivar Désirée 1 (SST18, SST20, SST36). The transgene codes for the enzyme sucrose:sucrose transferase (SST) which synthesises inulin-like fructans in potatoes, where they do not naturally occur.

Arabidopsis **-** This experiment comprises a heterogeneous set of 27 *Arabidopsis* lines comprising ecotypes, mutants or transgenic lines from the Landsberg (Ler0), Columbia (Col2) and C24 backgrounds. Details are given in Table 1.

Table 1. *Arabidopsis* dataset: genotype names and corresponding backgrounds

Line	Background	Line	Background	Line	Background
eds	Ler	Col2	Col2	lsd5	Col2
uvr21	Ler	amt11	Col2	35SNahG	Col2
Ler0	Ler	amt12	Col2	pgm1	Col2
Ak1	-	amt14	Col2	vtc12	Col2
Bch1	-	amt2	Col2	C24	C24
La0	-	axr31	Col2	AoPR1GFP	C24
Nd0	-	etr11	Col2	CPD2	C24
Ws0	-	fah12	Col2	CPD31	C24
Col0	Col2	lsd1	Col2	CPD9	C24

Data partitioning. FIE-MS data were collected in different analytical batches over a few weeks, each containing a randomised and representative selection of samples with good coverage of all genotypes. Training/test set partitioning was carried out on the basis of analytical batches. The number of training/test plant replicates per class is respectively 32/16 and 18/12 for the potato and the *Arabidopsis* datasets.

Data processing. Preprocessing of the data is not the scope of this paper. Briefly, we used a common approach that consists of combining sets of equally spaced m/z values into one mass unit bin. This results in good signal calibrations without the need of extensive spectrum alignment and reduces substantially the data dimension. Additionally, m/z signals may be linked directly to a specific metabolite, particularly in cases where a single, abundant metabolite dominates the intensity of a specific m/z value. The mass to charge ratios covered are respectively $m/z = [65, 997]$ and $m/z = [108, 1000]$ for the potato and *Arabidopsis* datasets (resulting in 933 and 893 variables). Data were log-transformed and normalised to total ion count before data analysis. Only results concerning the negative mode (*Arabidopsis*) and positive mode (potato) are presented here but similar conclusions were derived from both ionisation modes in data from both plant species.

3   Results

Initial investigations regarding RF model properties were carried out using all possible 351 pairwise comparisons between the 27 *Arabidopsis* genotypes. Despite the fact that we are using two machine learning methods which are able to cope with the multiclass problem, pairwise decomposition of the general problem has different advantages from both theoretical and biological angles. This includes a gain in interpretability, a lower complexity of the decision boundaries and a better association between relevant features and classes[6]. While the set of 27 genotypes is by nature very heterogeneous without any *a priori* biological hypothesis, this dataset covers a wide range of problems where some lines are very different whereas others are likely to be similar to other metabolic mutations and display little, if no, phenotype differences. Such coverage of weaker models to stronger models is of biological importance, as our aim is to provide tools for the detection of models with potentially sufficient explanatory power to guide deeper investigation of any significant metabolic phenotype.

The initial objective was to check whether the average margin $\overline{mg}(\mathbf{X}, Y)$ of the training samples for each classifier would provide sufficient information regarding generalisability of each model. Figure 1 illustrates the relationship between the area under curve (AUC) of each binary comparison and the average margin of the training set samples. The AUC is one way to summarize the relationship between sensitivity (true-positive rate) and specificity (false-positive rate) across all possible threshold values that define the decision boundary resulting in a better estimate than the model accuracy. Figure 1 can also be used to check the direct correlation between the confidence in the prediction votes of the training and those of the test data.

We also check how the margin can be linked to two class separability measures. On figure 1, the average margin for each model is compared to two widely used multivariate and univariate measures: 1) LDA canonical correlation R and 2) number of signals found significant by univariate analysis of the variance at different p-values (note that no correction for controlling the false discovery rate has been performed). A very good degree of agreement exists between the canonical correlation and the model margin. Despite the fact that LDA and RF differ

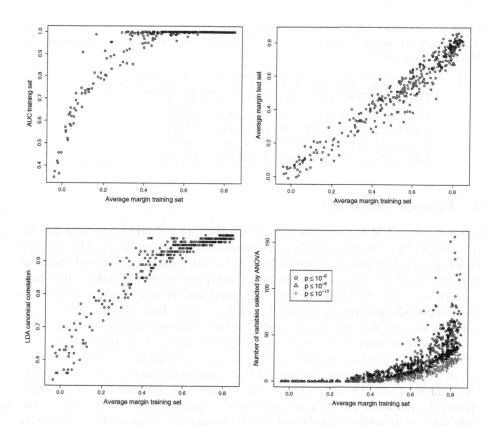

Fig. 1. Average margin of the training samples for each 351 classifier versus (i) AUC of the RF classifier(upper left); (ii) average margin of the unseen samples(upper right) (iii) LDA canonical correlation (lower left) and (iv) number of features found significant by univariate analysis of variance at different levels of significance(lower right)

by their principles, vote confidence is higher when the between-class/within-class ratio is maximised. The number of expected differences resulting from univariate testing is also reflected in the model average margins as this measure increases when discriminative features are more likely to exist. Deeper evaluation of potentially explanatory metabolome m/z signals generated by RF is detailed in the following section.

Statistics related to a selected set of comparisons is gathered in Table 2 in order to illustrate the biological relevance of the models. *Arabidopsis* genotype comparison involves the wild type Colombia Col-2 and two of its mutants (fah1-2 and pgm1) where strong metabolic changes were expected and lines which were not directly affected in their metabolism (ecotype C24 and its salt transporter lines CPD9 and CPD31). Model statistical measures are ordered according to the expected biological outcome; ecotype differences being the largest (C24 and Col2).

Table 2. Statistics related for five selected comparisons - [1] Values calculated on the training set and in brackets test set - [2] Number of features found significant by ANOVA at $p \leq 10^{-8}$

Comparison	R	$\overline{mg}(\mathbf{X}, Y)^{(1)}$	$\mathrm{AUC}^{(1)}$	$\mathrm{Accuracy}^{(1)}$	Num. feat.$^{(2)}$
C24-CPD9	0.71	0.07(0.11)	0.72(0.72)	83(72)	0
C24-CPD31	0.78	0.21(0.23)	0.88(0.85)	83(72)	0
Col2-fah12	0.86	0.37(0.40)	0.99(1.00)	97(96)	10
Col2-pgm1	0.93	0.56(0.51)	0.99(1.00)	92(92)	5
C24-Col2	0.98	0.81(0.80)	1.00(1.00)	100(100)	41

4   Feature Selection

One feature of ESI fingerprints is that signals from a single compound will be represented by a range of signals corresponding to different forms of the ionised molecule including combinations of Na^+, K^+ or NH_4^+ adducts or loss of chemical groups (CH_3COO^- or H_2O). Most importantly the problems under study may concern metabolic pathway changes and correlated signals could correspond to the compounds involved in the biochemical reaction. Thus identification of potentially relevant features would rarely lead to a unique solution [8] as opposed to biomarker discovery where one may be interested in a single signal that differentiate healthy to diseased patient for example. Analysis of redundant features are of importance to both the biologist to unravel metabolic mechanisms and the analytical chemist to interpret such signals and orientate complementary chemical identifications. As discussed before, Random Forests can cope naturally with multicolinear signals and the RF importance score is able to correctly highlight explanatory variables. However, there is no general (statistical) procedure to select and determine a cut off between the most and least relevant variables. Feature selection with Random Forest has been discussed elsewhere on transcriptomics data[7]. The basic approach (as in other algorithms such as SVM-RFE[8]) relies on ranking the initial pool of variables according to the importance score and subsequent iterative elimination of some of the variables is performed until a more parsimonious model with the lowest internal error is obtained. The final model contains a small number of "non-redundant" variables (note that the authors of [7] did not mention testing for orthogonality of features selected in the final subset) and exhibits similar or improved predictive abilities, but at the risk of excluding other meaningful biological information. To overcome this latter issue, we describe an alternative approach to define feature relevance in the context of RF. Rather than ranking each feature by their absolute importance score, we aim at assigning a level of significance to each signal, an approach which has more relevance for analytical chemists and biologists. As the importance score distribution is unknown, we test the null hypothesis that the importance score is not relevant to the model by means of a permutation test[9]: class labels are randomly permuted (preserving the fingerprint matrix structure) and importance scores are calculated for each permutation. The p-value is defined as the fraction of times one gets a score larger or equal to the original importance score of the un-permuted data. In the following, significance levels are determined by permuting the fingerprint matrix 2000 times.

We applied our feature selection technique to the same set of 351 pair-wise comparisons as above. Part of the results are illustrated in Figure 2. The number of significant feature highlighted by our approach is related to the average margin of classifier (Figure 2 - left). This matches previous observations where discrim-

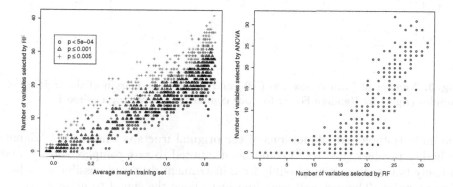

Fig. 2. Feature selection with Random Forests - Left: number of features selected by RF at different threshold versus the strength of the overall model for each classifier-Right: number of features by RF at $p \leq 0.001$ versus the number of significant signals by ANOVA at $p \leq 10^{-10}$ for each classifier

inatory signals were detected by straightforward univariate testing (Figure 1 - lower left). In order to check that RF feature selection does not forget "obvious" discriminative features, the number of variable found significant by both method is illustrated on Figure 2 at $p \leq 0.001$ (RF) and $p \leq 10^{-10}$ (ANOVA). Feature selection with Random Forests has several advantages over classical hypothesis testing: decision trees in general do not make assumptions regarding the shape of the data (mixture of categorical and continuous variable can be used); multi-variate interactions will be accounted for in the model; finally, importance score is related to the predictive abilities.

To illustrate in more details the aspects of our approach, we have applied identical procedures to compare four potato tuber lines. As the original aim of the project was to investigate unexpected effects in GM potato [3], each transgenic line containing novel fructans was compared to its progenitor (Désirée 1). In addition, we included the genotype Désirée 2 which is a near isogenic variant of Désirée 1 derived from tissue culture. The number of signals found significant at different p-values and the corresponding averaged margin of the training samples is given in Figure 3. The margin curve has a classic shape in a sense that removal of variables is translated into a more robust model, reaching a plateau for around 50 variables. These signals underwent a more comprehensive analysis by carbohydrate-specific LC-MS to determine which of these signals were fructans [3]. High ranked signals at the lowest p-value ($p \leq 5.10^{-4}$) corresponded to

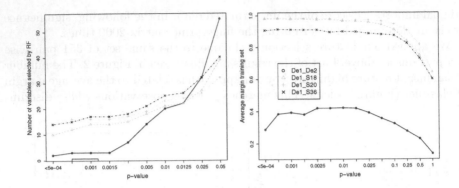

Fig. 3. Number of features selected (left) and the average margin of the training set samples (right) by Random Forest for four comparisons involving Désirée 1

the main isotope and adduct ions of the original fructans whereas m/z signals at higher p-values ($p \leq 10^{-3}$) could be attributed to rare complex adducts and unlikely isotopes resulting possibly from instrument drift and possible mass binning mistakes. These signals are often just above the signal to noise ratio and explain relative discrepancy between subsets identified in the three comparisons. For comparative purposes, variable selection as in [7] leads only to the following set of masses: $m/z = 544$ and 545 (S18), $m/z = 545$ and 705 (S20) and $m/z = 544, 545, 689, 690$ and 705 (S36). Those features can be positively interpreted, however one misses out on alternative explanatory solutions[3]. In Figure 3, we have compared Désirée 1 to its near isogenic line Désirée 2. Using the previous level of significance threshold ($p \leq 5.10^{-4}$), three signals ($m/z = 141, 316$ and 327) are still retained to explain differences between the two lines and could represent the effects of somaclonal variation on the metabolome. A significant increase in the average margin of the training set is not reflected in either accuracy and AUC of the training and test sets. Based on a maximum averaged margin, an optimal RF model would incorporate around 20 features ($p \leq 0.01$). Despite improving significantly prediction votes in both training and test sets, predictive accuracy remains low (75%). This is a common issue in feature selection when additions of extra features improve the model accuracy without carrying relevant biological knowledge[8].

5   Discussion

Apart from the obvious constraints imposed by natural variability one of the main problems facing a biologist using high dimensional omics technologies is data paucity (over fitting can be a problem) alongside a lack of direct access to variable relevance. Several key features of the Random Forest methodology we describe make important contributions to the model interpretation. By calculating (i) model margins and (ii) attributing p-values to each variable we can determine models with high predictive power and select only those that would lead to significant variables for further interpretation. Models utilising FIE-MS fingerprint data with lower predictive properties and generalisability rarely contain a high proportion of linked signals and are easily identified. In addition, we stress that the investigation of the relationship between correlated m/z signals can help to quickly identify potentially metabolites relevant to the study.

References

1. Weckwerth W.: Metabolomics in systems biology Annu. Rev. Plant Biol. **54** (2003) 66989
2. Allen J. et al.: High-throughput classification of yeast mutants for functional genomics using metabolic footprinting Nature Biotech. **21** (2003) 692-696
3. Catchpole, G.S. et al.: Hierarchical metabolomics demonstrates substantial compositional similarity between genetically modified and conventional potato crops. Proc. Natl. Acad. Sci. USA. **102** (2005) 14458-62
4. Breiman L.: Random Forests. Machine Learning **45(1)** (2001) 261-277
5. Thomaz C.E., Gillies D.F.: A maximum uncertainty LDA-based approach for limited sample size problems with application to face recognition. Technical Report 2004/1, Imperial College London (2004).
6. Tsujinishi D., Koshiba Y., Abe S.: Why Pairwise Is Better than One-against-All or All-at-Once, Proc. International Joint Conference on Neural Networks **1** (2004) 693-698
7. Diaz-Uriarte R., Alvarez de Andres S.: Gene selection and classification of microarray data using random forest BMC Bioinformatics **7** (2006) 3
8. Guyon I., Elisseeff A.: An Introduction to Variable and Feature Selection J. Machine Learning Res. **3** (2003) 1157-1182
9. Good P.: Permutation Tests: A Practical Guide to Resampling Methods for Testing Hypotheses, Springer Series in Statistics (2000).

Construction of Correlation Networks with Explicit Time-Slices Using Time-Lagged, Variable Interval Standard and Partial Correlation Coefficients

Wouter Meuleman[1,3], Monique C.M. Welten[1,2], and Fons J. Verbeek[1]

[1] Leiden Institute of Advanced Computer Science (LIACS), Leiden University,
Niels Bohrweg 1, 2333CA Leiden, The Netherlands
[2] Division of Molecular Cell Biology, Institute for Biology, Leiden University,
Wassenaarseweg 64, 2333AL Leiden, The Netherlands
{wmeulema, mwelten, fverbeek}@liacs.nl
[3] Current address W. Meuleman: Information and Communication Theory group,
Delft University of Technology, P.O. Box 5031, 2600GA Delft, The Netherlands

Abstract. The construction of gene regulatory models from microarray time-series data has received much attention. Here we propose a method that extends standard correlation networks to incorporate explicit time-slices. The method is applied to a time-series dataset of a study on gene expression in the developmental phase of zebrafish. Results show that the method is able to distinguish real relations between genes from the data. These relations are explicitly placed in time, allowing for a better understanding of gene regulation. The method and data normalisation procedure have been implemented using the R statistical language and are available from http://zebrafish.liacs.nl/supplements.html.

1 Introduction

Microarray data potentially disclose relations between genes. To that end a lot of research effort is spend on revealing networks of genes from microarray time-series experiments. Two established approaches for doing so are by using correlation networks ([1,2]) and dynamic Bayesian networks ([3,4,5]).

Correlation networks reveal only global associations between genes (via control mechanisms) over all time-points. No exact indication of time and interaction, i.e., interaction control, is given in these networks. The advantage of correlation networks is however, that they can be built in a deterministic manner, based on relatively simple calculations.

In general, dynamic Bayesian networks give more insight in the data by providing information on when certain genes are related, i.e., over which time-points. The major drawback of dynamic Bayesian networks however, is that there is no polynomial time algorithm known for finding the optimal network structure for a particular dataset, i.e., this problem is NP-hard ([6]). Network structure candidates have to be tried one by one and since the number of possibly suitable

M.R. Berthold, R. Glen, and I. Fischer (Eds.): CompLife 2006, LNBI 4216, pp. 236–246, 2006.

structures grows exponentially with the number of variables (genes) used, this quickly becomes computationally infeasible[1].

In this paper, we propose a method that combines these approaches so that correlation networks with explicit time-slices are obtained. This is achieved by using time-lagged, variable interval, standard and partial correlation coefficients. We build onto the notion of correlation networks. However, in order to avoid ambiguity with biological terms, from this point onwards the term 'correlation models' will be used.

2   Approach

The model building process involves the selection of strong correlations between gene expression profiles of which one profile is lagged in time and the interval over which the correlation is calculated is shortened. More specifically, correlations are calculated for all possible gene-pairs over transitions in time which are supported by the dataset. A t-test is used as a cutoff filter to obtain strong correlations only. These correlations are subsequently used to build an initial model incorporating explicit time-slices.

Indirect relations are removed from the model by calculating first order partial correlations for each strong correlation in the model. For two arbitrary genes g_1 and g_2 this works as follows. For each strong correlation between the profiles of g_1 and g_2 over time-points t_i and t_j, where $i < j$, partial correlations are calculated. This is done by controlling the influence of each of the genes in t_i, excluding g_1, on the correlation between g_1 and g_2.

Results are assessed for significance. Those below a certain significance threshold are removed from the model. The result of this is a correlation network with explicit time-slices, where connections between reporters indicate direct relations.

3   Methods

3.1   Correlations

In order to establish which genes portray a relation, we can calculate the correlation between their expression profiles. Standard correlation models are based on Pearson's correlation coefficient. However, these are not sufficient here, as we are primarily interested in correlations over time, that is, we would like to find out which expression profiles show a similar trend, with a certain time-lag.

An example of such correlations is given in Figure 1. Profiles 1 and 2 are strongly correlated with a time-lag of 2 units. A method for finding such correlations is used in [4] for grouping genes with possible common pathways. A drawback of this method is that the correlation calculation is always initiated

---

[1] Other recent approaches have acknowledged this problem and constructed ways of restricting the search space so that solutions could be generated in reasonable time ([7]).

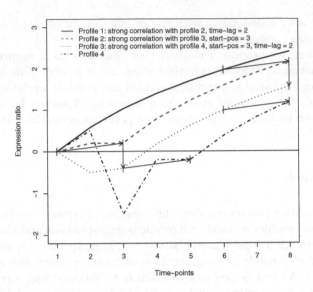

Fig. 1. Profiles with strong time-lagged and/or variable interval correlations. Arrows between profiles indicate the interval over which the correlation was calculated.

from the first time-point for one of the profiles. It is therefore not sufficient for finding the strong correlations present between profiles 2 and 3 and between 3 and 4. Profiles 2 and 3 correlate strongly from time-point 3 onwards, but not before that. Another example of this is the correlation between profile 3 and 4, which involves an additional time-lag of 2 units. These correlations must be considered, otherwise valuable information is underutilised. Therefore, we extended the standard formula for Pearson's correlation coefficient to incorporate, in addition to time-lags, a starting position s from which to look for correlations. This extension is formally described as:

$$\rho^*_{s,\ell}(x,y) = \frac{\displaystyle\sum_{i=s}^{|x|-\ell} (x_i - \mu_{x_{s,|x|-\ell}})\,(y_{i+\ell} - \mu_{y_{s+\ell,|x|}})}{\sqrt{\displaystyle\sum_{i=s}^{|x|-\ell} (x_i - \mu_{x_{s,|x|-\ell}})^2 \sum_{i=s}^{|x|-\ell} (y_{i+\ell} - \mu_{y_{s+\ell,|x|}})^2}}, \tag{1}$$

where $|x|\ (=|y|)$ denotes the number of time-points available and $\mu_{x_{s,|x|-\ell}}$ and $\mu_{y_{s+\ell,|x|}}$ are the means of variables x_s to $x_{|x|-\ell}$ and $y_{s+\ell}$ to $y_{|x|}$ respectively.

3.2   Significance of Correlations

The adjustment of intervals has consequences for the significance of correlations calculated. In order to test the significance of a correlation, a t-test with $n'-2$ degrees of freedom is used. Here, n' indicates the number of time-points used for calculating the correlation, given by

$$n' = n - (s - 1) - \ell \tag{2}$$

where n is the total number of time-points and ℓ and s the used time-lag and starting position respectively. Because we need at least one degree of freedom, we can define the allowed ranges for s and ℓ as

$$s = 1, \ldots, n - 3 \tag{3}$$
$$\ell = 1, \ldots, n - s - 2. \tag{4}$$

We do not look at zero-lag correlations, as we are interested in finding relations over time.

3.3 Partial Correlations

It may be the case that two strongly correlated variables (i.e., genes) have a common controller variable or that there is a moderator variable through which two variables are correlated. In both cases, two variables are strongly correlated but either one does not cause the other.

Partial correlations allow one to investigate the correlation between two variables while controlling (or, excluding the influence of) one or several other variables. That is, the partial correlation between variables x and y, controlling z, is the correlation between the parts of x and y that are uncorrelated with z. This restricts the results to only direct relations between variables, ruling out possible common controller or moderator variables.

The first order partial correlation between variables x and y, controlling z, is formally described by

$$\rho(x, y | z) = \frac{\rho(x, y) - \rho(x, z)\rho(y, z)}{\sqrt{\left(1 - \left(\rho(x, z)\right)^2\right)\left(1 - \left(\rho(y, z)\right)^2\right)}}. \tag{5}$$

Note that when testing first order partial correlations for significance, the number of degrees of freedom to be used is $n' - 3$.

3.4 Model Building

Using Eq. 1, pairwise correlations can be calculated between g selected genes, over ℓ time-lags and with variable start positions s. This results in a correlation matrix, given by

$$M_{s,\ell} = \begin{bmatrix} \rho^*_{s,\ell}(1, 1) & \cdots & \rho^*_{s,\ell}(1, g) \\ \vdots & \ddots & \vdots \\ \rho^*_{s,\ell}(g, 1) & \cdots & \rho^*_{s,\ell}(g, g) \end{bmatrix}. \tag{6}$$

For each combination of s and ℓ, supported by the data and allowing for enough degrees of freedom, such a correlation matrix is calculated. Analogous to Eq. 3 and 4, the number of valid correlation matrices m is given by

$$m = \sum_{s=1}^{n-3} \sum_{\ell=1}^{n-s-2} 1 \,. \tag{7}$$

Of each correlation in the correlation matrices, the significance is assessed using a t-test. All correlations with a p-value below a pre-defined value are selected. A p-value of 0.05 is commonly used as a cutoff point to distinguish strong correlations from weak ones. Using these correlations, an initial model can be built. An example of such a model is shown in Figure 2.

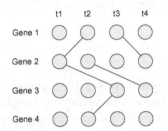

Fig. 2. Example of initial model, with $g = 4$ genes and $n = 4$ time-points   **Fig. 3.** Example of final model

The initial model consists of a matrix of nodes, made up of g rows (genes) and n columns (time-points). Each column can be seen as a *slice* of a time-series, a common notion in the field of dynamic Bayesian networks and time-series analysis in general. Each slice contains g nodes, depicting the genes at that time-point. Two nodes in different slices are connected if and only if the reporters they represent show a strong enough correlation.

After an initial model has been built, the correlations are verified using partial correlations. For each two reporters connected between time-points t_i and t_j in the initial model, where $i < j$, partial correlations are calculated. This is done in an iterative manner, controlling each of the remaining reporters in time-point t_i one by one.

The significance of each calculated partial correlation is calculated (using a t-test with $n' - 3$ degrees of freedom, as described before) and if one of them turns out not to be significant, this means the controlled reporter is responsible for a substantial part of the strong initial correlation. This could indicate a common controller or indirect relation.

We are particularly interested in direct relations. Therefore, in order to obtain strong direct relations only, if one or more partial correlations do not reach a pre-specified significance threshold, the initial correlation is removed from the model.

Inferring causation from correlation coefficients is discouraged ([8]), even after correction using partial correlations ([9]). However, we can extend the undirected connections in the model to directed ones when it is physically impossible for two-way relationships to occur. Because we are considering correlations between data from different time-points and time is generally directional, we can infer directions. Figure 3 shows an example of a final model, including the directed connections.

In the strict sense, the connections indicate which reporters have strongly correlating expression profiles over a certain interval. A life scientist may interpret these connections as *possible* regulations, or at least as hints towards them. After all, when a reporter shows a certain behaviour from one time-point onwards and another reporter shows similar behaviour from a later time-point onwards, the first could well be regulating the second.

4   Results

The model building process has been tested using the dataset resulting from experiments by Linney et al. ([10]). This dataset is the result of a zebrafish time-series microarray experiment and contains data for 8 equidistant time-points, ranging from 10 to 24 hours post fertilisation (hpf).

4.1   Data Preparation

The processed data as used in [10] is publicly available from the ArrayExpress[2] microarray data repository (accession number E-TIGR-17). However, to have full control over the dataset, we preferred to start from the "raw" readouts instead, provided by Renae Lin Malek ([10]). The normalisation procedure used consisted of spatial, dye and conditional normalisations ([11]). All procedures were implemented in the R programming language.

4.2   Variable Selection

Using ANOVA and Benjamini & Hochberg false discovery rate analysis ([12]) with a false positive level of 5%, genes significantly differentially expressed over the 8 time-points have been selected for further analysis.

We have focussed on one particular family of genes, so called *hox* genes. These have been recognised ([13,14]) as important factors in the developmental mechanism of vertebrates and are significant in research involving axial patterning during development. The dataset used contains data of early developmental stages, in this time-frame these genes are expected to show temporal differential expression. Because of this, the 28 *hox* genes present in the normalised dataset have been selected as a test case for further analysis.

4.3   Model Building

Following Eq. 7, for the 28 *hox* genes, $m = 15$ correlation matrices have been calculated. Out of the total amount of 11760 ($28 \times 28 \times 15$) correlations, 677 have been determined to be strong enough to be selected (p-value < 0.05).

---

[2] ArrayExpress: http://www.ebi.ac.uk/arrayexpress

Subsequently, partial correlations have been calculated, of which the significance was tested with p-value < 0.05 as well. Out of the initial 677 correlations, only 8 had strong enough partial correlations. This means that they suggest strong direct relations. In Table 1 the 8 selected correlations are listed with the lowest obtained partial correlation. The selected correlations have been used to build the final model as shown in Figure 4.

Table 1. Connections in final *hox* model. The 's' and '$s + \ell$' columns denote the starting points and time-lags of the connections as they are used in Eq. 1. The 'Acc. no.' column contains the accession numbers of reporters and the 'Corr.' and 'Part. corr.' columns contain the values of the standard and partial correlation coefficients respectively. Values have been rounded to 4 decimals for clarity. The last column contains references to literature showing evidence of found connections.

	Selected correlations						Lowest partial correlations			
s	Acc. no.	Gene	$s+\ell$	Acc. no.	Gene	Cor.	Acc. no.	Gene	Part. cor.	Evidence
1	U40995	*hoxb1b*	2	AF071261	*hoxc13a*	-0.9173	AF071258	*hoxc11a*	-0.8706	[15, 16]
1	U40995	*hoxb1b*	2	BI705747	*hoxb8a*	-0.9177	AF071258	*hoxc11a*	-0.8538	[15, 17–19]
1	AF071264	*hoxc4a*	3	AF071261	*hoxc13a*	-0.9697	AF071247	*hoxa5a*	-0.9092	[15, 16, 20]
1	AF071245	*hoxa3a*	3	BI705747	*hoxb8a*	0.9895	AF071252	*hoxb4a*	0.9456	[15, 17]
1	AF071241	*hoxa13a*	4	Y14530	*hoxb8a*	0.9880	BI705747	*hoxb8a*	0.9661	
2	AF071251	*hoxb1a*	3	BI705747	*hoxb8a*	-0.9841	AF071264	*hoxc4a*	-0.9121	[15, 17]
2	AF071264	*hoxc4a*	4	X17267	*hoxb6a*	-0.9972	AF071251	*hoxb1a*	-0.9850	[17]
3	AF071251	*hoxb1a*	4	AF071258	*hoxc11a*	-0.9908	AF071252	*hoxb4a*	-0.9669	[15, 17]

4.4   Result Verification

Hox genes are conserved throughout vertebrate evolution and possess a property commonly referred to as spatial and temporal co-linearity ([21,22,23]). This property is reflected in the names of the different *hox* genes. Lower numbered genes, e.g., *hoxb1a*, are typically expressed earlier and more anterior (closer to the head) than genes with a higher index, e.g., *hoxc11a*.

Thus, temporal co-linearity of *hox* genes already implicitly provides a method for verification of the model given in Figure 4. The exact origin of the order of expression of *hox* genes is still subject of research; moderators, regulation through other (*hox*) genes, might be involved ([24]).

Seven (7) out of the eight (8) relations present in the model do indeed directly correspond to the co-linearity pattern; the gene in the earlier time-point is lower-numbered than the gene in the later time-point. One relation, i.e. from *hoxa13a* to *hoxb8a* (cf. Figure 4) is an exception to the co-linearity rule and we have to reason as to whether this is a false positive result or another explanation clarifies this outcome.

The forward connections as computed by our method are confirmed by published experimental approaches ([15,16,17,18,19,20], cf. Table 1). Very pronounced examples from the recent literature are the connections between *hoxb1b* and *hoxb8* and between *hoxa3a* and *hoxb8a*.

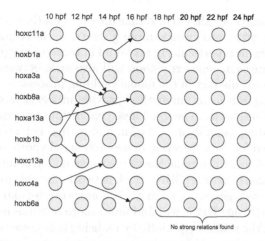

Fig. 4. Correlation model of *hox* genes. Genes are ordered in such a way to provide maximum graphical clarity. No strong relations were found between genes in time-points 18 to 24 hpf.

5   Discussion

We have elaborated on a methodology for developing correlation models to find good candidates for gene regulatory networks from microarray time-series data. In [25], a method is presented for inferring correlation networks using both standard and partial correlations. The principal difference between their method and the method presented in this paper is that we do incorporate explicit time-slices.

The principle of operation of our method was illustrated with a time-series dataset of zebrafish development and the results on this dataset provide insight in the usefulness of the method. We focussed on the connections between *hox* genes. The use of partial correlations helped to eliminate indirect relations from this subset of genes and thus focus on the more prominent ones over time. One should realise that indirect relations still might play a role, i.e., other genes/factors could influence the gene expression. Moreover, the connections found should not be interpreted as conclusive direct causations or regulations between genes. In our correlation model a connection indicates that gene expressions show a similar trend in behaviour over a certain time-span and with a certain time-lag.

Clearly, the more samples available the more distinct the results will be. Microarray data have some shortcomings that we should realise in the light of understanding the results. First, the samples used in Linney et al. consisted of crushed embryos. According to the co-linearity rule, in addition to a temporal order, *hox* genes also have a spatial order of expression. Using whole embryos complicates the possibility of detecting such arrangements which has an effect on the relations found by our method. Second, the accuracy with which relations are detected strongly depends on the sampling over time, i.e., time resolution. Given the time resolution of the dataset that we explored, consisting of samples taken with 2 hour intervals, a number of relations between genes might not have been detected. In considering our results one should also take into account that gene relation events do not necessarily adhere to given sampling intervals; the segmentation phase of zebrafish development is particularly profound in terms of morphological changes and therefore changes in gene expression will also arise between the time-points ([26]) and not appear distinct in the model building process.

Models resulting from using our method provide an overview of some of the more profound connections between genes. These connections are explicitly situated in time-slices, providing valuable information about their behaviour. The resulting models can be used for the generation of new research hypotheses, as well as for a starting point for heuristic approaches to the construction of further models, such as dynamic Bayesian networks (DBN). Such networks are able to model more complex relations between genes, which may yield more precise and complete models. On the other hand, using DBNs will find much more relations making it more difficult to interpret the results, especially by realising that noise is an intricate component of microarray data. Using microarrays in the onset of new experiments, a simple and straightforward outcome will help the researcher in a better way and this is exactly what we set out to accomplish with our method.

Spatio-temporal analysis approaches such as Whole Mount ISH ([27,28]) could complement the information gained using this method. Spatial, rather than temporal, information about the expression of genes yields information on co-location in addition to that of co-expression. The unresolved, rather differing, connection between *hoxa13a* and *hoxb8a* could be explained by understanding that these genes co-localize in distinct domains ([29,30]), which is something that can be shown using such co-locational analysis methods. Further research is therefore directed towards the combination of machine learning approaches on both spatial and time resolved gene expression patterns.

Our results were confirmed in the literature from analysis of wild types in zebrafish ([15,17]) and additional data from explant as well as knock-out experiments in other model systems (e.g., [20,19]). We have claimed that the dataset had some limitations; in addition we argue having data, i.e., microarrays, for more time points would enrich our findings so as to get more confirmation on the relations found and disclose other more transient relations. At the time of our experimentation these data were not available, however, the repositories that are currently being populated with microarrays can provide additional data necessary for such analysis.

We have shown that this method allows hypothesis generation and extraction of seed networks for computationally more intensive approaches. It makes use of multiple correlation matrices which can be calculated independently. As a consequence, the method allows for easy parallelization.

6 Conclusion

We have presented a method for building correlation networks with explicit time-slices which uses a novel way to find relations over time. The essential ingredients are time-lagged, variable interval, correlation coefficients.

The method was tested for genes of the zebrafish *hox*-family and has produced good results which are biologically meaningful and in correspondence with the literature.

Acknowledgements

The authors wish to acknowledge Renae Lin Malek for providing the dataset used to test our methodology. Many thanks to Veronica Vinciotti for providing invaluable support during normalisation of the used data and to Walter Kosters and Miranda Mandjes-van Uitert for helpful comments. This work has been partially supported by the ZF-Models consortium and the University Fund of Leiden University.

References

1. Wu, X., Ye, Y., Subramanian, K.: Interactive analysis of gene interactions using graphical Gaussian model. ACM SIGKDD Workshop on Data Mining in Bioinformatics **3** (2003) 63–69
2. Schfer, J., Strimmer, K.: An empirical Bayes approach to inferring large-scale gene association networks. Bioinformatics **21**(6) (2005) 754–764
3. Friedman, N.: Inferring cellular networks using probabilistic graphical models. Science **303** (2004) 799–805
4. Kellam, P., Liu, X., Martin, N., Orengo, C., Swift, S., Tucker, A.: A framework for modelling virus gene expression data. Intelligent Data Analysis **6**(3) (2002) 267–279
5. Kim, S.Y., Imoto, S., Miyano, S.: Inferring gene networks from time series microarray data using dynamic Bayesian networks. Brief Bioinform **4**(3) (2003) 228–235
6. Chickering, D., Geiger, D., Heckerman, D.: Learning Bayesian networks is NP-hard. Technical Report MSR-TR-94-17, Microsoft Research (1994)
7. Zou, M., Conzen, S.D.: A new dynamic Bayesian network (DBN) approach for identifying gene regulatory networks from time course microarray data. Bioinformatics **21**(1) (2005) 71–79
8. Hill, A.: The environment and disease: association or causation? In: Proc R Soc Med. Volume 58. (1965) 295–300
9. Phillips, C., Goodman, K.: The missed lessons of Sir Austin Bradford Hill. Epidemiologic Perspectives & Innovations **1**(3) (2004)
10. Linney, E., Dobbs-McAuliffe, B., Sajadi, H., Malek, R.L.: Microarray gene expression profiling during the segmentation phase of zebrafish development. Comp Biochem Physiol C Toxicol Pharmacol **138**(3) (2004) 351–362
11. Wit, E., McClure, J.: Statistics for microarrays: design, analysis and Inference. John Wiley & Sons Ltd, Chichester, West Sussex, U.K. (2004)
12. Pavlidis, P.: Using ANOVA for gene selection from microarray studies of the nervous system. Methods **31**(4) (2003) 282–289
13. Duboule, D.: The vertebrate limb: a model system to study the Hox/HOM gene network during development and evolution. Bioessays **14**(6) (1992) 375–384
14. Krumlauf, R.: Hox genes in vertebrate development. Cell **78**(2) (1994) 191–201

15. Prince, V.E., Moens, C.B., Kimmel, C.B., Ho, R.K.: Zebrafish hox genes: expression in the hindbrain region of wild-type and mutants of the segmentation gene, valentino. Development **125**(3) (1998) 393–406

16. Thummel, R., Li, L., Tanase, C., Sarras, M.P., Godwin, A.R.: Differences in expression pattern and function between zebrafish hoxc13 orthologs: recruitment of Hoxc13b into an early embryonic role. Dev Biol **274**(2) (2004) 318–333

17. Prince, V.E., Joly, L., Ekker, M., Ho, R.K.: Zebrafish hox genes: genomic organization and modified colinear expression patterns in the trunk. Development **125**(3) (1998) 407–420

18. Bel-Vialar, S., Itasaki, N., Krumlauf, R.: Initiating Hox gene expression: in the early chick neural tube differential sensitivity to FGF and RA signaling subdivides the HoxB genes in two distinct groups. Development **129**(22) (2002) 5103–5115

19. Forlani, S., Lawson, K.A., Deschamps, J.: Acquisition of Hox codes during gastrulation and axial elongation in the mouse embryo. Development **130**(16) (2003) 3807–3819

20. Gaunt, S.J., Strachan, L.: Temporal colinearity in expression of anterior Hox genes in developing chick embryos. Dev Dyn **207**(3) (1996) 270–280

21. Duboule, D., Doll, P.: The structural and functional organization of the murine HOX gene family resembles that of Drosophila homeotic genes. EMBO J **8**(5) (1989) 1497–1505

22. Graham, A., Papalopulu, N., Krumlauf, R.: The murine and Drosophila homeobox gene complexes have common features of organization and expression. Cell **57**(3) (1989) 367–378

23. Duboule, D.: Vertebrate hox gene regulation: clustering and/or colinearity? Curr Opin Genet Dev **8**(5) (1998) 514–518

24. Kmita, M., Duboule, D.: Organizing axes in time and space; 25 years of colinear tinkering. Science **301**(5631) (2003) 331–333

25. de la Fuente, A., Bing, N., Hoeschele, I., Mendes, P.: Discovery of meaningful associations in genomic data using partial correlation coefficients. Bioinformatics **20**(18) (2004) 3565–3574

26. Kimmel, C.B., Ballard, W.W., Kimmel, S.R., Ullmann, B., Schilling, T.F.: Stages of embryonic development of the zebrafish. Dev Dyn **203**(3) (1995) 253–310

27. Verbeek, F.J., Lawson, K.A., Bard, J.B.: Developmental bioinformatics: linking genetic data to virtual embryos. Int J Dev Biol **43**(7) (1999) 761–771

28. Welten, M., De Haan, S., Bertens, L., Noordermeer, J., Lamers, G., Spaink, H., Meijer, A., Verbeek, F.: ZebraFISH: Fluorescent *in situ* hybridization protocol and 3D images of gene expression patterns. Zebrafish (Accepted) (2006)

29. Knosp, W.M., Scott, V., Bchinger, H.P., Stadler, H.S.: HOXA13 regulates the expression of bone morphogenetic proteins 2 and 7 to control distal limb morphogenesis. Development **131**(18) (2004) 4581–4592

30. Sakaguchi, S., Nakatani, Y., Takamatsu, N., Hori, H., Kawakami, A., Inohaya, K., Kudo, A.: Medaka unextended-fin mutants suggest a role for Hoxb8a in cell migration and osteoblast differentiation during appendage formation. Dev Biol **293**(2) (2006) 426–438

The Language of Cortical Dynamics

Peter Andras

School of Computing Science
University of Newcastle
Newcastle upon Tyne
NE1 7RU, UK

Abstract. Cortical dynamics can be recorded in various ways. Theoretical works suggest that analyzing the dynamics of recorded activities might reveal the workings of the underlying neural system. Here we describe the extraction of an activity pattern language that characterizes the dynamics of high-resolution EEG data recorded. We show that the language can be formulated in terms of probabilistic continuation rules which predict reasonably well the dynamics of activity patterns in the data.

1 Introduction

Cortical dynamics has been analyzed since the availability of appropriate recording machinery and methodology (e.g., EEG, fMRI, etc.). There are several theories that explain some aspects of the observed dynamics [6,8,11].

Recent works have shown that observed cortical dynamics in awake animals in stimulus free environment resemble very much the cortical activity observed in stimulus rich environment [5,9]. It has been shown that the orientation column structure of cat V1 can be reconstructed by classification of activity patterns recorded from an animal in stimulus free environment [9]. It has also been shown that the activity patterns in the visual cortex of awake adult ferrets in stimulus free environment resemble to around 80% the activity of the visual cortex recorded in stimulus rich environment [5].

The above results suggest that the cortex performs continuous information processing, which is independent to a good extent from the actual environmental input to the nervous system. One way to test this hypothesis is to analyze high resolution cortical activity recordings and try to extract a probabilistic grammar describing the evolution of recorded activity patterns [2,3,4]. The hypothesis is confirmed if it is possible to extract a such grammar, which can predict the evolution of cortical activity sufficiently correctly. We note that similar approaches have been used before to analyse and describe multiple neural spike trains [12] and simultaneous recordings from single neurons [1,13].

Here we analyze high resolution EEG data recorded from cat cortex and we aim to extract regularities that can be used to predict the changes in the activity patterns. The results show that it is possible to extract high reliability rules that link together activity patterns. The set of such rules can be seen as an abstract probabilistic grammar of the cortical activity representing the language of cortical dynamics.

M.R. Berthold, R. Glen, and I. Fischer (Eds.): CompLife 2006, LNBI 4216, pp. 247–256, 2006.
© Springer-Verlag Berlin Heidelberg 2006

The paper is structured as follows. First we discuss the data, second we present the analysis methodology, third we show the analysis results and analyze the extracted language of cortical dynamics, and finally, we discuss the results and their implications.

2   The Data

The high resolution surface EEG (electro-encephalogram) data was recorded from cat cortex in the lab of WJ Freeman. The data consists of 20 sequences of 64-channel recordings, each sequence having 3000 recordings (i.e. the total data consists of 20 times 3000 items of 64-dimensional vectors = 60000 64-dimensional vectors). The recording was done in awake animals using high resolution recording from relatively small patches of exposed cortical surface. Further details and earlier analyses of the data in terms of similar models have been reported in other papers [7,11].

To reduce the variation in the data we consider the difference series of the original data, i.e., the 64 dimensional difference vectors between consecutive original recordings. A set of example recordings are shown in Figure 1. The 64 data values were organized into an 8 x 8 square (i.e. first 8 values in the first row, second 8 values in the second row, and so on).

3   Data Analysis: Methods and Results

Our aim is to find a set of typical activity patterns (i.e. activity patterns that occur with high frequency int he data) in the high resolution EEG data and to establish probabilistic regularities describing which activity pattern follows earlier activity patterns.

3.1   Searching for Typical Activity Patterns

We segmented the data into 2 x 2, 3 x 3 and 4 x 4 data matrices (i.e. the 8 x 8 data matrices are divided in smaller components, for example dividing the 8 x 8 matrices in 4 x 4 matrices means that the 8 x 8 matrices are divided in the middle horizontally and vertically resulting four 4 x 4 data matrices - upper-left, upper-right, lower-left, lower-right). We did the segmentation with and without overlap between the segments in order to capture many possible configurations of cortical activities that may organize into typical activity patterns. (Note that the segmentation does not induce a resolution hierarchy in a strict sense.)

For each such data matrix we analyzed the joint activity of the corresponding recordings with the aim of finding typical activity patterns. We consider an activity pattern typical if it appears regularly (i.e. with high frequency), possibly with minor random modifications.

We used Kohonen networks to find the typical activity patterns [10]. For each data matrix (i.e. a data matrix is defined by its size and position, for example

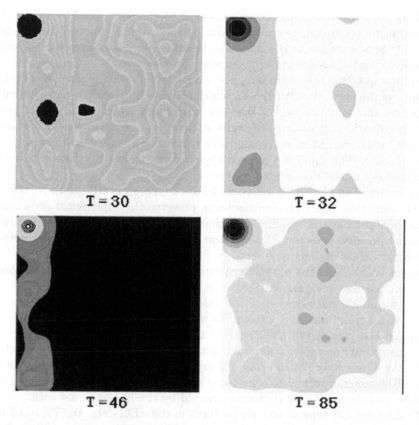

Fig. 1. High resolution EEG recordings from cat auditory cortex. The data is represented using pseudo-gray-scale colors. The T values indicate the sequential position of the data within the data sequence (each sequence contains 3000 data sets).

there are four 4 x 4 data matrices with no overlap) we built a separate Kohonen network with 25 nodes, containing a prototype vector and a position vector (note that the dimensionality of the prototype vector is equal with the dimensionality of the data vectors, i.e., the number of recordings within the data matrix; the dimensionality of the position vectors is 2, i.e., the Kohonen network nodes are arranged into a planar lattice). The Kohonen networks were trained with the corresponding data vectors (i.e., the 2 x 2, 3 x 3, 4 x 4 data matrices selected from each 64 electrode recordings, from 10 data sets considered for training of the networks). The training process is described by the following equations:

$$i = \arg\min_{j} ||x_t - w_j|| \tag{1}$$

$$I = \{j | \rho_t > ||v_j - v_i||\} \tag{2}$$

$$w_j = w_j + c_t \cdot (x_t - w_j), \quad j \in I \tag{3}$$

where x_t is the t-th data vector presented to the network. The values ρ_t, c_t are parameters that change gradually during the training process, starting from relatively large values towards zero. The training process stops when the values of these parameters reach zero (or get very close to zero).

An interpretation of the training process is the following: equation (1) selects the node of the network, with index i, which attracts the data vector x_t ; equation (2) defines the neighborhood of the node with index i within the planar lattice of nodes; equation (3) updates the value of prototype vectors contained in nodes belonging to the neighborhood of the node which attracts the currently presented data vector x_t The updating of prototype vectors means that they are moved slightly closer to the presented data vector.

The nodes of the Kohonen networks learn the representative data prototypes. At the same time the Kohonen networks preserve the topological structure of the original data space, which means that similar data vectors are attracted by the same prototype vector. Data vectors attracted by each Kohonen network node form a class of data vectors.

We analyzed each class of activity patterns identified by the trained Kohonen network to assess its compactness. Our underlying hypothesis is that if data vectors forming a class show little variation then they are likely to be representatives of a typical activity pattern. To measure the compactness of classes we computed the variance of the distances of data vectors belonging to the class from the prototype vector contained in the Kohonen network node, which attracts the data vectors of the class. Compact activity pattern classes identified by the Kohonen network were considered as typical activity patterns represented by their prototype activity pattern extracted by the Kohonen network.

We detected 437 typical activity patterns in the EEG data. In 77% of all data matrices we were able to detect at least one typical activity pattern. The number of typical activity patterns detected for each data matrix varied between 0 and 21. The average number of detected typical activity pattern for data matrices with at least one such pattern was 10.16, i.e., in average we detected around ten typical activity patterns for each data matrix.

The original data series were analyzed using the trained Kohonen networks. We classified the appropriately grouped data values and we generated a symbolic representation of the data sets using the detected typical activity patterns. If we found that they belong to a class corresponding to an identified typical activity pattern we included the associated symbol of the typical activity pattern in the symbolic translation of the data vector. In this way every 64 dimensional original data vector was translated into a set of symbols, each symbol corresponding to an identified typical activity pattern. An example of such translation is shown in Figure 2.

3.2   Searching for Language Rules

We analyzed the consecutiveness relationships between typical activity patterns with the aim of establishing probabilistic continuation rules (i.e. one-step ahead continuation rules). We note that in general many-steps ahead continuation rules

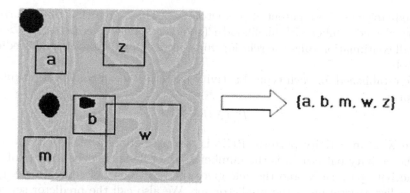

Fig. 2. Translation of original activity data into symbolic data. The rectangles surround the data matrices for which we are able to classify the data as belonging to a class associated with a typical activity pattern. The letters in the rectangles are the symbols associated with these typical activity patterns. On the right hand side of the arrow is the symbolic translation of the original data.

may play important role in determining the evolution of cortical dynamics. In our approach we approximate the complete set of continuation rules (which includes many-steps ahead rules as well) by a Markovian approximation consisting of one-step ahead rules.

For each typical activity pattern represented by a symbol in the translated data series, we considered all preceding sets of symbols. For example, if 'w' is symbol representing a typical activity pattern, which occurs in the symbolic translation of data vectors $x_{t_1}, x_{t_2}, x_{t_3}$, we consider as preceding symbol sets the symbol translations corresponding to the data vectors $x_{t_1-1}, x_{t_2-1}, x_{t_3-1}$. If the symbol set translations of these latter data vectors are $\{a, m, r\}$, $\{a, m, n, k, z\}$, $\{a, m, t, y\}$, then these symbol sets will be considered as preceding sets for the symbol 'w'.

Considering all preceding sets for a selected symbol we aim to identify subsets of these preceding sets, such that these subsets predict with relatively high likelihood the occurrence of the selected symbol in the symbol set translation of the next data vector. We call these subsets predictor sets of the selected symbol. Taking into account that the number of all possible combinations of symbols present in preceding sets can be extremely large, we adopt a simplified approach. We take all pairs of symbols in preceding sets and count the frequencies of such symbol pairs. We construct a graph of symbols found in preceding sets with weighted edges representing the frequency of the corresponding symbol pair. Cliques in this graph with all edges having a weight higher than an appropriately set lower limit are considered potential predictor sets of the selected symbol (in the case of our data we set this lower limit to be 20, i.e., we consider the cliques of the graph which contains edges between nodes which appear simultaneously in preceding sets more than 20 times). For all such potential predictor sets we count their occurrences in the whole data set and also the occurrences of the selected symbol in the symbol representation of the data vector following data vectors that contain in their symbol representation the potential predictor set. We accepted as predictor sets those potential predictor sets, which predicted the occurrence of the selected symbol with likelihood $P(selected\ symbol\ |\ predictor\ set) > 0.2$. We call continuation rules the relationships between predictor sets and predicted symbols.

We established for each typical activity pattern a set of continuation rules of the form

$$P_X^i(X|R^i(X)) \tag{4}$$

where X is an activity pattern, R(X) is the predictor set of activity patterns for the activity pattern X, is the number of the rule among the rules related to the activity pattern X, and the rule gives the probability of generating X given the earlier generation of the predictor set. We also call the predictor set of an activity patter X the reference set of this activity pattern [3]. The rules describing the probabilistic relationship between production of the pattern X after the preceding generation of patterns forming the reference set, are also called referencing rules of the pattern X [3]. Note that the same pattern may be generated by many rules, and the reference sets may overlap providing reference for many possible rules to generate new patterns. In general reference sets may contain earlier activity patterns with variable time distance between the predicted pattern and the pattern in the reference set. In the case of our analysis we restricted the reference sets to symbol sets generated immediately before the generation of the pattern X, implying that the rules that we extracted describe a Markovian language (i.e., symbols at time t+1 depend only on symbols that were present at time t).

Our analysis led to a set of rules describing which activity patterns may emerge after the presence of a set of activity patterns present at a given moment in the data. The set of these probabilistic rules constitute a probabilistic grammar describing the activity patterns emerging in the analyzed EEG data. In our view this rule set is an approximation of the true grammar governing the generation of activity patterns determining cortical dynamics. An example of such rules is presented in Figure 3.

3.3   Validation of Language Rules

To evaluate the correctness of predictions of established grammatical rules we analyzed another data set of similar EEG recordings that was not considered during the rule extraction phase.

The test data series were analyzed using the same Kohonen networks that were trained with the data used for rule extraction. The data vectors of test series were translated into series of symbol sets following the classification of the data matrices contained in the data vectors by the trained Kohonen networks. We searched for all occurrences of all established predictor sets for each of the 437 typical activity patterns. For each predictor set we calculated the probability of the predictor set being followed by the appropriate typical activity pattern. In this way we arrived to a set of probabilistic rules of the form

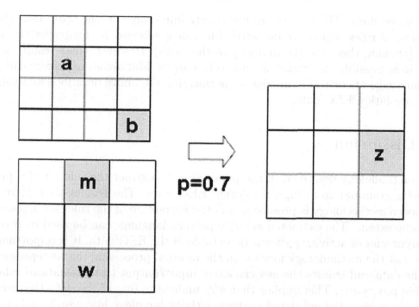

Fig. 3. An example of the language rules of the EEG data. The activity patterns on the left constitute a predictor set for the activity pattern on the right. The activity pattern on the right occurs with probability 0.7 after the occurrence of the predictor set. The letters represent identifiers of typical activity patterns in the corresponding data matrix.

$$\hat{P}_X^i\,(X|R^i(X))\tag{5}$$

a test rule being calculated for each of the established reference rules.

If the calculated reference rules are correct and they describe the dynamics of cortical activity, we expect that the probability values of rules in equations (4) and (5) do not differ significantly. Figure 4 shows the results for one 2 x 2 data matrix and for typical activity patterns associated with this data matrix. We calculated the differences between the calculated reference rule probabilities and the probabilities calculated for the same reference rule using the test data (i.e., $P_X^i(X|R^i(X)) - \hat{P}_X^i\,(X|R^i(X)))$. We averaged these differences for each typical activity pattern across all referencing rules of the given typical activity pattern. For each averaged difference we calculated the measure of how significantly it differs from zero using the t-test. The average difference does not differ significantly from zero if the significance measure (t value) in absolute value is below 1.96, which is the case in 5 out of 11 cases, and it differs very significantly from zero if the absolute value is above 2.57, which is the situation in 4 out of 11 cases, the remaining 2 cases having intermediary values of difference.

The results show that in half of the cases the rules are valid for the unseen data, and there are fewer cases in which there is a statistically very significant difference between the predictions of the extracted rules and the dynamics of the unseen data. The results are not overly impressive, in the sense that there are several rules which are detected, but not confirmed by analyzing the test data (though, these are the minority of the cases). However, these results show that it is possible to extract a relatively correct Markovian approximation of the language of cortical dynamics by performing the above described analysis of high resolution EEG data.

4   Discussion

The methodology described in the paper aims to extract the rule set of a probabilistic grammar from high resolution EEG data. The results show that the proposed methodology is promising and the extraction of the rules set is possible to some extent. The extracted activity pattern language can be used to predict the dynamics of activity patterns detectable in the EEG data. It is important to note that the methodology focuses on the internal processing that is represented by the data and ignores the association of input/output transformations related to this processing. This implies that any understanding of the data that might be derived from the extracted pattern activity language may not be related in any direct way to input/output processing and may not reveal explicitly how inputs are represented and processed within the system, and how outputs of the system under study are generated. Earlier theoretical analysis [2,3] suggests that such internal computations should be critical for computations performed by neural systems, and without understanding them it might not be possible to understand how action-perception cycles of the system are generated in the context of the system's experience and processing structure.

We note that the data analysis that we applied to determine the abstract language of the EEG data is equivalent to learning an abstract language from positive examples only. Such learning problems as data analysis problems are ill defined in principle, and can be solved correctly under appropriate constraints, which regularize the search space of possible language rules. The implication of this for the presented work is that the rules that we extract form the data might be side effects of the actual rules. At the same time another factor that contributes to establishment of incorrect rules is that we search for a Markovian approximation (i.e., current state depends only on the previous state) of the full language that might contain referencing rules including in their reference sets activity patterns from the longer past.

We show here that it is possible to extract an internal language from EEG data and that this language can be seen as the language of cortical dynamics. This language of cortical dynamics includes in a generic sense the experience of the system and its own structure and other constraints that determine how the system processes information. The presented approach opens new avenues for research on cortical dynamics and suggests that we should focus on understanding of the inner language of cortical dynamics instead of focusing on input/output transformation properties of cortical neural activity. This focus on internal processing is in agreement with earlier theoretical works [4,3] and also

1:	2:	3:	4:	5:
X	X	X	X	X
6:	7:	8:	9:	10:
X	X	2.0924	3.6264	1.5205
11:	12:	13:	14:	15:
3.0546	2.4206	7.1962	3.0181	X
16:	17:	18:	19:	20:
1.6241	0.5750	X	-0.4839	X
21:	22:	23:	24:	25:
X	-0.4544	X	X	X

Fig. 4. The significance of differences between probabilities associated with language rules and corresponding probabilities determined from the test data. Out of the 25 possible (corresponding to the 25 nodes of the Kohonen network) typical activity patterns 11 were recognized (i.e., these formed sufficiently compact classes around the prototype activity pattern). The differences between the probabilities of corresponding referencing rules were calculated, and we calculated how significantly the average of these differences differs from zero using the t-test. The significance measures (t-values) are shown in the lower parts of the boxes.

with experimental findings [5,9], which show that internal processing resembles the general processing in presence and absence of external stimuli.

Finally we note that previous works on extracting similar probabilistic language descriptions of neural activity focused primarily on single neurons [1] or on small sets of neurons [13]. In a previous paper [4] we described the conceptual background of activity pattern computation in neural systems and we have shown that it can be applied to large scale neural systems through the analysis of emerging activity patterns and establishing of probabilistic continuation rules. In the present paper we continue this work and focus on the data analysis methodology and the derivation of the continuation rule representation of the language of cortical dynamics.

Acknowledgement. The author would like to thank Professor Walter J Freeman for providing the high resolution EEG data.

References

1. Abeles, M., Bergman, H., Gat, I., Meilijson, I., Seidemann, E., Tishby, N., & Vaadia, E.: Cortical activity flips among quasi-stationary states. PNAS 92 (1995) 8161-8620.
2. Andras, P.: A model for emergent complex order in small neural networks. Journal of Integrative Neuroscience 2 (2003) 55-70.
3. Andras, P.: Pattern languages: A new paradigm for neurocomputation. Neurocomputing 58 (2004) 223-228.
4. Andras, P.: Computation with chaotic patterns. Biological Cybernetics 92 (2005) 452-460.
5. Fiser, J., Chiu, C., & Weliky, M.: Small modulation of ongoing cortical dynamics by sensory input during natural vision. Nature 431 (2004) 703-718.
6. Freeman, W.J.: Role of chaotic dynamics in neural plasticity. Progress in Brain Research 102 (1994) 319-333.
7. Freeman, W.J. & Burke, B.C.: A neurobiological theory of meaning in perception. Part IV: Multicortical patterns of amplitude modulation in gamma EEG. International Journal of Bifurcation and Chaos 13 (2003) 2857-2866.
8. Kay, L.M., Lancaster, L.R., & Freeman, W.J.: Reafference and attractors in the olfactory system during odor recognition. International Journal of Neural Systems 7 (1996) 489-495.
9. Kenet, T., Bibitchkov, D., Tsodyks, M., Grinvald, A., & Arieli, A.: Spontaneously emerging cortical representations of visual attributes. Nature 425 (2003) 954-956.
10. Kohonen, T: Self-Organizing Maps. Springer-Verlag, Heidelberg (1995).
11. Ohl, F.W., Scheich, H., & Freeman, W.J.: Change in pattern of ongoing cortical activity with auditory category learning. Nature 412 (2001) 733-736.
12. Radons, G., Becker, J.D., Dulfer, B., & Kruger, J.: Analysis, classification, and coding of multielectrode spike trains with hidden Markov models. Biological Cybernetics 71 (1994) 359-373.
13. Seidemann, E., Meilijson, I., Abeles, M., Bergman, H., & Vaadia, E.: Simultaneously recorded single units in the frontal cortex go through sequences of discrete and stable states in monkeys performing a delayed localization task. The Journal of Neuroscience 16 (1996) 752-768.

A Simple Method to Simultaneously Track the Numbers of Expressed Channel Proteins in a Neuron

A. Aldo Faisal[1] and Jeremy E. Niven[2]

[1] Dept. of Zoology, Cambridge University, CB2 3EJ Cambridge, UK
a.faisal@zoo.cam.ac.uk
[2] Smithsonian Tropical Research Inst., Roosevelt Ave., Tupper Building 401 Balboa,
Ancón Panamá, República de Panamá

Abstract. Neurons express particular combinations of ion channels that confer specific membrane properties. Although many ion channels have been characterized the functional implications of particular combinations and the regulatory mechanisms controlling their expression are often difficult to assess in vivo and remain unclear. We introduce a method, Reverse Channel Identification (RCI), which enables the numbers and mixture of active ion channels to be determined. We devised a current-clamp stimulus that allows each channels characteristics to be determined. We test our method on simulated data from a computational model of squid giant axons and from fly photoreceptors to identify both the numbers of ion channels and their specific ratios. Our simulations suggest that RCI is a robust method that will allow identification of ion channel number and mixture in vivo.

1   Introduction

Ion channels confer specific membrane properties on neurons, affecting their input-output relationships. Although numerous studies of neurons have shown that they contain characteristic sets of ion channels it is increasingly clear that the specific number of a particular type of ion channel and the ratio of particular ion channel types within neurons varies substantially. This variation is likely to exist not only within homologous neurons between individuals but also within populations of homologous neurons from the same individual. The specific variation in the numbers of particular ion channels and their ratios may have important functional implications not just on neural processing but also on neural energy requirements. Previous studies have suggested that neurons use activity sensors, such as calcium, to regulate the number and type of ion channels in their membranes [1,2]. This tuning of expressed and active channel composition may be beneficial for improving signaling characteristics of neurons [3,4,5,6,7,8,9,10]. Furthermore, the regulation of channel density ratios can improve metabolic efficiency [11,12,13] and reduce the effects of channel noise on action potential initiation and propagation[14,15]. The potential implications of

M.R. Berthold, R. Glen, and I. Fischer (Eds.): CompLife 2006, LNBI 4216, pp. 257–267, 2006.
© Springer-Verlag Berlin Heidelberg 2006

changes in the number of expressed ion channels and ratios on neural function suggest that knowledge of how these vary is essential for understanding neural processing, plasticity and adaptability. These values are, however, difficult to assess in vivo. Here we set out to develop a simple and quick method to identify these values in such a way that the same experimental preparation can be used for future experiments.

Our method is designed to identify the number of ion channels, N_i, given that the presence of each type ion channel identified by genetic,immunological or pharmacological methods in the membrane is accounted for and that each channel's kinetics $p_i(V, t)$ and conductance are known. The dynamics of a neurons membrane potential V can be described for an iso-potential patch of membrane (or equivalently a spherical neuron) by an inhomogeneous first-order differential equation [16]

$$\dot{V} = \frac{1}{C_m}\left[g_{\text{Leak}}(E_{\text{Leak}} - V) + \sum_i g_i(V, t)(E_i - V) + I(t)\right] \qquad (1)$$

where C_m is the specific membrane capacitance, g_{Leak} is the membrane leak (the transmembrane conductance without the voltage-gated conductances), E_{leak} is the leak reversal potential, $g_i(V, t)$ is the total voltage-dependent time-varying conductance for a set i of ion channels, E_i is the respective ion's reversal potential, and finally $I(t)$ is some time-varying input (be it synaptic or injected through an electrode) [16]. The solution of Eq. 1 is non-trivial, even for a well behaved input $I(t)$, as it involves products of V with a time-varying function $(g_i(V, t))$ and, therefore, usually requires numerical computation. The ionic conductances $g_i(V, t)$ are a non-linear and time-varying function of the channel open probability p_i times the number of channels N_i present in the membrane and its single channel conductance γ_i.

$$g_i(V, t) = p_i(V, t)\, N_i\, \gamma_i \qquad (2)$$

In the following we will drop the index i as we look in more detail at the parameters characterizing a channel.

Following [17,18] an ion channel will be open with probability p if it binds k gating particles (these particles are conceptual not physical). Each gating particle has a voltage-dependent probability $n(V, t)$ of being bound to the channel. The probability of the channel being open is, thus,

$$p_i(V, t) = n(V, t)^k \qquad (3)$$

n is defined by a first order ordinary differential equation (linear chemical kinetics), which depends on the forward and backward transition rate functions $\alpha(V)$ and $\beta(V)$.

$$\frac{d}{dt}n = \alpha(V)(1 - n(t)) - \beta(V)n(t) \qquad (4)$$

The transition rate function in turn is typically described by exponential or sigmoidal functions (which reflect the transition rates thermodynamic heritage)

[18]. The activation time constant τ of a channel is defined by $\tau = \frac{1}{\alpha+\beta}$. Similarly the steady state (voltage-clamped) channel open probability is given by

$$p(V, t \to \infty) = p_\infty(V) = \left(\frac{\alpha}{\alpha+\beta} \right)^k = (\alpha \, \tau)^k \tag{5}$$

Assume a neuron and its channels to be in their resting (steady) state. An up-shift in membrane potential followed by a down-shift in membrane potential, would thus first drive channels towards their new steady open probability and then, after the down-shift they would return to their previous steady state. Two types of channels with different α and β will therefore approach their steady state open probabilities to a differing degree depending on the duration of the membrane potential shift. Note, that such potential shifts would require a more difficult and complicated (sometimes infeasible) experimental approach to voltage-clamp a neuron (typically *in vitro*). *In vivo* the terms governing the

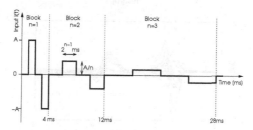

Fig. 1. Structure of the CRI input stimulus sequence (see text for details)

membrane potential $V(t)$ in response to a current input I(t) in Eq. 1 reflect the voltage-dependence of the ion channels involved:

$$\sum_i p_i(V, t) N_i \gamma_i (E_i - V) \tag{6}$$

where we have expanded the term g_i in terms of the probabilities p_i. The membrane response $V(t)$ is thus the solution of an integral-differential equation which is not analytically solvable, (without approximating away the specific kinetic properties of ion channels [16]). The voltage response to a stimulus will thus be non-linear (voltage-dependent) convolution of the contributions of each channel type's open probabilities $p_i(V, t)$ times its frequency of occurency N_i.

2   Result

The aim of our method is to yield quickly and accurately the mix and density of each ion channel species present in the cell, such that the results can be known while the experimental preparation can be used for other types of experiments.

We therefore use short stimulus protocol based on current-clamp injections (stimulus) and simultaneous membrane potential measurements (response),

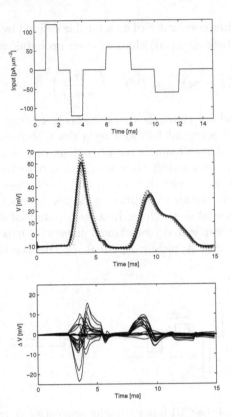

Fig. 2. CRI processing example (Top) The CRI stimulus $I(t)$. (Middle) Test neuron response (thick, solid line) and responses stored in the cross-reference (thin, dotted lines). (Bottom) Pairwise differences between the test neuron response and each stored response (thin, solid lines, for reference a thick solid line $\Delta V(t) = 0$ is drawn).

which can be easily measured intra-cellularly (sharp electrodes) or via whole cell patch-clamp.

The CRI method has the following prerequisites

- The total conductance for each ionic species e.g. potassium (K).
- The identity of the voltage-gated ion channels present in the membrane and their kinetics and single channel conductance have to characterized. In general the number of different types of ion channels per cells is limited to about two to twelve channels.
- The total membrane capacitance and resistance, which is used to calculate the membrane area to obtain absolute channel numbers (otherwise RCI will only yield channel mixture ratios).
- Reasonable space-clamp for the neuron recording (or if the ion channels are uniformly embedded in the membrane a sufficiently large area of membrane ($> 1 \ \mu m^2$) in an membrane-attached patch-clamp pipette).

The CRI method relies on capturing the relative contribution of each ion channel by using a suitable stimulus $I(t)$.

Varying the duration and amplitude of the input current will vary the path (rate of rise and duration at peak) of the membrane potential, thus probing regions of an ion channels rate functions α and β for different amount of times. Thus, one should be able to distinguish the differing underlying rate functions of the involved ion channels by probing them with different steps.

We conceived a stimulus sequence composed of blocks, each block consisting of a positive and negative step with the steps logarithmically increasing in their duration and linearly decreasing in amplitude (see Fig. 1 for illustration). Thus, the n-th block in the sequence is structured as follows: We give no current input for n ms, then a current step of size $+\frac{A}{n}$ of n ms duration, after a pause with no input of n ms duration, we give a step amplitude $-\frac{A}{n}$ of n ms duration. The purpose of this stimulus is to elicit membrane responses at varying combinations of: (1) the rate of change of membrane potential $\dot{V}$, (2) at different membrane potentials V and (3) over different periods of time.

The structure of the stimulus prunes the individual channel time constants and voltage activation characteristics apart, and thus exposing sufficient information such that the amount of each ion channel species' contribution to the total neuron response can be measured.

We represent the neuron's voltage response as follows. At $t = 0$ the current stimulus $I(t)$ is injected into a neuron (regardless whether *in silicio* or *in vitro/vivo*) and its membrane potential response $V(t)$ is digitized with a sampling rate f_{sample}. A response vector $\mathbf{v} = (\mathbf{v_1} \ldots \mathbf{v_n})$ is constructed, which elements v_i are the membrane potential $V((i-1)\frac{1}{f_{\text{sample}}})$. A response $\mathbf{v}$ can now be quantitatively compared to the similarity of any other response $\mathbf{v}^*$ by computing the Euclidean distance $d = \sqrt{\sum_{i=1}^{n}(v_i - v_i^*)^2}$. With this prerequisites we can outline the general method to process data with the CRI method.

CRI is divided into two steps contingent steps. The first step consists of constructing a cross-reference table of simulated neuron responses, each entry corresponding to a specific combination of ion channels and densities (Channel Response). I.e. if only two types of ion channels A and B are known to exist, then the simulated responses can be arranged in a 2-dimensional array and a simulated responses in a give row column pair thus specifies a specific mix and density of ion channels A and B. Successive rows contain simulated responses for increasing densities of channel A and each successive column containing simulated responses for an increasing density of channel B. This cross-reference table has to be constructed only once for each type of neuron. This allows the second step to be computationally light-weight and fast as it avoids to compute the simulated responses on-the-fly for each neuron recording.

In the second step, the experimentally recorded neuron responses are compared with the stored neuron responses in the cross-reference table. The simulated response which is closest to the neuron response is selected and the channel mix and density looked up (reverse Identification). A neuron response may be close to several simulated responses (which will be typically adjacent to each other) in the table and thus the channel mix and density can be interpolated between the individual reverse looked up entries. Given the membrane area and single channel conductance the individual channel numbers can be straightforward computed, while channel mixture ratios can also be computed without this information. The duration of the RCI protocol, from the first stimulus injection to the processed answer requires less then a second. In the following presentation of our results we used $f_{sample} = 10$ kHz and made our stimulus $n = 2$ blocks long.

2.1 The Squid Potassium Channel and Derived Virtual Channels Are Used to Test CRI Sensitivity

We test the robustness of our method by applying it to a virtual dataset. We simulated an iso-potential membrane patch, corresponding to a spherical cell soma, with membrane properties based on squid axon membrane (membrane resistance $g_{Leak} = 0.3$ mS cm^{-2} and membrane capacitance $C_m = 1$ μF cm^{-2}) [16]. To test the sensitivity we embedded the standard squid delayed rectifier potassium (K) channel and a second, slightly modified version of the K channel (virtual channel) in our model membrane.

2.2 The Reverse Channel Response Identification (CRI) Method

The squid delayed rectifier K channel is described by a 5 state discrete state model of gating [18]

$$c_1 \underset{\beta_n}{\overset{4\alpha_n}{\rightleftharpoons}} c_2 \underset{2\beta_n}{\overset{3\alpha_n}{\rightleftharpoons}} c_3 \underset{3\beta_n}{\overset{2\alpha_n}{\rightleftharpoons}} c_4 \underset{4\beta_n}{\overset{\alpha_n}{\rightleftharpoons}} o \quad .$$

The horizontal sequence of states c_1 to o corresponds to 0 to 4 bound gating particles and $k = 4$. Closed states, therefore, are c_1 to c_4 and the only open state is o. We define parameterized versions of the voltage-dependent forward transition rate functions $\alpha_n(V)$ and backward transition rate function $\beta(V)$

$$\alpha_K(V) = A_\alpha \frac{V_\alpha - V}{\left(\exp\left[1 - \frac{V}{V_\alpha}\right] - 1 \right)} \tag{7}$$

$$\beta_K(V) = A_\beta \, \exp\left[\frac{-V}{V_\beta}\right] \tag{8}$$

Following common convention the rate functions have units of ms^{-1} and the membrane potential V is applied "module" units [mV]. The indexed constants depend on the channel properties and in the case of the squid delayed rectifier K channel their values are $A_\alpha = \frac{1}{100}, A_\beta = \frac{1}{8}, V_\alpha = 10, V_\beta = 80$ [17] (without loss of generality the resting potential is 0 mV).

The parameterized rate equations allow us to vary the microscopic properties of channels that account for mesoscopic properties of channel behavior such as p or τ. Parameter variations will allow us to create virtual channels derived from

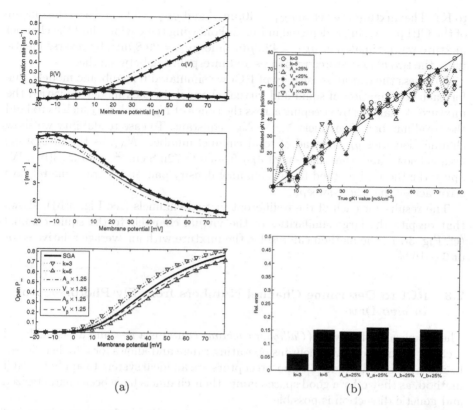

(a)   (b)

Fig. 3. (a) Overview of squid and six derived virtual channels used to test the acuity of the CRI method. (Top) Plot of forward (α) and backward (β) rate functions versus membrane potential. (Middle) Channel time constant ($\tau = \frac{1}{\alpha+\beta}$) versus membrane potential. (Bottom) Steady state channel open probability (p_{∞}) versus voltage-clamped membrane potential. **(b)** CRI channel density mixture detection acuity for mixtures of squid K and virtual K channels. Simulated membrane responses for a mix of squid K channel (gK1) and each of the six virtual channels (labeled here gK2) to test the acuity of the CRI method. Channel mixtures were varied using the relationship $g_{K1} + g_{K2} = 72$ mS cm^{-2}. (Top) Estimated versus true channel mixture (plotted as density of g_{K1}). (Bottom) Rel. error of the CRI method in detecting the channel mixture for each for the six mixtures (averaged over the all test mixtures).

the squid K channel with very similar response properties. Combining a virtual channel with the squid channel in a model membrane, allows us to test if the CRI method is sensitive enough to detect the mixture of two very similar channels. We created 4 virtual channels by varying each channel parameter by +25%, while holding all other parameters constant. Furthermore, we also increased and decreased the number k of gating particles ($k = 3$, $k = 5$), while keeping all other parameters constant (Fig. 3(a)), yielding two more virtual channels We illustrate this procedure with an example of mixing the squid K channel (here labeled K1) with a virtual channel (here labeled K2), where A_{β} was increased by 25% with respect to K1. The mixture was set to $g_{K1} = 40\mu m^{-2}$ and $g_{K2} = 32\mu m^{-2}$ and the output of the CRI processing is depicted in Fig. 2. Following the CRI method the channel mixture was estimated as $g_{K1} = 42.2\mu m^2$ and $g_{K2} = 29.8\mu m^2$, by reverse identifying the two closest stored responses and interpolating between them.

To assess the overall sensitivity of RCI we simulated 6 membrane models, one for each combination of squid and virtual channel. In each model we varied the mixture of channels by sweeping across the range of N_{K1} and N_{K2} such as to hold the total number of channels $N_{K1} + N_{K2}$ constant. To ease readability we show channel densities g_{Ki} instead of total channel numbers N_{Ki}, as the membrane area is known and, thus, increment g_{K1} from 0 to 72mS cm^{-2} (see Fig. 3(b)). We apply the the RCI method to each channel density pair and average the relative error across the whole sweep.

The results for each of the 6 different virtual channels (see Fig. 3(b)), shows that despite the large similarities of the virtual channels to the squid channel (cf. Fig. 3(a)), the method can resolve the mixture with an average relative error of $6 - 16$ %.

2.3 RCI to Determine Channel Numbers from Fly Photoreceptor *in vivo* Data

The membrane of blowfly (*Calliphora vicina*) contains a fast and a slow delayed rectifier K channel which neurons operating range and allows for a higher signal-to-noise ratio in daylight [3]. Photoreceptors are an ideal system to apply the RCI method, as they offer a good space-clamp, their channels have been characterized and genetic dissection is possible.

Membrane properties derived from current- and voltage-clamp experiments of blow fly photo receptors [3] yield for typical cells $g_{\text{Leak}} = 0.24$mS cm^{-2} and a membrane area of $A = 1.310^4$ μm^2 with $C_m = 1$ μF cm^{-2} (reversal potentials of $E_K = -24.6$ mV,$E_{\text{Leak}} = 0$ mV).

In contrast to squid-like K channels, the blowfly K channels are described by two exponential rate functions.

$$\alpha(V) = A_\alpha \exp\left(\frac{V + V_{\text{off}}}{V_\alpha}\right) \tag{9}$$

$$\beta(V) = A_\beta \exp\left(\frac{-V - V_{\text{off}}}{V_\beta}\right) \tag{10}$$

The slow K channel has parameters $A_\alpha = 0.9, A_\beta = 0.0037, V_\alpha = 13, V_\beta = 33.8, V_{\text{off}} = -60$ and $k = 1$ [3]. The fast K channel has an unsuitable kinetic description for the purpose of this study. Therefore, we refitted the published data [3] with a more suitable set of exponentials using a simplex method to fit the parameters for our two-exponential model, resulting in $A_\alpha = 0.0472, A_\beta = 0.0262, V_\alpha = 13.4, V_\beta = 23.2, V_{\text{off}} = -60$ and $k = 1$ (see Fig. 4(a)). Sweeping across the ratio of possible channel densities, using the previous relationship $g_{\text{Kfast}} + g_{\text{Kslow}} = 72$ mS cm^{-2} (see Fig. 4(b)), produced very accurate estimates with an average accuracy error of only 4% (see Fig. 4(c)).

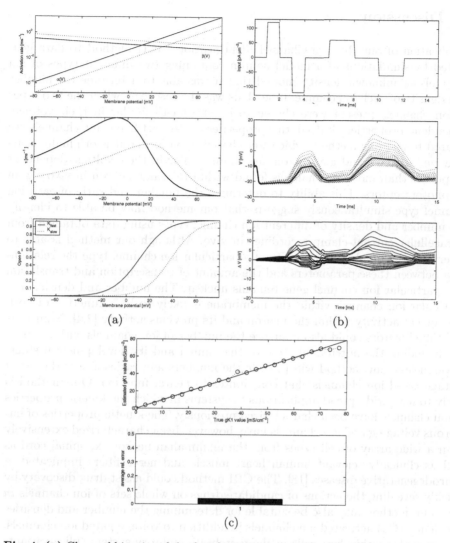

Fig. 4. (a) Channel kinetics of the fast (solid line) and slow (dotted line) delayed rectifier K channel in blow fly photoreceptors. (Top) Plot of forward (α) and backward (β)
rate functions versus membrane potential. (Middle) Channel time constant ($\tau = \frac{1}{\alpha+\beta}$)
versus membrane potential. (Bottom) Steady state channel open probability (p_∞) versus
voltage-clamped membrane potential. **(b)** CRI processing of fly photoreceptor responses.
(Top) Logarithmic CRI stimulus $I(t)$. (Middle) Test neuron response (thick, solid line)
and responses stored in the cross-reference (thin, dotted lines). (Bottom) Pairwise differences between the test neuron response and each stored response (thin, solid lines). **(c)**
CRI channel density mixture detection acuity for fly photoreceptors. Simulated membrane responses for a mix of squid K channel (gK1) and each of the six virtual channels
(labeled here gK2) to test the acuity of the CRI method. Channel mixtures were varied
using the relationship $g_{K1} + g_{K2} = 72$ mS cm^{-2}. (Top) Estimated versus true channel
mixture (plotted as density of g_{K1}). (Bottom) Rel. error of the CRI method in detecting
the channel mixture for each of the six mixtures (averaged over the all test mixtures).

3   Discussion

Application of our Reverse Channel Identification (RCI) method to data generated by simulations of a model neuron containing two different types of ion channels at unknown density showed that it was able to determine the number and density of each ion channel type. This was the case even when the two types of ion channels present were chosen to be extremely similar in their voltage-dependent properties. Indeed the properties of the artificial ion channels we created to test our method, which were based upon K channels originally characterized in the squid giant axon, are more similar in their voltage-dependent properties than most ion channels found within the same neuron in vivo e.g. in fly photoreceptors. This ability to determine the number and ratio of each ion channel type simultaneously suggests that our method may be able to identify the number and density of different ion channel types using data obtained from intracellular current-clamp recordings in vivo. Although our method is able to determine the number and density of a particular ion channel type the relationship between these parameters and the amount of transcription and translation of a particular ion channel gene remains unclear. The number and density of a particular ion channel within the membrane is likely to be a function of both the current activity within the neuron and its previous activity [2,4]. Numerous additional factors could also influence the number of ion channels within a neuron including the nutritional state of the animal and its developmental stage. Nevertheless, our method will provide the numbers and ratios across the set of voltage-gated ion channels that contributes to neural function. Our method is likely to have widespread applications to systems in which the kinetic properties of ion channels have been characterized previously. The kinetic properties of numerous voltage-gated ion channels have, however, been characterized extensively from a wide array of cell types from the mammalian neocortex, spinal cord as well as clinically relevant human heart muscle and nerve fibers implicated in neurodegenerative diseases [18]. The CRI method could assist drug discovery by quickly revealing the actions of candidate drugs on whole sets of ion channels *in vivo*. Our method may also be suitable for determining the number and densities of calcium (Ca) activated ion channels in addition to voltage-gated ion channels. For example, within hair cells in the vertebrate auditory system splice-variants of Ca-activated K channels differ more in their kinetic properties than our artificially created variations of squid voltage-gated K channels [19]. We are currently working on an extension of the CRI method to fit experimental data where not only the number and density ratios of ion channels are unknown but also the specific kinetic properties of the ion channels are not precisely known.

Acknowledgments

The authors would like to thank Simon Laughlin for his advice. AAF was supported by a Junior Research Fellowship by Wolfson College, Cambridge University. JEN was supported by the Smithsonian Institute.

References

1. Marder, E., Prinz, A.A.: Modeling stability in neuron and network function: the role of activity in homeostasis. Bioessays **24**(12) (2002) 1145–54
2. Zhang, W., Linden, D.J.: The other side of the engram: experience-driven changes in neuronal intrinsic excitability. Nat Rev Neurosci **4**(11) (2003) 885–900
3. Weckstrom, M., Hardie, R., Laughlin, S.: Voltage-activated potassium channels in blowfly photoreceptors and their role in light adaptation. J Physiol **440** (1991) 635–57
4. Stemmler, M., Koch, C.: How voltage-dependent conductances can adapt to maximize the information encoded by neuronal firing rate. Nature Neurosci. **2**(6) (1999) 521–527
5. McAnelly, M.L., Zakon, H.H.: Coregulation of voltage-dependent kinetics of na(+) and k(+) currents in electric organ. J Neurosci **20**(9) (2000) 3408–14
6. Brickley, S.G., Revilla, V., Cull-Candy, S.G., Wisden, W., Farrant, M.: Adaptive regulation of neuronal excitability by a voltage-independent potassium conductance. Nature **409**(6816) (2001) 88–92
7. Niven, J.E., Vahasoyrinki, M., Juusola, M.: Shaker K(+)-channels are predicted to reduce the metabolic cost of neural information in Drosophila photoreceptors. Proc Biol Sci **270 Suppl 1** (2003) 58–61
8. Aizenman, C.D., Akerman, C.J., Jensen, K.R., Cline, H.T.: Visually driven regulation of intrinsic neuronal excitability improves stimulus detection in vivo. Neuron **39**(5) (2003) 831–42
9. Niven, J.E.: Channelling evolution: canalization and the nervous system. PLoS Biol **2**(1) (2004) E19
10. Vahasoyrinki, M., Niven, J.E., Hardie, R.C., Weckstrom, M., Juusola, M.: Robustness of neural coding in Drosophila photoreceptors in the absence of slow delayed rectifier K+ channels. J Neurosci **26**(10) (2006) 2652–2660
11. Attwell, D., Laughlin, S.B.: An energy budget for signalling the the grey matter of the brain. J. Cereb. Blood Flow and Metabolism **21** (2001) 1133–1145
12. Laughlin, S.B.: Energy as a constraint on the coding and processing of sensory information. Curr Opin Neurobiol **11**(4) (2001) 475–480
13. Niven, J.E., Vahasoyrinki, M., Kauranen, M., Hardie, R.C., Juusola, M., Weckstrom, M.: The contribution of Shaker K+ channels to the information capacity of Drosophila photoreceptors. Nature **421**(6923) (2003) 630–634
14. Faisal, A., Laughlin, S., White, J.: How reliable is the connectivity in cortical neural networks? In Wunsch, D., ed.: Proceedings of the IEEE Intl. Joint. Conf. Neural Networks 2002, INNS (2002) 1–6
15. Faisal, A.A., White, J.A., Laughlin, S.B.: Ion-channel noise places limits on the miniaturization of the brain's wiring. Curr Biol **15**(12) (2005) 1143–1149
16. Koch, C.: Biophysics of computation. Computational neuroscience. Oxford University Press, Oxford (1999)
17. Hodgkin, A., Huxley, A.: Quantitative description of membrane current and its application to conduction and excitation in nerve. J. Physiol. (London) **117** (1952) 500–544
18. Hille, B.: Ion channels of excitable membranes. 3rd edn. Sinauer Associates, Sunderland, MA (2001) 814 pp.
19. Jones, E.M., Gray-Keller, M., Fettiplace, R.: The role of ca2+-activated k+ channel spliced variants in the tonotopic organization of the turtle cochlea. J Physiol **518** (1999) 653–665

Author Index

Lecture Notes in Bioinformatics